AF327115

Catalan Numbers with Applications

Catalan Numbers with Applications

Thomas Koshy
Framingham State College

2009

Oxford University Press, Inc., publishes works that further
Oxford University's objective of excellence
in research, scholarship, and education.

Oxford   New York
Auckland   Cape Town   Dar es Salaam   Hong Kong   Karachi
Kuala Lumpur   Madrid   Melbourne   Mexico City   Nairobi
New Delhi   Shanghai   Taipei   Toronto

With offices in
Argentina   Austria   Brazil   Chile   Czech Republic   France   Greece
Guatemala   Hungary   Italy   Japan   Poland   Portugal   Singapore
South Korea   Switzerland   Thailand   Turkey   Ukraine   Vietnam

Published by Oxford University Press, Inc.
198 Madison Avenue, New York, New York 10016

www.oup.com

Oxford is a registered trademark of Oxford University Press

Library of Congress Cataloging-in-Publication Data
Koshy, Thomas.
Catalan numbers with applications / Thomas Koshy.
    p. cm.
Includes bibliographical references and index.
ISBN 978-0-19-533454-8
1. Catalan numbers.   I. Title.
QA431.K726 2008
511′.6—dc22   2008033798

9 8 7 6 5 4 3 2 1

Printed in the United States of America
on acid-free paper

*To my grandson
Nathan Malayil Koshy*

Preface

Man has the faculty of becoming completely absorbed in one subject,
no matter how trivial, and no subject is so trivial that it will not assume
infinite proportions if one's entire attention is devoted to it.

— Tolstoy, *War and Peace*

A North Star on the Mathematical Sky

In *Fibonacci and Lucas Numbers with Applications*, I called the Fibonacci and Lucas sequences "two shining stars in the vast array of integer sequences," because of their ubiquitousness, tendency to appear in quite unexpected and unrelated places, abundant applications, intriguing properties, and the fascination and easy accesibility that have attracted amateurs and mathematicians alike. However, Catalan numbers are even more fascinating. Like the North Star in the evening sky, they are a beautiful and bright light in the mathematical heavens. They continue to provide a fertile ground for number theorists, especially Catalan enthusiasts and computer scientists.

Since the publication of Euler's triangulation problem (1751) and Catalan's parenthesization problem (1838), nearly 400 articles and problems on Catalan numbers have appeared in various periodicals. They exude the beauty and ubiquity of Catalan numbers. "They have the same delightful propensity for popping up unexpectedly, particularly in combinatorial problems," Martin Gardner wrote in 1976 in his popular column *Mathematical Games* in *Scientific American*. "Indeed," he adds, "the Catalan sequence is probably the most frequently encountered sequence that is still obscure enough to cause mathematicians lacking access to

N. J. A. Sloane's *A Handbook of Integer Sequences* to expend inordinate amounts of energy re-discovering formulas that were worked out long ago."

As Gardner noted, many amateurs and mathematicians may know the *ABC*'s of Catalan sequence, but may not be familar with their myriad unexpected occurrences, delightful applications, properties, or the beautiful and surprising relationships among numerous examples. This book is the first to collect and present the various occurrences, applications, and properties of Catalan numbers from a multitude of sources in an orderly and enjoyable fashion. It presents a clear and comprehensive introduction to one of the most delightful topics in mathematics. This book, as one reviewer observed, reflects my "intense labor of love" and fascination with the central theme.

Audience

This book is intended for a broad audience: mathematical amateurs, high school students and teachers, as well as undergraduate and graduate students in both mathematics and computer science. Catalan numbers provide invaluable opportunities for projects, group discussions, talks, and capstone experiences, such as senior and master's theses. In the process, they will be able to explore new applications and discoveries, and advance the frontiers of mathematical knowledge even further.

Like Fibonacci and Lucas numbers, Catalan numbers are also an excellent source of fun and excitement. They can be used to generate interesting dividends for students, such as intellectual curiosity, experimentation, pattern recognition, and conjecturing. For example, I have presented several Catalan examples for experimentation, pattern recognition, and conjecturing in *Mathematics for Liberal Arts* courses in the past several years. My students loved the examples and became intrigued with Catalan numbers. To my great surprise, one student, a Spanish major, even conjectured Segner's recursive formula for the nth Catalan number (see formula [5.6] in Chapter 5), a rewarding experience indeed.

Organization

In the interest of clarity, the book is divided into seventeen chapters, most of them of approximately of equal length. The first four chapters, which are short, present the fundamental properties of binomial coefficients, since Catalan numbers involve binomial coefficients. My goal is to bring the reader from the familiar to the less well known, as the reviewer noted.

The central character in the nth Catalan number $\frac{1}{n+1}\binom{2n}{n}$ is the central binomial coefficient $\binom{2n}{n}$; so two of the four chapters pursue the properties of the central binomial coefficients.

Prerequisites

A number of examples are within reach of amateurs. However, a good background in precalculus mathematics, and a reasonable amount of mathematical maturity are desirable. An exposure to bijections, binomial coefficients, Pascal's triangle, combinatorics, recursion, generating functions, congruences, principle of mathematical induction, and extended binomial coefficients is highly recommended for enjoying the beauty and fascination of this bright star.

Historical Flavor

Every effort has been made to present material with its historical context. I have tried to incorporate the name and affiliation of every contributor, as well as the year of contribution. My apologies for any inadvertent omissions; I would gratefully appreciate hearing about any omissions by e-mail.

Pascal's Triangle

Since the explicit formula for the nth Catalan number involves the central binomial coefficient, it is obvious that Catalan numbers can be extracted from Pascal's triangle. In fact, there are a number of ways they can be read from Pascal's triangle; every one of them is described and exemplified in Chapter 12. This brings Catalan numbers a step closer to number theory enthusiasts, especially high school students and teachers.

Appendix

Throughout the book, the discussion hinges on a number of elementary mathematical tools. They include the *fundamental counting principles, bijection, recursion, generating functions, mathematical induction, congruences*, and *extended binomial coefficients* $\binom{n}{r}$, where n can be a negative integer $-m$, a positive rational number p/q, or a negative rational number $-p/q$. A brief introduction to each topic is presented in Appendix A for a quick review or reference. For a detailed discussion of each, you may want to consult the author's *Elementary Number Theory with Applications* or *Discrete Mathematics with Applications*, as needed.

Symbols and Biographical Sketches

A key to symbols used in the book and their definitions appear on the inside of the front cover for easy reference. They are all standard symbols in mathematics.

For a quick reference, an index appears inside the back cover of biographical sketches of about twenty-five mathematicians who have made significant contributions to the development of the subject matter.

References

Six sources have played a major role in the development of this unique project: M. J. Kuchinski's master's thesis, *Catalan Structures and Correspondences*; R. P. Stanley's *Enumerative Combinatorics*, vols. 1 and 2, and *Catalan Addendum*; H. W. Gould's extensive bibliography on Catalan numbers, *Bell and Catalan Numbers: Research Bibliography of Two Special Number Sequences,* revised edition; and J. A. S. Growney's doctoral dissertation, *Finitely Generated Free Groups*. The discourses and references in these works were invaluable in my own journey. The list of references at the end of the book includes nearly all resources in English on central binomial coefficients and Catalan numbers, and it continues to grow.

Acknowledgments

I am grateful to a number of people for their cooperation, encouragement, and support in the development of the book.

To begin with, I am indebted to the following reviewers for their constructive suggestions and overwhelming enthusiasm for the project: Joyce Cutler at Framingham State College; Thomas E. Moore at Bridgewater State College; Ralph P. Grimaldi at Rose-Hulman Institute of Technology; Arthur T. Benjamin at Harvey Mudd College; and Henry W. Gould at West Virginia University.

Thanks also to Jeff Gao for preparing the lists of matrices in Examples 11.3 and 11.4, and the list of Catalan numbers C_1 through C_{100} in the Appendix; to Margarite Roumas for her superb editorial assistance; to Michael Penn, senior editor in mathematics at Oxford University Press, for his boundless enthusiasm for the project; to Phyllis Cohen, editor of math and physics at Oxford University Press; and to Ned Sears, Dayne Poshusta, and Stephanie Attia for cooperation and support.

Framingham, Massachusetts
August 21, 2008

Thomas Koshy
tkoshy@frc.mass.edu

Contents

Symbols

$\{x, y, z\}$	set consisting of the elements x, y, z
$\mathbf{N}$	set of positive integers $1, 2, 3, \ldots$
$\mathbf{W}$	set of whole numbers $0, 1, 2, \ldots$
$\{s_n\}_1^\infty = \{s_n\}$	sequence with general term s_n
$\displaystyle\sum_{i=k}^{i=m} a_i = \sum_{i=k}^{m} a_i$	$a_k + a_{k+1} + \cdots + a_m$
$\displaystyle\sum_{i \in I} a_i$	sum of the values of a_i as i runs over the values in I
$\displaystyle\sum_{P} a_i$	sum of the vaues of a_i, where i has properties P
$\displaystyle\sum_i \sum_j a_{ij}$	$\displaystyle\sum_i \left(\sum_j a_{ij} \right)$
$\displaystyle\prod_{i=k}^{i=m} a_i = \prod_{i=k}^{m} a_i$	$a_k a_{k+1} \ldots a_m$
$n!$	$n \cdot (n-1) \cdots 3 \cdot 2 \cdot 1$, where $0! = 1$
$n!!$	$n!(n-1)! \cdots 3!2!1!$
$|x|$	absolute value of x
$\lfloor x \rfloor$	greatest integer $\leq x$
$\iff$	if and only if
LHS	left-hand side
RHS	right-hand side
PMI	principle of mathematical induction

$a \prec b$	a precedes b
$a \mid b$	a is a factor of b
$a \nmid b$	a is not a factor of b
(a, b)	greatest common factor of a and b
$a \in A$	a belongs to the set A
$\lvert A \rvert$	number of elements in set A
$A \cup B$	union of the sets A and B
$A \cap B$	intersection of the sets A and B
$A = (a_{ij})_{m \times n}$	$m \times n$ matrix A whose ijth element is a_{ij}
$a \approx b$	a is approximately equal to b
$a \equiv b \pmod{m}$	a is congruent to b modulo m
$a \bmod b$	remainder when a is divided by b
∞	infinity
$\overline{AB}$	line segment AB
$\overleftrightarrow{AB}$	line containing the points A and B
$v - w$	edge between vertices v and w
$\dbinom{n}{r}$	$\dfrac{n!}{r!(n-r)!}$
$\dbinom{-n}{r}$	$(-1)^r \dbinom{n+r-1}{r}$
$\dbinom{p/q}{r}$	$\dfrac{p(p-q)(p-2q)\cdots[p-(r-1)q]}{q^r\, r!}$
$\dbinom{-p/q}{r}$	$(-1)^r \dfrac{p(p+q)(p+2q)\cdots[p+(r-1)q]}{q^r\, r!}$
$(a_1, a_2, \ldots, a_n)$	n-tuple
$f = \begin{pmatrix} a_1 \ a_2 \ \ldots \ a_n \\ b_1 \ b_2 \ \ldots \ b_n \end{pmatrix}$	permutation f such that $f(a_i) = b_i$
$\Gamma(z)$	$\int_0^z t^{z-1} e^{-t}\, dt$
$\mathbf{Z}_n$	additive group of integers modulo n
t_m	$\dfrac{m(m+1)}{2}$
t_m^*	$t_m t_{m-1} \cdots t_2 t_1$
$\begin{bmatrix} n \\ r \end{bmatrix}$	$\dfrac{t_n^*}{t_r^* t_{n-r}^*}$
■	end of an example, proof, theorem, or corollary

Catalan Numbers with Applications

1

Binomial Coefficients

Binomials, which are sums of two terms, occur often in mathematics. There is a systematic way of expanding the positive integral powers of binomials. This is the essence of the binomial theorem, which we will study in Chapter 2. First, we begin with a brief discussion of *binomial coefficients*, which are coefficients that occur in binomial expansions; binomial coefficients form the cornerstone of all discussions in this and other chapters.

Binomial Coefficients

Let n and r be nonnegative integers. The *binomial coefficient* $\binom{n}{r}$ is defined as

$$\binom{n}{r} = \frac{n!}{r!(n-r)!}$$

where $0 \leq r \leq n$. If $r > n$, then $\binom{n}{r}$ is defined as 0.

For example, $\binom{8}{3} = \frac{8!}{3!(8-3)!} = \frac{8!}{3!5!} = 56$ and $\binom{7}{9} = 0$.

The binomial coefficient $\binom{n}{r}$ is also denoted by C(n,r) and nCr. For example, the graphing calculator TI-89 employs the notation nCr for computing binomial coefficients.

Those familiar with combinatorics will recall that the binomial coefficient $\binom{n}{r}$ denotes the number of combinations (unordered arrangements) of n distinct things taken r at a time. For example, we can form $\binom{8}{3} = 56$ subcommittees of 3 people

The term *binomial coefficient* was introduced by German mathematician **Michel Stifel** (1486–1567), considered the greatest German algebraist of the sixteenth century. Stifel was born in Eslingen and was educated in the monastry there. In 1555, he became professor at the University of Jena. He is best known for his mathematical work, *Arithmetica Integra*, published in 1544. The book is divided into three parts: rational numbers, irrational numbers, and algebra. In the first part, Stifel gives the binomial coefficients for $n \leq 17$.

Stifel is one of the strangest personalities in the history of mathematics. Originally an Augustinian monk, he converted to Protestantism in 1523 under Martin Luther's influence, and became a fanatical reformer. He believed in numerical mysticism. For instance, analyzing biblical writings and using astrological computations, Stifel predicted the end of the world on October 3, 1533. He was forced to seek asylum in a prison after ruining the lives of many trusting peasants who abandoned their work and property to follow him to heaven. He also employed arithmography to "prove" that Pope Leo X was the "beast" mentioned in the *Book of Revelation*. He claimed that this discovery was inspired by God.

The bilevel parentheses notation for binomial coefficient was introduced by German mathematician and physicist **Baron Andreas von Ettinghausen** (1796–1878).

Von Ettinghausen was born in Heidelberg, Germany. He attended the University of Vienna in Austria, where he worked for two years as an assistant in mathematics and physics. In 1821, von Ettinghausen became professor of mathematics, and in 1835, professor of physics and director of the Physics Institute. Thirteen years later, he became the director of the Mathematical Studies and Engineering Academy in Vienna.

A pioneer in mathematical physics, von Ettinghausen worked in algebra, analysis, differential geometry, electromagnetism, mechanics, and optics.

each from a committee of 8 people; we can draw exactly $\binom{8}{2} = 28$ distinct lines passing through 8 points in a plane, no 3 of which are collinear; and we can draw $\binom{10}{3} = 120$ distinct triangles using 10 points in a plane, no 3 of which are collinear.

A regular polygon with n sides has $\binom{n}{2} - n = \frac{n(n-1)}{2} - n = \frac{n(n-3)}{2}$ diagonals. Three diagonals intersect in $\binom{n}{4}$ points, not necessarily all distinct.

It is interesting to note that n points can divide a line into $1 + n = \binom{n}{0} + \binom{n}{1}$ parts; n lines (cuts) can divide a plane (pizza) into $1 + n + \binom{n}{2} = \binom{n}{0} + \binom{n}{1} + \binom{n}{2}$ parts; n planes (cuts) can divide space (a three-dimensional pizza) into $1 + n + \binom{n}{2} + \binom{n}{3} = \binom{n}{0} + \binom{n}{1} + \binom{n}{2} + \binom{n}{3}$ parts. More generally, n cuts can divide an m-dimensional pizza into $\binom{n}{0} + \binom{n}{1} + \binom{n}{2} + \binom{n}{3} + \cdots + \binom{n}{m}$ pieces, as L. Schläfli proved in 1901.

It follows from the definition that $\binom{n}{0} = 1 = \binom{n}{n}$. You can confirm both.

Triangular and Tetrahedral Numbers

Interestingly enough, the triangular number $t_n = \frac{n(n+1)}{2}$ and the tetrahedral number $T_n = \frac{n(n+1)(n+2)}{6}$ can be rewritten as binomial coefficients:

$$t_n = \frac{n(n+1)}{2} = \binom{n+1}{2}$$

and

$$T_n = \frac{n(n+1)(n+2)}{6} = \binom{n+2}{3}$$

Since $t_n = t_{n-1} + n$, it follows that $\binom{n+1}{2} = \binom{n}{2} + n = \binom{n}{2} + \binom{n}{1}$. This is a special case of Pascal's identity in Theorem 1.2.

Next, we take a look at "The Twelve Days of Christmas," a well-known holiday carol, and see how it is related to triangular numbers.

Twelve Days of Christmas

The carol reads: "On the first day of Christmas, my true love sent me a partridge in a pear tree. On the second day of Christmas, my true love sent me two turtle doves and a partridge in a pear tree. On the third day, my true love sent me three French hens, two turtle doves, and a partridge in a pear tree." This pattern continues until the twelfth day; on that day my true love sent me "twelve drummers drumming, eleven pipers piping, ten lords a-leaping, nine ladies dancing, eight maids a-milking, seven swans a-swimming, six geese a-laying, five gold rings, four calling birds, three french hens, two turtle doves, and a partridge in a pear tree."

This carol raises two interesting questions:

- If the given pattern continues for n days, how many gifts g_n would be sent on the nth day?
- What is the total number of gifts s_n sent in n days?

First, notice that the number of gifts sent on the nth day equals n more than the number of gifts g_{n-1} sent on the previous day; so $g_n = g_{n-1} + n$, where $g_1 = 1$. Consequently,

$$g_n = t_n = \frac{n(n+1)}{2} = \binom{n+1}{2}$$

It follows from this that

$$
\begin{aligned}
s_n &= \sum_{i=1}^{n} t_i \\
&= \sum_{i=1}^{n} \frac{i(i+1)}{2} \\
&= \frac{1}{2}\left(\sum_{i=1}^{n} i^2 + \sum_{i=1}^{n} i\right) \\
&= \frac{1}{2}\left[\frac{n(n+1)(2n+1)}{6} + \frac{n(n+1)}{2}\right] \\
&= \frac{n(n+1)}{12}[(2n+1) + 3] \\
&= \frac{n(n+1)(n+2)}{6} \\
&= \binom{n+2}{3}
\end{aligned}
$$

Very often we will need to compute the binomial coefficients $\binom{n}{r}$ and $\binom{n}{n-r}$ in the same problem. The next theorem shows that we do not need to compute each separately.

Theorem 1.1 Let n and r be nonnegative integers. Then

$$\binom{n}{r} = \binom{n}{n-r}$$

Proof

$$
\begin{aligned}
\binom{n}{n-r} &= \frac{n!}{(n-r)![n-(n-r)]!} \\
&= \frac{n!}{(n-r)!r!}
\end{aligned}
$$

$$= \frac{n!}{r!(n-r)!}$$

$$= \binom{n}{r}$$

For example, $\binom{23}{15} = \binom{23}{8} = 490{,}314$.

Combinatorial Identities

The following combinatorial identities can be verified algebraically:

$$\binom{n}{r} = \frac{n}{r}\binom{n-1}{r-1} \tag{1.1}$$

$$\binom{n}{r} = \frac{n}{n-r}\binom{n-1}{r} \tag{1.2}$$

$$\binom{n}{r} = \frac{n-r+1}{r}\binom{n}{r-1} \quad \text{(Cardano's Rule of Succession)} \tag{1.3}$$

$$\binom{n}{k}\binom{k}{r} = \binom{n}{r}\binom{n-r}{k-r} \quad \text{(Newton's Identity)} \tag{1.4}$$

$$\binom{n}{2} + \binom{n-1}{2} = (n-1)^2 \tag{1.5}$$

$$\frac{1}{n}\binom{nm}{m} = \binom{nm-1}{m-1} \tag{1.6}$$

$$n\binom{n}{r} = (r+1)\binom{n}{r+1} + r\binom{n}{r} \tag{1.7}$$

where $0 \le r \le k \le n$.

Binomial coefficients satisfy an interesting and useful recurrence relation, as the next theorem shows. It is called *Pascal's identity*, in honor of French mathematician Blaise Pascal.

Theorem 1.1 (*Pascal's identity*) Let n and r be any positive integers, where $r \le n$. Then

$$\binom{n}{r} = \binom{n-1}{r-1} + \binom{n-1}{r}$$

Blaise Pascal (1623–1662) was born in Clermont-Ferrand, France. Although he showed remarkable mathematical ability at an early age, his father, Étienne Pascal (1588–1640), himself an able mathematician, encouraged him to pursue other subjects, such as ancient languages. His father even refused to teach young Blaise any sciences, but relented when he found that young Pascal by age twelve had discovered many results in elementary geometry. At fourteen, Pascal attended weekly meetings of a group of French mathematicians, which became the French Academy in 1666. At sixteen, he developed many results in conic sections and even wrote a book on them in 1648.

Observing that his father would spend countless hours auditing laborious government accounts, and feeling that intelligent people should not waste their time doing mundane things, Pascal, at the age of nineteen, invented the first mechanical calculating machine.

Pascal died in Paris at the age of thirty-nine. His short but very productive life was racked with both physical pain and religious torment. He is often described as the greatest "might-have-been" in the history of mathematics. With his astounding gifts and remarkable geometric intuition, Pascal likely would have accomplished much more.

The programming language Pascal is named after him.

Proof   We shall simplify the RHS[†] and show that it equals the LHS.

$$\binom{n-1}{r-1} + \binom{n-1}{r} = \frac{(n-1)!}{(r-1)!(n-r)!} + \frac{(n-1)!}{r!(n-r-1)!}$$

$$= \frac{r(n-1)!}{r(r-1)!(n-r)!} + \frac{(n-r)(n-1)!}{r!(n-r)(n-r-1)!}$$

$$= \frac{r(n-1)!}{r!(n-r)!} + \frac{(n-r)(n-1)!}{r!(n-r)!}$$

$$= \frac{(n-1)![r+(n-r)]}{r!(n-r)!}$$

$$= \frac{(n-1)!n}{r!(n-r)!}$$

$$= \frac{n!}{r!(n-r)!}$$

$$= \binom{n}{r} \qquad \blacksquare$$

For example, $\binom{23}{10} = \binom{22}{9} + \binom{22}{10}$.

Runyon Numbers

It is easy to verify that

$$\frac{1}{n+1}\binom{n-1}{m}\binom{n+1}{m+1} = \frac{1}{m+1}\binom{n-1}{m}\binom{n}{m}$$

These numbers are called *Runyon numbers*, after J. P. Runyon of Bell Telephone Laboratories, Murray Hill, New Jersey. He encountered them in his study of a telephone traffic system with inputs from two sources.[*]

The definition of the binomial coefficient $\binom{n}{r}$ seems to display a fractional appearance, so we might be tempted to assume that its value can be a fraction. Fortunately, this is far from the truth, as the following theorem confirms. We shall take advantage of this fact on numerous occasions.

Theorem 1.3   The binomial coefficient $\binom{n}{r}$ is an integer for every integer $n \geq 0$.

[†] RHS and LHS are abbreviations of right-hand side and left-hand side, respectively.
[*] See J. A. Morrison, "A Certain Functional-Difference Equation," *Duke Mathematical Journal* **31** (1964), 445–448.

Proof (*by PMI*[†])   Since $\binom{0}{r} = 1$ is an integer, the result is true when $n = 0$.

Now assume that it is true for all nonnegative integers $< k$. Then $\binom{k-1}{r-1}$ and $\binom{k-1}{r}$ are integers by the inductive hypothesis. So is their sum $\binom{k-1}{r-1} + \binom{k-1}{r}$; that is, $\binom{n}{r}$ is an integer by Pascal's identity.

Thus, by PMI, every binomial coefficient is an integer.   ∎

Theorem 1.3 has an interesting byproduct, as the following corollary reveals.

Corollary 1.1   The product of r consecutive integers is divisible by $r!$.

Proof   Clearly, it suffices to consider the product of r consecutive positive integers. Let n be the least of them. Since

$$\frac{n(n+1)\cdots(n+r-1)}{r!} = \frac{(n+r-1)!}{r!(n-1)!} = \binom{n+r-1}{r}$$

by Theorem 1.3, $\dfrac{n(n+1)\cdots(n+r-1)}{r!}$ is an integer; so $n(n+1)\cdots(n+r-1)$ is divisible by $r!$.   ∎

For example, $7! \mid 5 \cdot 6 \cdot 7 \cdot 8 \cdot 9 \cdot 10 \cdot 11$ and $5! \mid (-6)(-7)(-8)(-9)(-10)$.

Pascal's Triangle

The various binomial coefficents $\binom{n}{r}$, where $0 \le r \le n$, can be arranged in the form a triangular array, as Figures 1.1 and 1.2 show. The triangular array is named after Pascal, who wrote a book about it, *Treatise on the Arithmetic Triangle* in 1653 (published posthumously in 1665). In his book, Pascal shows the relationship between binomial coefficients and polynomials. Although the array is named after him, it had appeared on the title page of a 1529 arithmetic book by German astronomer Petrus Apianus (1495–1552), which was the first appearance of the triangular array in the Western world.

Hermite's Divisibility Properties

We now turn to two divisibility properties of binomial coefficients, discovered by French mathematician Charles Hermite (1822–1901). They were also proved by E. C. Catalan and G. B. Mathews, among others.

[†] PMI is an abbreviation of the Principle of Mathematical Induction.

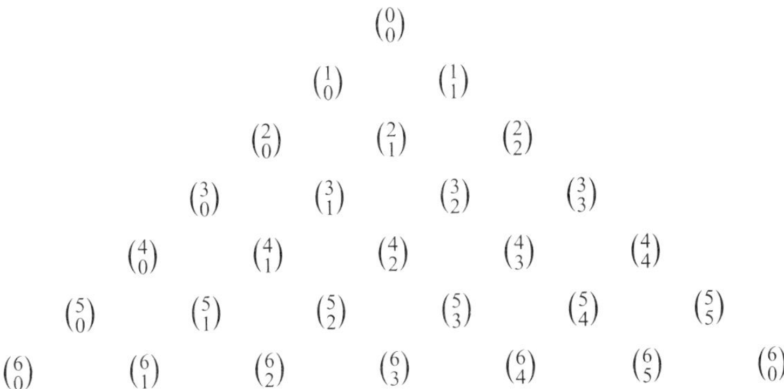

Figure 1.1 Pascal's Triangle

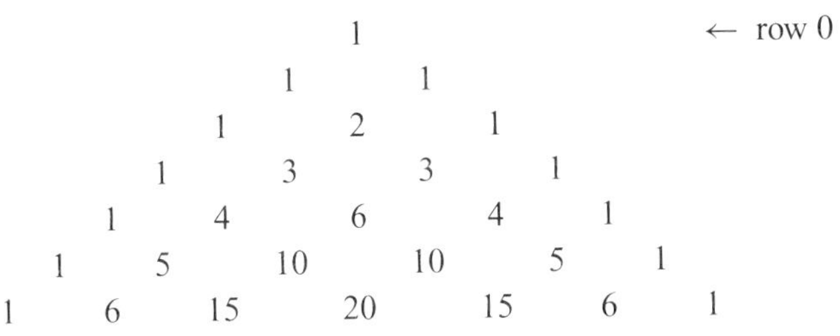

Figure 1.2 Pascal's Triangle

Theorem 1.4 (*Hermite*)   Let $m, n \geq 1$. Then

$$\frac{m}{(m, n)} \left| \binom{m}{n} \right. \tag{1.8}$$

$$\frac{m - n + 1}{(m + 1, n)} \left| \binom{m}{n} \right. \tag{1.9}$$

where (a, b) denotes the greatest common divisor (gcd) of the positive integers a and b.

Proof

1. Let $d = (m, n)$. Then, by the euclidean algorithm, there exist integers A and B such that

$$d = Am + Bn$$

Charles Hermite, a prominent algebraist of the nineteenth century, was born in Lorraine, France. After attending the Collège of Nancy and the Collège Henri IV, he transferred to the Collège Louise-le-Grand, where he was taught by the same professor who had taught E. Galois fifteen years earlier. While there, he read the works of Euler, Gauss's *Disquisitiones arithmeticae*, and Lagrange's *Traité sur la résolution des équations numériques*. During this time, he tried to establish the impossibility of solving a fifth-degree equation by radicals. Hermite continued his studies at the École Polytechnique in Paris, where he was taught by Catalan.

Hermite became familiar with the work of A. L. Cauchy and J. Liouville on the theory of functions, and the work of C. G. J. Jacobi on elliptic and hyperbolic functions. In 1843, when he was only twenty years old, he communicated his discovery on elliptic functions to Jacobi, followed by his discoveries on number theory. Throughout his life, Hermite exercised profound influence through his correspondence with prominent mathematicians.

In 1856, Hermite was elected a member of the Academy of Sciences. Six years later, through Louis Pasteur's influence, a position was created for him at the École Polytechnique. In 1869, he was appointed professor of analysis, resigning in 1876.

Hermite's work spans the theory of abelian and elliptic functions, theory of quadratic forms, theory of invariants, number theory, and integral calculus. His study of algebraic continued fractions led in 1873 to his celebrated theorem that e is transcendental. In 1883, F. Lindemann, adapting Hermite's proof, established the transcendence of π. Both proofs were subsequently refined by K. Weierstrass, D. Hilbert, and F. C. Klein.

Multiplying both sides by $\binom{m}{n}$, this yields

$$d\binom{m}{n} = Am\binom{m}{n} + Bn\binom{m}{n}$$

$$= m\left[A\binom{m}{n} + B\binom{m-1}{n-1} \right]$$

$$= mC$$

where C is an integer. So $\frac{m}{d} \mid \binom{m}{n}$. That is, $\frac{m}{(m,n)} \mid \binom{m}{n}$.

2. Let $d = (m+1, n)$. As before, there exist integers P and Q such that

$$d = P(m+1) + Qn$$

$$= (m-n+1)P + n(P+Q)$$

$$d \cdot \frac{m!}{n!(m-n+1)!} = \binom{m}{n}P + \binom{m}{n-1}(P+Q)$$

$$= R \quad \text{(say)}$$

$$d\binom{m}{n} = (m-n+1)R$$

So

$$\frac{m-n+1}{d} \, \bigg| \, \binom{m}{n}$$

That is,

$$\frac{m-n+1}{(m+1,n)} \, \bigg| \, \binom{m}{n} \qquad \blacksquare$$

Hermite's divisibility properties have interesting byproducts, as the next corollary shows. We will use them in later chapters.

Corollary 1.2

1. The binomial coefficient $\binom{2n}{n}$ is an even integer, where $n \geq 1$.
2. Let p be a prime. Then $\binom{p}{r} \equiv 0 \pmod{p}$, where $1 \leq r \leq p-1$.
3. $n+1 \mid \binom{2n}{n}$

Proof

1. It follows from property (1.8) that $\binom{2n}{n}$ is an even integer, where $n \geq 1$.
2. Since $1 \leq r \leq p-1, (p,r) = 1$. So, again by property (1.8), $p \mid \binom{p}{r}$; that is, $\binom{p}{r} \equiv 0 \pmod{p}$.
3. Let $m = 2n$ in (1.9). Since $(2n+1,n) = 1$, it follows that $n+1 \mid \binom{2n}{n}$. $\blacksquare$

It follows from this corollary that $\frac{1}{n+1}\binom{2n}{n}$ is an integer. Integers $C_n = \frac{1}{n+1}\binom{2n}{n}$ are called *Catalan numbers*. We will encounter them numerous times in later chapters.

2

The Central Binomial Coefficient

This chapter focuses on the popular and seemingly ubiquitous central binomial coefficient (CBC) $\binom{2n}{n}$. The CBC occurs in many unexpected places and therefore deserves special attention. It has been studied extensively, and we will explore some important occurrences of the coefficient in the next several chapters.

The CBCs $\binom{2n}{n}$ are centrally located in even-numbered rows in Pascal's triangle, as Figure 2.1 shows.

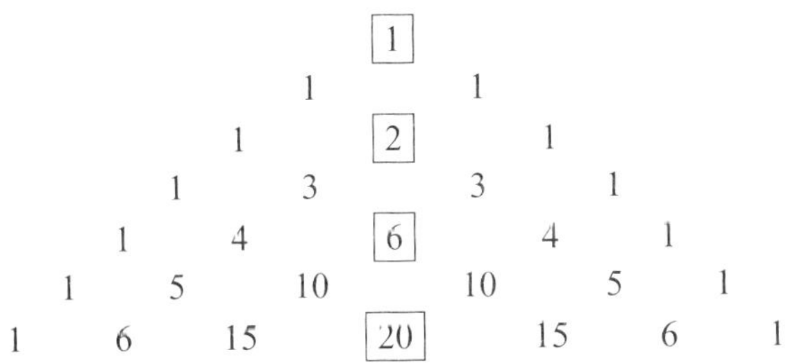

Figure 2.1 Pascal's Triangle

Parity of the CBC

In Corollary 1.2, we found that the CBC $\binom{2n}{n}$ is an even integer, when $n \geq 1$. In 1988, Richard K. Guy of the University of Calgary, Alberta, Canada, gave two

Richard Kenneth Guy (1916–) was born in Nuneaton, Warwickshire, England. After graduating from Warwick School, he entered the University of Cambridge, and received his B.A. in 1938 and M.A. in 1941. From 1939–1947, he taught at Stockport Grammar School, and from 1948–1952 at Goldsmiths' College, London. Then he joined the faculty at the University of Malaya, Singapore, where he stayed for four years. In 1959, he was appointed senior lecturer and then professor of Mathematics at the Indian Institute of Technology, New Delhi. In 1969, he was appointed professor of Mathematics at the University of Calgary. A professor emeritus since 1982, Guy received an honorary doctorate from the university nine years later.

A gifted problem-poser and problem-solver, Guy has published over 250 articles, and authored or coauthored 10 books. His research interests include combinatorial games, combinatorial geometry, enumerative combinatorics, graph theory, and number theory. He is best known for coauthorship (with John H. Conway and E. R. Berlekamp) of the four-volume *Winning Ways for Your Mathematical Plays*, "the best-seller that first described the theory that underlies games ranging from the simplest pencil-and-paper-amusements to those, like *Go*, that have challenged us for centuries."

An expert on endgame theory, Guy was endgame editor for *British Chess Magazine* from 1948 to 1951. He received the Lester R. Ford Award in 1989 and has served on the editorial boards of numerous mathematical periodicals. He is an ardent mountain climber.

elementary proofs that $\binom{2n}{n}$ is an even integer:

$$\binom{2n}{n} = \frac{2(2n-1)(2n-2)!}{n!(n-1)!}$$

$$= 2(2n-1)\frac{(2n-2)!}{n!(n-1)!}[2n-(2n-1)]$$

$$= 2(2n-1)\left[2\binom{2n-2}{n-1} - \binom{2n-1}{n-1}\right]$$

Because each binomial coefficient is an integer, it follows that $\binom{2n}{n}$ is an even integer.

A slight change in line 2 of this proof yields the second proof:

$$\binom{2n}{n} = 2(2n-1)\frac{(2n-2)!}{n!(n-1)!}[n-(n-1)]$$

$$= 2(2n-1)\left[\binom{2n-2}{n-1} - \binom{2n-2}{n-2}\right]$$

We present an equally elementary proof using Pascal's identity:

$$\binom{2n}{n} = \binom{2n-1}{n-1} + \binom{2n-1}{n}$$

$$= \binom{2n-1}{n} + \binom{2n-1}{n}$$

$$= 2\binom{2n-1}{n}$$

The following combinatorial identities also yield the same result:

$$\binom{n}{r} = \frac{n}{r}\binom{n-1}{r-1}$$

$$\binom{n}{r} = \frac{n}{n-r}\binom{n-1}{r}$$

More generally, $\frac{(nr)!}{(r!)^n}$ is an integer. This follows from the fact that $\frac{(nr)!}{n!(r!)^n}$ is an integer; the latter was, in fact, proposed as a problem in 1910 by H. C. Feemster of York College, Nebraska, and solved in the same year using induction by G.B.M. Zerr of Temple College (now Temple University), Philadelphia. Its proof employs Theorem 1.3.

Example 2.1 Prove that $\frac{(nr)!}{n!(r!)^n}$ is an integer for every $n \geq 0$.

Proof *(by PMI)*   Since $\frac{(0)!}{0!(r!)^0} = 1$ and $\frac{(r)!}{1!r!} = 1$ are both integers, the result is true when $n = 0$ and $n = 1$.

Now, assume that is is true for an arbitrary integer $k \geq 0$; that is, $\frac{(kr)!}{k!(r!)^k}$ is an integer. Then

$$\frac{[(k+1)r]!}{(k+1)!(r!)^{k+1}} \div \frac{(kr)!}{k!(r!)^k} = \frac{[(k+1)r]!}{(k+1)!(r!)^{k+1}} \cdot \frac{k!(r!)^k}{(kr)!}$$

$$= \frac{(kr+r)(kr+r-1)\cdots(kr+1)}{(k+1)r!}$$

$$= \frac{(kr+r-1)\cdots(kr+1)}{(r-1)!}$$

$$= \binom{kr+r-1}{r-1}$$

This is an integer, so $\frac{[(k+1)r]!}{(k+1)!(r!)^{k+1}}$ is also an integer.
Thus, by PMI, the result is true for every $n \geq 0$.   ∎

It follows from this example that $\frac{(nr)!}{(r!)^n}$ is also an integer. This was proposed as a problem in 1902 by J. W. Young, a graduate student at Cornell University.

Next we will show that each CBC is twice its northwestern (or northeastern) neighbor. We have

$$\binom{2n}{n} = \frac{(2n)!}{n!n!}$$

$$= \frac{2n}{n} \cdot \frac{(2n-1)!}{(n-1)!n!}$$

$$= 2\binom{2n-1}{n-1}$$

where $n \geq 1$.

This is essentially an application of the identity $\binom{n}{r} = \frac{n}{r}\binom{n-1}{r-1}$.

For example,

$$\binom{10}{5} = 252$$

$$= 2 \cdot 126$$

$$= 2\binom{9}{4}$$

The next example shows that $n + 1 \mid \binom{2n}{n}$, a property already established in Corollary 1.2. We will use this fact throughout our discussion.

Example 2.2 Prove that $(n+1) \mid \binom{2n}{n}$, where $n \geq 0$.

Proof   Notice that

$$\binom{\binom{2n}{n} - 2n}{n-1} = \frac{(2n)!}{(n!)^2} - \frac{(2n)!}{(n-1)!(n+1)!}$$

$$= \frac{(2n)!}{n!(n+1)!}$$

$$= \frac{1}{n+1}\binom{2n}{n}$$

By Theorem 1.3, the LHS is an integer, so $\frac{1}{n+1}\binom{2n}{n} = C_n$ is an integer. In other words, $n + 1 \mid \binom{2n}{n}$. ∎

For example, let $n = 12$. Then

$$\binom{2n}{n} = \binom{24}{12} = 2{,}704{,}156$$

Notice that $13 \mid 2{,}704{,}156$.

Alternate Proofs

We now present an alternate proof of this result, based on the one given in 1988 by R. J. Clarke of the University of Adelaide.

We have

$$\binom{2n+1}{n} = \frac{2n+1}{n+1}\binom{2n}{n}$$

Since $n+1$ and $2n+1$ are relatively prime, it follows that $n+1 \mid \binom{2n}{n}$.

The next alternate proof is due to L. Takács of Case Western Reserve University, Cleveland, Ohio:

$$\binom{2n}{n} = \frac{n+1}{2(2n+1)}\binom{2n+2}{n+1}$$

Because $(n+1, 2n+1) = 1$ and $\binom{2n+2}{n+1}$ is even, it follows that $n+1 \mid \binom{2n}{n}$.

Here is yet another proof, given in 1988 by Guy. Let $C_n = \frac{1}{n+1}\binom{2n}{n}$. Then

$$C_n = (2n+1)C_n - 2nC_n$$

$$= \frac{2n+1}{n+1}\binom{2n}{n} - \frac{2n}{n+1}\binom{2n}{n}$$

$$= \frac{(2n+1)!}{(n+1)!n!} - 2 \cdot \frac{(2n)!}{(n+1)!(n-1)!}$$

$$= \binom{2n+1}{n+1} - 2\binom{2n}{n+1}$$

Since the RHS is an integer, it follows that every Catalan number C_n is also an integer; in other words, $n+1 \mid \binom{2n}{n}$.

Wahlin's Proof

In 1911, G. E. Wahlin of the University of Illinois presented a proof that $n+1 \mid \binom{2n}{n}$. His proof, basically the same as Clarke's, appeared as a solution to the problem proposed in 1910 by Feemster. It invokes Corollary 1.1 twice:

$$\frac{1}{n+1}\binom{2n}{n} = \frac{(2n)!}{(n+1)!n!}$$

$$= \frac{(2n)(2n-1)(2n-2)\cdots(n+1)}{(n+1)!}$$

By Corollary 1.1, $N = \frac{(2n)(2n-1)\cdots(n+1)}{n!}$ is an integer. For the same reason, $N' = \frac{(2n+1)(2n)(2n-1)\cdots(n+1)}{(n+1)!}$ is also an integer.

But $N' = \frac{2n+1}{n+1}N$. Since $(2n+1, n+1) = 1$, it follows that $\frac{N}{n+1} = \frac{1}{n+1}\binom{2n}{n}$ is an integer.

In 1910, Feemster established a generalization of this result: $\frac{(2m)!(2n)!}{m!n!(m+n)!}$ is an integer. Wahlin provided a proof of this also in 1991. He added that a proof can be found in *Die Elemente der Zahlen Theorie* by German mathematician Paul Gustav Heinrich Bachmann (1837–1920).

Interestingly, in 2000 D. Callan of the University of Wisconsin at Madison discovered three similar binomial coefficients. They are:

$$\frac{3}{n}\binom{2n}{n-3}, \frac{3}{n}\binom{3n}{n+1}, \text{ and } \frac{4}{(3n+1)(3n+2)}\binom{3n+2}{n}$$

Largest Exponent of a Prime Factor of the CBC

Next we would like to find the largest exponent e of a prime p in the prime factorization of the CBC. First, we make a definition.

The *floor* of a real number x, denoted by $\lfloor x \rfloor$, is the largest integer $\leq x$. For example, $\lfloor -3.45 \rfloor = -4, \lfloor -3 \rfloor = -3$, and $\lfloor 3.45 \rfloor = 3$.

Now, we make an important observation. Let p be a prime $\leq m$. Then

$$m! = \cdots p \cdots (2p) \cdots (p \cdot p) \cdots m$$

Clearly, p is a factor of every pth number on the RHS, so p occurs $\left\lfloor \frac{m}{p} \right\rfloor$ times in the prime factorization of $m!$. But, each of the factors $p^2, p^3, \cdots$ contributes an additional factor. There are $\left\lfloor \frac{m}{p^2} \right\rfloor$ such factors. Likewise, there are $\left\lfloor \frac{m}{p^3} \right\rfloor$ factors that contribute an additional factor p in the prime factorization. Continuing like this, the total number of times p is a factor of $m!$ is $\sum_{i \geq 1} \left\lfloor \frac{m}{p^i} \right\rfloor$. That is,

$$e = \sum_{i \geq 1} \left\lfloor \frac{m}{p^i} \right\rfloor \tag{2.1}$$

Let k be the largest exponent of p such that $p^k \leq m$, so $p^{k+1} > m$. Then $\left\lfloor \frac{m}{p^{k+1}} \right\rfloor = 0$. Consequently, the sum (2.1) is finite. Thus

$$e = \sum_{i=1}^{k} \left\lfloor \frac{m}{p^i} \right\rfloor \tag{2.2}$$

where $k = \lfloor \log_p m \rfloor$. We will employ this formula in Lemma 2.2.

We are now ready to compute the largest exponent e such that $p^e \mid \binom{2n}{n}$.

Since $\binom{2n}{n} = \frac{(2n)!}{n!n!}$, the largest power of p that divides $(2n)!$ is $\sum_{i=1}^{k} \left\lfloor \frac{2n}{p^i} \right\rfloor$ and that of p that divides $n!$ is $\sum_{i=1}^{k} \left\lfloor \frac{n}{p^i} \right\rfloor$, where $k = \lfloor \log_p 2n \rfloor$. Consequently, the largest power of p that divides $\frac{(2n)!}{n!n!}$ is

$$e = \sum_{i=1}^{k} \left\lfloor \frac{2n}{p^i} \right\rfloor - \sum_{i=1}^{k} \left\lfloor \frac{n}{p^i} \right\rfloor - \sum_{i=1}^{k} \left\lfloor \frac{n}{p^i} \right\rfloor$$

$$= \sum_{i=1}^{k} \left\lfloor \frac{2n}{p^i} \right\rfloor - 2 \sum_{i=1}^{k} \left\lfloor \frac{n}{p^i} \right\rfloor$$

For example, let $n = 12$ and $p = 7$. Then

$$\binom{2n}{n} = \binom{24}{12} = 2{,}704{,}156$$

and $k = \lfloor \log_7 24 \rfloor = 1$. The largest exponent e of 7 such that $7^e \mid \binom{24}{12}$ is given by

$$e = \sum_{i=1}^{1} \left\lfloor \frac{24}{7^i} \right\rfloor - 2 \sum_{i=1}^{1} \left\lfloor \frac{12}{7^i} \right\rfloor = 3 - 2 = 1$$

Notice that $7 \mid 2{,}704{,}156$, but $7^2 \nmid 2{,}704{,}156$.

When $n = 14$ and $p = 5$, $k = \lfloor \log_5 28 \rfloor = 2$. So, the largest exponent e of 5 such that $5^e | \binom{28}{14}$ is given by

$$
\begin{aligned}
e &= \sum_{i=1}^{2} \left\lfloor \frac{28}{5^i} \right\rfloor - 2 \sum_{i=1}^{2} \left\lfloor \frac{14}{5^i} \right\rfloor \\
&= (5 + 1) - 2(2 + 0) \\
&= 2
\end{aligned}
$$

Again, notice that $5^2 | 40{,}116{,}600$, but $5^3 \nmid 40{,}116{,}600$, where $40{,}116{,}600 = \binom{28}{14}$.

The next theorem, inspired by the fact that $C_n = \frac{(2n)!}{n!(n+1)!}$ is an integer for $n \geq 0$, was proposed as a problem in 1986 by H. D. Ruderman of Lehman College, The Bronx, New York. The proof is based on the one given in the following year by B. Poonen, then a student at Harvard University.

The proof requires two simple lemmas, and they both employ the *floor function*. We begin with the two lemmas.

Lemma 2.1 Let x be a real number ≥ 0. Then

$$
\lfloor 3x \rfloor = \lfloor x \rfloor + \lfloor x + 1/3 \rfloor + \lfloor x + 2/3 \rfloor
$$

Proof   Since $\lfloor x + n \rfloor = \lfloor x \rfloor + n$, it suffices to prove the result for $0 \leq x < 1$, where n is an integer.

Case 1   Let $0 \leq x < 1/3$, so $0 \leq 3x < 1$. Then

$$
\text{RHS} = 0 + 0 + 0 = 0 = \text{LHS}
$$

Case 2   Let $1/3 \leq x < 2/3$, so $1 \leq 3x < 2$. Then

$$
\text{RHS} = 0 + 0 + 1 = 1 = \text{LHS}
$$

Case 3   Let $2/3 \leq x < 1$, so $2 \leq 3x < 3$. Then

$$
\text{RHS} = 0 + 1 + 1 = 2 = \text{LHS}
$$

Thus, the result is true for every real number $x \geq 0$.   ∎

Lemma 2.2 Let p be a prime and m a positive integer such that $p^m \geq 3$. Then

$$
\left\lfloor \frac{3n}{p^m} \right\rfloor \geq \left\lfloor \frac{n}{p^m} \right\rfloor + \left\lfloor \frac{n+1}{p^m} \right\rfloor + \left\lfloor \frac{n+2}{p^m} \right\rfloor
$$

Proof   This follows by letting $x = \frac{n}{p^m}$ in Lemma 2.1.   ∎

We are now ready to present the theorem.

Theorem 2.1 Let $n \geq 3$. Then

$$R_n = \frac{(3n)!}{n!(n+1)!(n+2)!}$$

is an integer.

Proof   Let e be the largest power of an arbitrary prime p such that $p^e | R_n$. Then

$$e = \sum_{m=1}^{\infty} \left(\lfloor 3n/p^m \rfloor - \lfloor n/p^m \rfloor - \lfloor (n+1)/p^m \rfloor - \lfloor (n+2)/p^m \rfloor \right)$$

It suffices to show that $e \geq 0$.

If $p \neq 2$, then $e \geq 0$ by Lemma 2.2.

Suppose $p = 2$. Let k be an integer such that $n + 2 < 2^k \leq 3n$, where $n \geq 3$. Then, by Lemma 2.2, we have

$$e = \sum_{m=1}^{\infty} \left(\lfloor 3n/2^m \rfloor - \lfloor n/2^m \rfloor - \lfloor (n+1)/2^m \rfloor - \lfloor (n+2)/2^m \rfloor \right)$$

$$\geq \lfloor 3n/2 \rfloor - \lfloor n/2 \rfloor - \lfloor (n+1)/2 \rfloor - \lfloor (n+2)/2 \rfloor + \lfloor 3n/2^k \rfloor$$

$$\geq f(n)$$

where

$$f(n) = \lfloor 3n/2 \rfloor - \lfloor n/2 \rfloor - \lfloor (n+1)/2 \rfloor - \lfloor (n+2)/2 \rfloor + 1$$

$$= \lfloor 3n/2 \rfloor - \lfloor n/2 \rfloor - \lfloor (n+1)/2 \rfloor - \lfloor n/2 \rfloor$$

$$= \lfloor 3n/2 \rfloor - 2 \lfloor n/2 \rfloor - \lfloor (n+1)/2 \rfloor$$

Notice that $f(0) = 0 - 2 \cdot 0 - 0 = 0$ and $f(1) = 1 - 2 \cdot 0 - 0 - 1 - 0$, so $f(0) = 0 = f(1)$. Furthermore,

$$f(n+2) = \lfloor 3(n+2)/2 \rfloor - 2 \lfloor (n+2)/2 \rfloor - \lfloor (n+3)/2 \rfloor$$

$$= (\lfloor 3n/2 \rfloor + 3) - 2 (\lfloor n/2 \rfloor + 1) - (\lfloor (n+1)/2 \rfloor + 1)$$

$$= \lfloor 3n/2 \rfloor - 2 \lfloor n/2 \rfloor - \lfloor (n+1)/2 \rfloor$$

$$= f(n)$$

Thus, by induction, $f(n) = 0$ for every integer $n \geq 0$.

Consequently, $e \geq 0$ always. Hence R_n is an integer, as desired.   ∎

Paul Erdös (pronounced *air-dish*) (1913–1996), the greatest Hungarian mathematician and one of the greatest mathematicians of the twentieth century, was born in Bucharest. His parents taught high school mathematics and physics; his father had spent six years in a Siberian prison. Young Erdös was home-taught, mostly by his father, except for three years in school.

A child prodigy, Erdös discovered negative numbers for himself at the age of three. At seventeen, he entered Eötvös University. Three years later, he wrote a delightful proof of Chebychev's theorem that there is a prime number between the positive integers n and $2n$. Erdös received his Ph.D. at the age of twenty-one.

One of the most prolific writers in mathematics, Erdös authored about 1500 articles and coauthored about 500. In a tribute in 1983, E. Straus described Erdös as "the prince of problem-solvers and the absolute monarch of problem-posers" Also called the "Euler of our time," Erdös wrote extensively in number theory, combinatorics, the theory of functions, complex analysis, set theory, group theory, and probability. Number theory and combinatorics were, by far, his favorites.

(continued)

> Erdös regarded worldly possessions "as a nuisance." Straus writes that he never owned a home, car, or a bank account, and never had a family or an address. Always searching for mathematical truths, he traveled from meeting to meeting carrying a half-empty suitcase. He stayed with mathematicians wherever he went, and donated his honorariums as prizes for students.
>
> A receipient of numerous honors, Erdös died of a massive heart attack while attending a mathematics conference in Warsaw, Poland.

For example,

$$R_5 = \frac{15!}{5!6!7!} = 3{,}003$$

is an integer.

Although the Catalan number C_n counts a number of things, as we shall see in Chapters 4–11, it is not yet known if R_n counts anything.

For the curious-minded, we add that $\frac{(4n)!}{n!(n+1)!(n+2)!(n+3)!}$ is an integer, except when n is a *Mersenne number*, which is an integer of the form $2^k - 1$. This was established in 1987 by John O. Kiltinen of Northern Michigan University, Marquette.

Returning to the CBC, although $n + 1 \mid \binom{2n}{n}$, we find that the number of occurrences of n dividing $\binom{2n}{n}$ seems to be quite rare, according to Hungarian mathematician P. Erdös, and American mathematician R. L. Graham of AT&T Research Labs and now of the University of San Diego. For example, $4 \mid \binom{6}{3}$, but $3 \nmid \binom{6}{3}$; and $3 \mid \binom{4}{2}$ and $2 \mid \binom{4}{2}$. In 1947, Erdös established that for every positive integer k, there are infinitely many ns such that $\frac{(2n)!}{n!(n+k)!}$ is an integer.[†]

Let p be a prime such that $n < p < 2n$. Then $p \mid (2n)!$, but $p \nmid n!$. So $p \mid \binom{2n}{n}$; in other words, $\binom{2n}{n}$ is divisible by every prime in the open interval $(n, 2n)$; so it is divisible by the product of all primes in the same range.

For example, let $n = 10$. Then $\binom{2n}{n} = \binom{20}{10} = 184{,}756$. There are four primes in the range $10 < p < 20$, namely, 11, 13, 17, and 19. Notice that $11! \mid 184756$, $13 \mid 184756$, $17 \mid 184756$, $19 \mid 184756$, and $11 \cdot 13 \cdot 17 \cdot 19 \mid 184756$. Notice also that $184{,}756 = 4 \cdot 11 \cdot 13 \cdot 17 \cdot 19$.

[†] See *American Mathematical Monthly* **56** (Jan. 1949), 42–43.

Ronald L. Graham, chief scientist at AT&T Research Labs, also an expert juggler, and a master ping-pong player, was born in Taft, California, where his father worked in the oil fields. Since his father moved many times, young Graham never stayed longer than two years at any school; nonetheless, he developed a prodigious aptitude for mathematics and science.

At the age of fifteen, Graham received a Ford Foundation scholarship to pursue a program for gifted students at the University of Chicago. After three years, his father, claiming that the university was too leftist, convinced the son to transfer to the University of California at Berkeley, where he majored in electrical enginnering. After a year, he enlisted in the air force to avoid a military draft and was sent to Alaska, where he worked as a communications specialist at night and attended classes full-time during the day at the University of Alaska at Fairbanks. After his discharge, Graham returned to Berkeley and received his Ph.D. in number theory.

In 1962, Graham joined the Bell Telephone Laboratories, Murray Hill, New Jersey, and pursued his own research. His outstanding contributions to number theory and other areas of mathematics earned him the coveted Polya Prize in 1972 and membership in the National Academy of Sciences in 1985. A member of the editorial board of forty mathematics and computer science journals, he served on the high-profile National Research Council committee on cryptography during the years 1994–1996.

(continued)

Graham met Erdös in 1963, and Erdös played a significant role in Graham's mathematical thinking. The two collaborated on numerous articles, the first one appearing in 1972. Graham kept an "Erdös room" in his house for his mentor.

In 1991, the late Gian-Carlo Rota of Massachusetts Institute of Technology nominated Graham for president of the American Mathematical Society. Rota characterized Graham as "one of the charismatic figures in contemporary mathematics, as well as the leading problem-solver of his generation. For the last twenty-five years, he has been the central figure in the development of discrete mathematics. His seminal work has led to the birth of at least three new branches of mathematics: Ramsey theory, computational geometry, and worst-case analysis of multiprocessing algorithms."

Conway's Generalization

In 1988, John Conway, an outstanding English mathematician at Princeton University, communicated to Guy the following generalization of Example 2.2:

$$m!n! \mid (m,n)(m+n-1)!$$

where (m,n) denotes the greatest common divisor (gcd) m and n. Its proof employs the following facts:

- The gcd of two positive integers is a positive integer.
- Every binomial coefficient is an integer.
- $\frac{m(m+n-1)!}{m!n!} = \binom{m+n-1}{m-1}$
- $\frac{n(m+n-1)!}{m!n!} = \binom{m+n-1}{n-1}$
- $(ab, ac) = a(b,c)$

So

$$\left(\frac{m(m+n-1)!}{m!n!}, \frac{n(m+n-1)!}{m!n!} \right) = \left(\binom{m+n-1}{m-1}, \binom{m+n-1}{n-1} \right)$$

Since the RHS is an integer, it follows that $\frac{(m+n-1)!}{m!n!}(m,n)$ is also an integer. In other words, $m!n! \mid (m,n)(m+n-1)!$. ∎

For example, let $m = 6$ and $n = 9$; so $(m,n) = (6,9) = 3$. Then

$$\frac{(m,n)(m+n-1)!}{m!n!} = \frac{3 \cdot 14!}{6!9!}$$
$$= 1,001$$

is an integer, as expected.

The next example[†] provides an interesting application of Feemster's result.

Example 2.3 Let m and n be positive integers. Show that

$$\frac{\left[\binom{2m}{m}\binom{2n}{n}\right]^2}{\binom{m+n}{m}}$$

is an integer.

Proof   We have

$$\frac{\left[\binom{2m}{m}\binom{2n}{n}\right]^2}{\binom{m+n}{m}} = \frac{(2m)!(2m)!(2n)!(2n)!m!n!}{m!m!m!m!n!n!n!n!(m+n)!}$$

$$= \binom{2m}{m}\binom{2n}{n}\frac{(2m)!(2n)!}{m!n!(m+n)!}$$

So the LHS is an integer if and only if $\frac{(2m)!(2n)!}{m!n!(m+n)!}$ is an integer. By Feemster's result, $\frac{(2m)!(2n)!}{m!n!(m+n)!}$ is an integer. Thus, the value of the given expression is an integer.   ∎

We add that $\frac{(2a)!(2b)!(2c)!}{a!b!c!(a+b+c)!}$ need *not* be an integer.

A Recurrence Relation for the CBC

Next, we will derive a recurrence relation for the CBC. To this end, let $A(n) = \binom{2n}{n}$. Then

$$A_{n+1} = \binom{2n+2}{n+1}$$

$$= \frac{(2n+2)!}{(n+1)!(n+1)!}$$

$$= \frac{2(2n+1)}{n+1} \cdot \frac{(2n)!}{n!n!}$$

$$= \frac{2(2n+1)}{n+1} A_n$$

Thus we have the recurrence relation

$$(n+1)A_{n+1} = 2(2n+1)A_n$$

where $A_0 = 1$.

[†] G. Klambauer, *Problems and Propositions in Analysis*, Marcel Dekker, New York, 1979.

A Generating Function for the CBC

We are now in a position to construct a generating function for the CBC A_n, where $n \geq 0$. We will let differential and integral calculus do the job for us, as J. L. Brown Jr. of Pennsylvania State University and V. E. Hoggatt Jr. of San Jose State University did in 1978.

Let

$$A(x) = \sum_{n=0}^{\infty} A_n x^n$$

$$= A_0 + \sum_{n=0}^{\infty} A_{n+1} x^{n+1}$$

Differentiating both sides with respect to x,

$$A'(x) = \sum_{n=0}^{\infty} (n+1) A_{n+1} x^n$$

$$x A'(x) = \sum_{n=0}^{\infty} (n+1) A_{n+1} x^{n+1}$$

$$= \sum_{n=0}^{\infty} n A_n x^n$$

Using the above recurrence relation,

$$A'(x) = \sum_{n=0}^{\infty} (n+1) A_{n+1} x^n$$

$$= \sum_{n=0}^{\infty} 2(2n+1) A_n x^n$$

$$= 2 \left[2 \sum_{n=0}^{\infty} n A_n x^n + \sum_{n=0}^{\infty} A_n x^n \right]$$

$$= 2[2x A'(x) + A(x)]$$

$$= 4x A'(x) + 2A(x)$$

$$(1 - 4x) A'(x) = 2A(x)$$

So

$$\frac{A'(x)}{A(x)} = \frac{2}{1 - 4x}$$

Integrating both sides with respect to x, this yields

$$\ln A(x) = \ln \frac{1}{\sqrt{1 - 4x}} + \ln C$$

$$A(x) = \frac{C}{\sqrt{1 - 4x}}$$

Since $A(0) = A_0 = 1$, $C = 1$. Thus

$$A(x) = \frac{1}{\sqrt{1 - 4x}}$$

That is,

$$\frac{1}{\sqrt{1 - 4x}} = \sum_{n=0}^{\infty} \binom{2n}{n} x^n$$

as desired.

We will return to this generating function in Chapter 5, which deals with Catalan numbers.

The Binomial Theorem

Binomial coefficients and Pascal's identity play a significant role in the development of the binomial theorem, as we will see shortly. The binomial theorem can be employed to develop a vast array of combinatorial identities.

The binomial expansion for $n = 2$ can be found in Euclid's work (c. 300 B.C). There is strong evidence that the prominent Indian mathematician-astronmer Aryabhatta was familiar with the binomial theorem for $n = 2$ and $n = 3$. The binomial theorem for positive integral exponents was known to Persian poet and mathematician Omar Khayyám.

Outstanding English astronomer, mathematician, and physicist Sir Isaac Newton showed how to expand $(1 + x)^n$ directly without knowing the expansion of $(1 + x)^{n-1}$. He accomplished this by showing that each coefficient can be computed from the preceding one using the formula $\binom{n}{r} = \frac{n-r+1}{r}\binom{n}{r-1}$, where $r \geq 1$. This formula is called *Cardano's rule of succession*, after Italian mathematician Girolano Cardano, a doctor by profession and "one of the most extraordinary characters in the history of mathematics," according to mathematical historian Howard Eves of the University of Maine.

Aryabhatta (c. 476–c. 550), the first prominent Indian astronomer-mathematician, was born in Kusumapura, near Patna on the Ganges. He studied at Nalanda University, and later became its head. Aryabhatta used mathematics to solve astronomical problems and was interested in Diophantus's work on indeterminate equations, and Indian astronomer Parasar's work on comets and planetary motion. He described the Earth as spherical and computed its diameter as 7980 miles. He understood the nature of eclipses and that the sun was the source of moonlight, both ideas unknown to the West until Copernicus and Galileo, a thousand years later. Aryabhatta's accurate astronomical calculations led to the development of a calendar in India. He developed the expansions of $(x+y)^2$ and $(x+y)^3$, and formulas for computing square roots and cube roots. Around 500, he calculated the value of π as $62832/20000 = 3.1416$, more accurate than previously known.

His masterpiece, *The Aryabhatia*, written in 499, deals with astronomy, plane and spherical geometry, algebra, quadratic equations, sums of powers of the first n natural numbers, and a table of sine values. It was translated into Arabic around 800 and into Latin in the thirteenth century.

India's first space satellite was named *Aryabhata*, in recognition of his contributions to astronomy and mathematics.

Theorem 2.1 (*The Binomial Theorem*) Let x and y be arbitrary real numbers, and n an arbitrary nonnegative integer. Then

$$(x+y)^n = \sum_{r=0}^{n} \binom{n}{r} x^{n-r} y^r$$

Proof (*by PMI*) Let $P(n)$ denote the given statement. When $n = 0$, LHS $= (x+y)^0 = 1$ and RHS $= \sum_{r=0}^{0} \binom{0}{r} x^{0-r} y^r = x^0 y^0 = 1$, so LHS = RHS. Thus the formula works when $n = 0$.

Assume $P(k)$ is true for an arbitrary integer $k \geq 0$:

$$(x+y)^k = \sum_{r=0}^{k} \binom{k}{r} x^{k-r} y^r$$

Then

$$(x+y)^{k+1} = (x+y)^k (x+y)$$

$$= \left[\sum_{r=0}^{k} \binom{k}{r} x^{k-r} y^r \right] (x+y)$$

Omar Khayyám (1048–1131) was born in Nishapur, Persia (now Iran). His name literally translated means "tent maker,"indicating that tent-making was the trade of his father Ibrahim or other ancestors. The political chaos of the time played a significant role in Khayyám's life. He describes the difficulties learned men faced during this uncertain period in his *Treatise on Demonstration of Problems of Algebra*:

> I was unable to devote myself to the learning of this algebra and the continued concentration upon it, because of obstacles in the vagaries of time which hindered me; for we have been deprived of all the people of knowledge save for a group, small in number, with many troubles, whose concern in life is to snatch the opportunity, when time is asleep, to devote themselves meanwhile to the investigation and perfection of a science; for the majority of people who imitate philosophers confuse the true with the false; and they do nothing but deceive and pretend knowledge, and they do not use what they know of the sciences except for base and material purposes; and if they see a certain person seeking for the right and preferring the truth, doing his best to refute the false and untrue and leaving aside hypocrisy and deceit, they make a fool of him and mock him.

Despite these unfavorable circumstances, Khayyám became an outstanding mathematician and astronomer. He wrote several books, including *Problems of Arithmetic*, a book on music, and one on algebra, before he was twenty-five years old.

About 1070, Khayyám moved to Samarkand, Uzbekistan. There he was supported by the chief justice of Samarkand, which enabled him write his famous algebra book, *Treatise on Demonstration of Problems of Algebra*. This book contains geometric solutions of cubic equations using conic sections. Khayyám realized that a cubic equation can have more than one solution; in fact, he demonstrated the existence of equations with two solutions.

Khayyám was invited to set up an observatory in Isfahan, where he stayed for eighteen years. This was the most peaceful period of his life, when he made substantial contributions to astronomy. For example, he computed the number of days in a year as 365.24219858156, very close to the number of days in the Gregorian calendar.

In 1077, he wrote his commentaries on the postulates in Euclid's *Elements*, contributing to noneuclidean geometry. He discusses Pascal's triangle in one of his books, although al-Karaji had discussed it earlier. He had a method for extracting the nth roots based on the binomial theorem, and was familiar with binomial coefficients.

(continued)

In 1092, political events halted funding for the observatory, ending Khayyám's serene life and his calendar reform of 1079. Orthodox Muslims attacked his faith and his religious free thinking.

In addition to his scientific achievements, Khayyám is well known for the *Rubaiyat*, a collection of 600 four-line poems.

$$= \sum_{r=0}^{k} \binom{k}{r} x^{k+1-r} y^r + \sum_{r=0}^{k} \binom{k}{r} x^{k-r} y^{r+1}$$

$$= \left[\binom{k}{0} x^{k+1} + \sum_{r=1}^{k} \binom{k}{r} x^{k+1-r} y^r \right] + \left[\sum_{r=0}^{k-1} \binom{k}{r} x^{k-r} y^{r+1} + \binom{k}{k} y^{k+1} \right]$$

$$= \binom{k+1}{0} x^{k+1} + \sum_{r=1}^{k} \binom{k}{r} x^{k+1-r} y^r + \sum_{r=1}^{k} \binom{k}{r-1} x^{k+1-r} y^r + \binom{k+1}{k+1} y^{k+1}$$

$$= \binom{k+1}{0} x^{k+1} + \sum_{r=1}^{k} \left[\binom{k}{r} + \binom{k}{r-1} \right] x^{k+1-r} y^r + \binom{k}{r} y^{k+1}$$

$$= \binom{k+1}{0} x^{k+1} + \sum_{r=1}^{k} \binom{k+1}{r} x^{k+1-r} y^r + \binom{k+1}{k+1} y^{k+1}$$

$$= \sum_{r=0}^{k+1} \binom{k+1}{r} x^{k+1-r} y^r$$

Thus, by induction, the formula is true for every integer $n \geq 0$. ∎

For example,

$$(3a - 2b)^4 = \binom{4}{0}(3a)^4(-2b)^0 + \binom{4}{1}(3a)^3(-2b)^1 + \binom{4}{2}(3a)^2(-2b)^2$$

$$+ \binom{4}{3}(3a)^1(-2b)^3 + \binom{4}{4}(3a)^0(-2b)^4$$

$$= (3a)^4 + 4(3a)^3(-2b) + 6(3a)^2(-2b)^2 + 4(3a)(-2b)^3 + (-2b)^4$$

$$= 81a^4 - 216a^3 b + 216a^2 b^2 - 96ab^3 + 16b^4$$

Isaac Newton (1642–1727) was born into a farming family in Woolsthorpe, England, on Christmas day, 1642. His father had died a few weeks earlier.

His early education was at the dame schools at Skillington and Stoke and at King's School in Grantham. In 1656, his mother called him home to learn farming. However, uninterested in farm chores, he showed great skill and passion for solving problems, conducting experiments, and developing mechanical models. For example, he constructed a model of a mill, powered by a mouse, that ground wheat into flour and a wooden clock run by water.

In June 1661, Newton entered Trinity College, Cambridge, an environment that encouraged his interests and skills. In October, he happened to pick up a book on astrology at the Stourbridge Fair, which he could not understand because of references to geometry and trigonometry. He read Euclid's *Elements*, which he found to be obvious; and then René Descartes's *La Géométrie*, which he found a bit difficult. He also read William Oughtred's *Clavis*, Johann Kepler's *Optics*, François Viète's works, Frans van Schoonten's *Miscellanies*, and John Wallis's *Arithmetica Infinitorum*. These readings moved him to pursue mathematics rather than chemistry. He received his B.A. from Cambridge in 1665.

The same year, Newton discovered the binomial theorem and laid the foundation for the theory of fluxions, now called differential calculus. Since the university was closed for part of 1665 and 1666 due to the plague, Newton lived at home, where he made several spectacular discoveries. He developed his calculus and applied it to the theory of equations; formulated the fundamental principles of his theory of gravitation; and performed his first experiments in optics. Interestingly, three important discoveries—the

(*continued*)

theory of gravitation, the nature of light and color, and calculus—occurred in a span of eighteen months, from 1665 to 1666.

Although Newton shared his results with his friends and students, they were not published until many years later. The delay resulted in an unfortunate dispute with great German mathematician Gottfried Wilhelm Leibniz (1646–1716) about who had first invented the calculus.

In 1669, at the age of twenty-six, Newton was appointed Lucasian professor, succeeding Isaac Barrow (1630–1677) who had resigned his professorship for Newton. He continued his work in optics. In 1675, Newton communicated his corpuscular theory of light to the Royal Society, where he had been a fellow since 1672.

Newton's greatest work, *Philosophiae Naturalis Principia Mathematica* (Mathematical Principles of Natural Philosophy), popularly known as *Principia*, continues to be the most influential work in the history of science. It includes Newton's laws of motion and the theory of gravitation, and provides the first unified system of scientific principles about terrestrial and celestial motion. Newton's spectacular discoveries in optics appear in *Opticks*, published in 1704. This book includes the discovery that white light is made up of all colors. His *Arithmetica Universalis*, published in 1707, contains his lectures from 1673 to 1683 and many important contributions to the theory of equations. *The Method of Fluxions*, although written in 1671, was not published until 1736. In fact, all his works, except *Principia*, were published years after he developed the theories in them; nearly all of them appeared in print only because his friends insisted.

In 1703, Newton was elected president of the Royal Society, a position he held until his death. In 1705, he was knighted by Queen Anne.

A superb analyst and experimentalist, with remarkable powers of concentration, Newton is ranked among the greatest mathematical geniuses. His insight into physical problems and his ability to analyze them mathematically remain unparalleled. On one occasion when he was praised for his greatness, he remarked that if he had seen farther than other men, it was only because he was able to stand on the shoulders of giants.

Newton died at the age of eighty-four after a long and painful illness. He was buried in Westminister Abbey.

The binomial theorem can be employed to derive a host of *combinatorial identities*, that is, identities involving binomial coefficients, as the next two corollaries demonstrate.

Corollary 2.1

$$\sum_{r=0}^{n} \binom{n}{r} = 2^n$$

That is, the sum of the binomial coefficients is 2^n.

Girolamo Cardano (1501–1576), a man of extraordinary gifts and versatility, as well as great contradictions, was born in Pavis, Italy, the illegitimate son of a Milanese lawyer. He studied at the Universities of Pavia and Padua. A doctor by profession, he practiced medicine at Sacco and Milan from 1524 to 1550; during this period, he studied, taught, and wrote his works.

(continued)

After spending about a year in France, Scotland, and England, he returned to Milan and became professor of science at the University of Pavia and then at the University of Bologna in 1562. Eight years later, he was imprisoned briefly for publishing a horoscope of Christ. The following year, he resigned his chair at Bologna, moved to Rome, and became the most distinguished astrologer at the papal court, receiving pensions for his services. Cardano took his own life in 1576 to fulfill his astrological prediction of the date of his death.

Cardano authored a number of works on arithmetic, astronomy, physics, and medicine. His principal mathematical work is *Ars Magna*, published at Nuremberg in 1545. It was the first great Latin book devoted solely to algebra.

Proof   The proof follows by letting $x = 1 = y$ in the binomial theorem.   ∎

For example,

$$\sum_{r=0}^{5} \binom{5}{r} = \binom{5}{0} + \binom{5}{1} + \binom{5}{2} + \binom{5}{3} + \binom{5}{4} + \binom{5}{5}$$

$$= 1 + 5 + 10 + 10 + 5 + 1$$

$$= 2^5$$

Corollary 2.1 has a fascinating set-theoretic interpretation. Since an n-element set has $\binom{n}{r}$ r-element subsets, where $0 \leq r \leq n$, it follows that an n-element has a total of $\sum_{r=0}^{n} \binom{n}{r} = 2^n$ subsets.

Next we present an interesting byproduct of Corollary 2.1.

Parity of the CBC Revisited

Earlier in the chapter, we established that the CBC $\binom{2n}{n}$ is even when $n \geq 1$. We now reconfirm it using the elementary proof given in 1988 by L. Takács of Case Western Reserve University.

By Corollary 2.1, we have

$$\sum_{r=0}^{2n} \binom{2n}{r} = 2^{2n}$$

That is,

$$\sum_{r=0}^{n-1}\binom{2n}{r} + \binom{2n}{n} + \sum_{r=n+1}^{2n}\binom{2n}{r} = 2^{2n}$$

$$\sum_{r=0}^{n-1}\binom{2n}{r} + \binom{2n}{n} + \sum_{r=0}^{n-1}\binom{2n}{2n-r} = 2^{2n}$$

$$\sum_{r=0}^{n-1}\binom{2n}{r} + \binom{2n}{n} + \sum_{r=0}^{n-1}\binom{2n}{r} = 2^{2n}$$

So

$$\binom{2n}{n} = 2^{2n} - 2\sum_{r=0}^{n-1}\binom{2n}{r}$$

Because the RHS is an even integer, it follows that $\binom{2n}{n}$ is also an even integer, when $n \geq 1$.

Corollary 2.1 yields yet another result:

$$\sum_{n \geq 0}\frac{1}{\displaystyle\sum_{r=0}^{n}\binom{n}{r}} = \sum_{n=0}^{\infty}\frac{1}{2^n}$$

$$= \frac{1}{1 - 1/2}$$

$$= 2$$

The next corollary also follows from the binomial theorem.

Corollary 2.2 Let x be any real number. Then

$$(1 + x)^n = \sum_{r=0}^{n}\binom{n}{r}x^r$$

$\blacksquare$

Klamkin's Formula

The next example, an application of Corollary 2.2 and both differential and integral calculus, was proposed in 1954 by M. S. Klamkin of the Polytechnic Institute of

Brooklyn, New York. The solution is due to D. C. Russell of Birkbeck College, University of London, and once again contains the ubiquitous CBC.

Example 2.4 Evaluate the sum

$$\sum_{r=0}^{n}(-1)^r\left(\frac{1}{r+1}+\frac{1}{r+2}+\cdots+\frac{1}{r+n}\right)\binom{n}{r}$$

Solution   Let S denote the given sum. Then

$$S=\sum_{r=0}^{n}\sum_{s=1}^{n}\frac{(-1)^r\binom{n}{r}}{r+s}$$

$$=\sum_{s=1}^{n}\sum_{r=0}^{n}\frac{(-1)^r\binom{n}{r}}{r+s} \tag{2.3}$$

Let

$$f_s(x)=\sum_{r=0}^{n}\frac{(-1)^r\binom{n}{r}x^{r+s}}{r+s}$$

Then

$$f_s'(x)=\sum_{r=0}^{n}(-1)^r\binom{n}{r}x^{r+s-1}$$

$$=x^{s-1}\sum_{r=0}^{n}\binom{n}{r}(-x)^r$$

$$=x^{s-1}(1-x)^n \tag{2.4}$$

So

$$f_s(x)=\int_0^x t^{s-1}(1-t)^n\,dt$$

Thus, by equations (2.3) and (2.4),

$$S = \sum_{s=1}^{n} f_s(1)$$

$$= \sum_{s=1}^{n} \int_0^1 t^{s-1}(1-t)^n \, dt$$

$$= \int_0^1 (1-t)^n \sum_{s=1}^{n} t^{s-1} \, dt$$

$$= \int_0^1 (1-t)^n \left(\frac{1-t^n}{1-t} \right) dt$$

$$= \int_0^1 (1-t)^{n-1} dt - \int_0^1 t^n (1-t)^{n-1} \, dt$$

$$= \frac{1}{n} - \frac{n!(n-1)!}{(2n)!}$$

$$= \frac{1}{n} \left[1 - \binom{2n}{n}^{-1} \right]$$

That is,

$$\sum_{r=0}^{n} (-1)^r \left(\frac{1}{r+1} + \frac{1}{r+2} + \cdots + \frac{1}{r+n} \right) \binom{n}{r} = \frac{1}{n} \left[1 - \binom{2n}{n}^{-1} \right] \tag{2.5}$$

In particular, let $n = 3$. Then

$$\text{LHS} = \sum_{r=0}^{3} (-1)^r \left[\sum_{s=1}^{3} \frac{1}{r+s} \binom{3}{r} \right]$$

$$= \frac{19}{60}$$

$$= \frac{1}{3} \left[1 - \binom{6}{3}^{-1} \right]$$

$$= \text{RHS} \qquad \blacksquare$$

We add that L. Carlitz of Duke University, Durham, North Carolina, established identity (2.5) using the technique of finite differences.

The next example, proposed by E. M. Gibson of Drury College, Missouri, is related to Corollary 2.2, as we will see shortly. The solution uses PMI and is based on the one given in 1914 by Feemster. It exhibits yet another occurrence of the CBC.

Example 2.5 Evaluate the sum

$$\binom{2n-1}{0} - \binom{2n-1}{1} + \binom{2n-1}{2} - \cdots + (-1)^{n-1}\binom{2n-1}{n-1}$$

Solution   Let S_r denote the sum of the first r terms of this sum. Then:

$$S_1 = \binom{2n-1}{0} = (-1)^0\binom{2n-2}{0}$$

$$S_2 = \binom{2n-1}{0} - \binom{2n-1}{1} = (-1)^1\binom{2n-2}{1}$$

$$S_3 = \binom{2n-1}{0} - \binom{2n-1}{1} + \binom{2n-1}{2} = (-1)^2\binom{2n-2}{2}$$

More generally, we conjecture that

$$S_r = (-1)^{r-1}\binom{2n-2}{r-1} \tag{2.6}$$

Then

$$S_{r+1} = S_r + (-1)^r\binom{2n-1}{r}$$

$$= (-1)^{r-1}\binom{2n-2}{r-1} + (-1)^r\binom{2n-2}{r}$$

$$= (-1)^r\left[\binom{2n-1}{r} - \binom{2n-2}{r-1}\right]$$

$$= (-1)^r\left[\frac{(2n-1)!}{r!(2n-r-1)!} - \frac{(2n-2)!}{(r-1)!(2n-r-1)!}\right]$$

$$= (-1)^r\frac{(2n-2)!(2n-r-1)}{r!(2n-r-1)!}$$

$$= (-1)^r\binom{2n-2}{r}$$

Thus, by induction, formula (2.6) works for every partial sum. In other words,

$$S_n = (-1)^{n-1}\binom{2n-2}{n-1}$$

as required.   ∎

Notice that $S_n = (-1)^{n-1} nC_{n-1}$, where C_k denotes the kth Catalan number.

To see how this example is related to Corollary 2.2, notice that the given sum is just one-half of the terms in the expansion of $(1-1)^{2n-1}$:

$$(1-1)^{2n-1} = \sum_{r=0}^{2n-1} (-1)^r \binom{2n-1}{r}$$

That is,

$$0 = S_n + \sum_{r=n}^{2n-1} (-1)^r \binom{2n-1}{r}$$

So

$$\sum_{r=n}^{2n-1} (-1)^r \binom{2n-1}{r} = -S_n$$

$$= (-1)^n \binom{2n-2}{n-1}$$

The following example provides a delightful confluence of Corollary 2.2, generating functions, and differential calculus. It was proposed in 1989 by K. L. McAvaney of Deakin University, Victoria, Australia. The solution, based on the one given the following year by the Chico Problem Group, California State University at Chico, uses the power series expansion of the hyperbolic sine function. As a byproduct, we will see that a special case of the example marks a surprising appearance of the CBC.

Example 2.6 Let $n \geq 2$ and $k = 2, 4, \ldots, 2n - 2$. Prove that

$$\sum_{r=1}^{n} (-1)^r \binom{2n}{n+r} r^k = 0 \tag{2.7}$$

Proof   We begin with the generating function

$$g(x) = (1 - e^x)^{2n}$$

$$= \sum_{j=0}^{2n} (-1)^j \binom{2n}{j} e^{jx}$$

$$= \sum_{r=-n}^{n} (-1)^{n+r} \binom{2n}{n+r} e^{(n+r)x}$$

where we have substituted $j = n + r$. Let

$$F(x) = e^{-nx}g(x)$$
$$= e^{-nx}(1 - e^x)^{2n}$$
$$= (-1)^n \sum_{r=-n}^{n} (-1)^r \binom{2n}{n+r} e^{rx}$$

Differentiating this n times with respect to x, this yields

$$F^{(n)}(x) = (-1)^n \sum_{r=-n}^{n} (-1)^r \binom{2n}{n+r} r^k e^{rx}$$

Then

$$F^{(n)}(0) = (-1)^n (S' + S)$$

where

$$S' = \sum_{r=-n}^{n} (-1)^r \binom{2n}{n+r} r^k \quad \text{and} \quad S = \sum_{r=1}^{n} (-1)^r \binom{2n}{n+r} r^k$$

Notice that k is even, in which case $S = S'$. But

$$F(x) = e^{-nx}(1 - e^x)^{2n}$$
$$= e^{-nx}(1 - e^x)^n(1 - e^x)^n$$
$$= (e^{-x} - 1)^n(1 - e^x)^n$$
$$= (-1)^n(1 - e^{-x})^n(1 - e^x)^n$$
$$= (-1)^n(2 - \cosh^p x)$$
$$= (-1)^n 4^n \sinh^{2n}(x/2)$$

Since

$$\sinh x = x + \frac{x^3}{3!} + \frac{x^5}{5!} + \frac{x^7}{7!} + \cdots$$

it follows that the least power of x in the power series expansion of $F(x)$ is x^{2n}; consequently, $F^{(k)}(0) = 0$ for all k, where $1 \leq k \leq 2n-1$. In particular, when k is even, $2(-1)^n S = 0$; so the given sum S is zero, as desired. ∎

Suppose we let $k = 0$ in formula (2.7). Then the LHS becomes

$$S = \sum_{r=1}^{n} (-1)^r \binom{2n}{n+r}$$

We would like to compute the value of S. To this end, notice by Corollary 2.2 that

$$S = \sum_{r=0}^{2n} (-1)^r \binom{2n}{r} = 0$$

That is,

$$\sum_{r=0}^{n-1} (-1)^r \binom{2n}{r} + (-1)^n \binom{2n}{n} + \sum_{r=n+1}^{2n} (-1)^r \binom{2n}{r} = 0$$

$$\sum_{r=0}^{n-1} (-1)^r \binom{2n}{r} + (-1)^n \binom{2n}{n} + \sum_{s=1}^{n} (-1)^{n+s} \binom{2n}{n+s} = 0$$

$$(-1)^n \sum_{s=1}^{n} (-1)^s \binom{2n}{n-s} + (-1)^n \binom{2n}{n} + (-1)^n \sum_{s=1}^{n} (-1)^s \binom{2n}{n+s} = 0$$

$$(-1)^n S + (-1)^n \binom{2n}{n} + (-1)^n S = 0$$

Therefore,

$$S = \sum_{r=1}^{n} (-1)^r \binom{2n}{n+r} = -\frac{1}{2} \binom{2n}{n} \tag{2.8}$$

This result was noted in 1990 by D. Callan.

For example,

$$\sum_{r=1}^{4} (-1)^r \binom{8}{4+r} = -35 = -\frac{1}{2} \binom{8}{4}$$

and

$$\sum_{r=1}^{3} (-1)^r \binom{6}{3+r} = -10 = -\frac{1}{2} \binom{6}{3}$$

As yet another byproduct, D. Callan, R. Doucette, and H. Seiffert (Germany) proved in 1990 that

$$\sum_{r=1}^{n} (-1)^r \binom{2n}{n+r} r^{2n} = \frac{(-1)^n}{2} (2n)! \tag{2.9}$$

For example, let $n = 3$. Then

$$\sum_{r=1}^{3}(-1)^r \binom{6}{3+r} r^6 = -360$$

$$= \frac{(-1)^3}{2} 6!$$

There are several formulas that involve the CBC. The best-known formula is *Lagrange's identity*:

$$\sum_{r=0}^{n} \binom{n}{r}^2 = \binom{2n}{n}$$

We will establish this in Chapter 4.

Using Lagrange's identity, we can establish the following formula:

$$\sum_{r=0}^{n} \frac{(2n)!}{(r!)^2 (n-r)!^2} = \binom{2n}{n}^2$$

Lagrange's Identity and Random Walks

The next example is an application of Lagrange's identity to random walks. It is based on a problem proposed in 1992 by M. Klamkin and A. Liu of the University of Alberta, Edmonton, Canada. The solution is based on the one given in the following year by J. S. Sumner and K. L. Dove of the University of Tampa, Florida.

Example 2.7 A *lattice point* (x, y) on the cartesian plane is a point with integer coordinates x and y. Consider a cop, initially at the origin, and a robber, initially at the lattice point (n, n), whether $n \geq 0$; see Figure 2.2. At each lattice point (x, y), each flips a coin and decides the next move; the cop moves east or north to the next lattice point, whereas the robber moves west or south to the next point. They stay within the square lattice. Find the probability that they will meet on the square lattice.

Solution If the cop and the robber meet, they must meet on the diagonal at some lattice point $P(i, n - i)$, where $0 \leq i \leq n$. The number of paths the cop can take to each the point P from the origin is $\binom{n}{i}$ and the robber also can take $\binom{n}{n-i} = \binom{n}{i}$ different paths to reach the point P. By the addition and multiplication principles

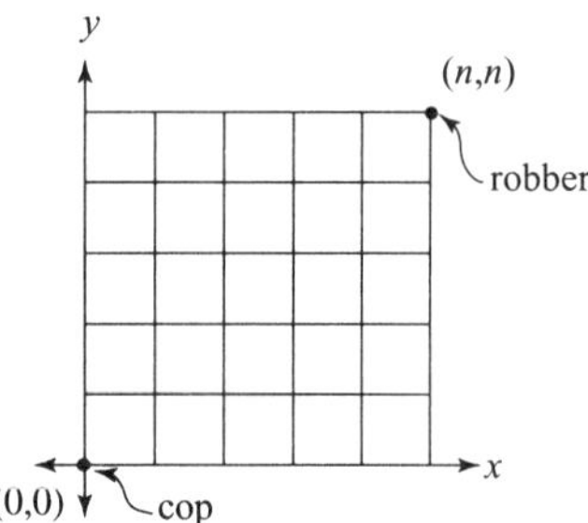

Figure 2.2 Initial Position of the
Cop and the Robber

in combinatorics, the total number of ways the two can meet is given by

$$\sum_{i=0}^{n}\binom{n}{i}\binom{n}{i} = \sum_{i=0}^{n}\binom{n}{i}^2 = \binom{2n}{n}$$

But the total number of pairs of paths equals $2^n \cdot 2^n = 2^{2n}$. So the probability that the cop and the robber will meet is given by $\frac{1}{2^{2n}}\binom{2n}{n}$ ∎

Remainder When $\binom{2p}{p}$ Is Divided by an Odd Prime p

Suppose we let p be a prime. Then, by Lagrange's identity,

$$\binom{p}{0}^2 + \sum_{r=1}^{p-1}\binom{p}{r}^2 + \binom{p}{p}^2 = \binom{2p}{p}$$

That is,

$$\sum_{r=1}^{p-1}\binom{p}{r}^2 + 2 = \binom{2p}{p}$$

But $\binom{p}{r} \equiv 0 \pmod{p}$, where $1 \le r < p$, by Corollary 1.2. Therefore,

$$\binom{2p}{p} \equiv 2 \pmod{p}$$

Thus, when $\binom{2p}{p}$ is divided by an odd prime p, the remainder is 2.

For example, let $p = 7$. Then

$$\binom{2p}{p} = \binom{14}{7}$$

$$= 3,432$$

and 3432 mod 7 = 2, as expected.

Bounds on the CBC

There are several double inequalities satisfied by the CBC $\binom{2n}{n}$. The simplest of them[†] is

$$2^n < \binom{2n}{n} < 2^{2n} \tag{2.10}$$

where $n \geq 2$. This can be established fairly easily using induction as follows.

Since $4 < \binom{4}{2} < 16$, the result is true when $n = 2$. Now, assume that it is true for an arbitrary integer $k \geq 2$:

$$2^k < \binom{2k}{k} < 2^{2k}$$

Then

$$\binom{2k+2}{k+1} = \binom{2k}{k} \frac{(2k+1)(2k+2)}{(k+1)(k+1)}$$

$$= 2 \binom{2k}{k} \frac{2k+1}{k+1}$$

Since $1 < \frac{2k+1}{k+1} < 2$ when $k \geq 2$, this implies that

$$2 \binom{2k}{k} < \binom{2k+2}{k+1} < 2 \cdot 2 \binom{2k}{k}$$

Using the inductive hypothesis, this yields

$$2 \cdot 2^k < \binom{2k+2}{k+1} < 2^2 \cdot 2^{2k}$$

[†] See D. I. A. Cohen, *Basic Techniques of Combinatorial Theory*, Wiley, New York, 1978.

That is,

$$2^{k+1} < \binom{2k+2}{k+1} < 2^{2k+2}$$

Thus, by induction, inequality (2.10) holds for every $n \geq 2$. ∎

For example, let $n = 6$. Then $2^n = 2^6 = 64, \binom{2n}{n} = \binom{12}{6} = 924$, and $2^{2n} = 2^{12} = 4{,}096$. Clearly, $2^6 < \binom{12}{6} < 2^{12}$.

Another Pair of Bounds

Let $M = \binom{2n}{n}$. Since M is the largest binomial coefficient in the binomial expansion of $(1+1)^{2n}$, that is, in row $2n$ of Pascal's triangle, and since the sum of the binomial coefficients in that row equals 2^{2n}, it follows that $M < 2^{2n}$. But there are $2n + 1$ binomial coefficients in row $2n$, so $(2n + 1)M > 2^{2n}$; that is, $\frac{2^{2n}}{2n+1} < m$. Thus

$$\frac{2^{2n}}{2n + 1} < \binom{2n}{n} < 2^{2n} \tag{2.11}$$

For example, let $n = 5$. Then

$$\frac{2^{2n}}{2n + 1} = \frac{2^{10}}{11} \approx 93.090909$$
$$\binom{2n}{n} = \binom{10}{5} = 252$$
$$2^{2n} = 2^{10} = 1{,}024$$

Clearly,

$$\frac{2^{10}}{11} < \binom{10}{5} < 2^{10}$$ ∎

Wang's Bounds

In 1990, E. T. H. Wang of Wilfrid Laurier University, Ontario, Canada, proposed a stronger double inequality for the CBC:

$$\frac{2^{2n-1}}{\sqrt{n}} < \binom{2n}{n} < \frac{2^{2n}}{\sqrt{2n}} \tag{2.12}$$

where $n \geq 4$. We will employ the technique used by H. O. Pollak of Summit, New Jersey, to establish these inequalities.

Let $a_n = \frac{1}{2^{2n}} \binom{2n}{n}$. Then it suffices to show that

$$\frac{1}{2\sqrt{n}} < a_n < \frac{1}{\sqrt{2n}}$$

We have

$$
\begin{aligned}
a_n &= \frac{(2n)!}{2^{2n}n!n!} \\
&= \frac{1 \cdot 2 \cdot 3 \cdot 4 \cdots (2n-1)(2n)}{2^{2n}n!n!} \\
&= \frac{2^n n![1 \cdot 3 \cdot 5 \cdots (2n-1)]}{2^{2n}n!n!} \\
&= \frac{1 \cdot 3 \cdot 5 \cdots (2n-1)}{2 \cdot 4 \cdot 6 \cdots (2n)}
\end{aligned}
$$

So

$$
\begin{aligned}
a_n^2 &= \frac{1}{2} \cdot \frac{1}{2} \cdot \frac{3}{4} \cdot \frac{3}{4} \cdot \frac{5}{6} \cdot \frac{5}{6} \cdots \frac{2n-1}{2n} \cdot \frac{2n-1}{2n} \\
&> \frac{1}{2} \cdot \frac{1}{2} \cdot \frac{2}{3} \cdot \frac{3}{4} \cdot \frac{4}{5} \cdot \frac{5}{6} \cdots \frac{2n-2}{2n-1} \cdot \frac{2n-1}{2n} \\
&= \frac{1}{4n}
\end{aligned}
$$

Thus

$$a_n > \frac{1}{2\sqrt{n}} \tag{2.13}$$

Similarly,

$$
\begin{aligned}
a_n^2 &< \frac{1}{2} \cdot \frac{2}{3} \cdot \frac{3}{4} \cdot \frac{4}{5} \cdots \frac{2n-3}{2n-2} \cdot \frac{2n-2}{2n-1} \cdot \frac{2n-1}{2n} \cdot \frac{2n}{2n} \\
&= \frac{1}{2n}
\end{aligned}
$$

So

$$a_n < \frac{1}{\sqrt{2n}} \tag{2.14}$$

Combining inequalities (2.13) and (2.14), we get

$$\frac{2^{2n-1}}{\sqrt{n}} < \binom{2n}{n} < \frac{2^{2n}}{\sqrt{2n}} \tag{2.15}$$

This yields the desired double inequality (2.12) with a smaller upper bound.

For example, let $n = 5$. Then

$$\frac{2^{2n-1}}{\sqrt{n}} = \frac{2^9}{\sqrt{5}} \approx 228.973360896$$

$$\binom{2n}{n} = \binom{10}{5} = 252$$

$$\frac{2^{2n}}{\sqrt{2n}} = \frac{2^{10}}{\sqrt{10}} \approx 323.817232401$$

Clearly,

$$\frac{2^9}{\sqrt{5}} < \binom{10}{5} < \frac{2^{10}}{\sqrt{10}}$$

It is worth noting that the double inequality (2.15) holds for $n \geq 2$.

Interestingly, inequality (2.13) can be established slightly differently also. To see this, we have:

$$a_n = \frac{1 \cdot 3 \cdot 5 \cdots (2n-1)}{2 \cdot 4 \cdot 6 \cdots (2n)} \tag{2.16}$$

$$= \frac{1}{2} \cdot \frac{3}{4} \cdot \frac{5}{6} \cdots \frac{2n-1}{2n}$$

$$> \frac{1}{2} \cdot \frac{2}{3} \cdot \frac{4}{5} \cdots \frac{2n-2}{2n-1} \tag{2.17}$$

Multiplying (2.16) and (2.17), we get

$$a_n^2 > \frac{1}{4n}$$

So

$$a_n > \frac{1}{2\sqrt{n}}$$

A Lower Upperbound

In fact, the upperbound in inequality (2.15) can slightly be improved:

$$\frac{2^{2n-1}}{\sqrt{n}} < \binom{2n}{n} < \frac{2^{2n}}{\sqrt{2n+1}} \tag{2.18}$$

where $n \geq 2$. To establish this, as before, we have:

$$a_n = \frac{1 \cdot 3 \cdot 5 \cdots (2n - 1)}{2 \cdot 4 \cdot 6 \cdots (2n)}$$

$$= \frac{1}{2} \cdot \frac{3}{4} \cdot \frac{5}{6} \cdots \frac{2n - 1}{2n}$$

$$< \frac{2}{3} \cdot \frac{4}{5} \cdot \frac{6}{7} \cdots \frac{2n}{2n + 1}$$

$$= \frac{2 \cdot 4 \cdot 6 \cdots (2n)}{1 \cdot 3 \cdot 5 \cdots (2n - 1)} \cdot \frac{1}{2n + 1}$$

$$= \frac{1}{a_n} \cdot \frac{1}{2n + 1}$$

Thus

$$a_n^2 < \frac{1}{2n + 1}$$

$$a_n < \frac{1}{\sqrt{2n + 1}}$$

So

$$\binom{2n}{n} < \frac{2^{2n}}{\sqrt{2n + 1}}$$

This coupled with inequality (2.15) yields inequality (2.18).

For example, let $n = 5$. Then:

$$\frac{2^{2n-1}}{\sqrt{n}} = \frac{2^9}{\sqrt{5}} \approx 228.973360896$$

$$\binom{2n}{n} = \binom{10}{5} = 252$$

$$\frac{2^{2n}}{\sqrt{2n + 1}} = \frac{2^{10}}{\sqrt{11}} \approx 308.747616848$$

Clearly,

$$\frac{2^9}{\sqrt{5}} < \binom{10}{5} < \frac{2^{10}}{\sqrt{11}}$$

It follows from the above discussions that

$$\frac{1}{2\sqrt{n}} < \frac{1 \cdot 3 \cdot 5 \cdots (2n - 1)}{2 \cdot 4 \cdot 6 \cdots (2n)} < \frac{1}{\sqrt{2n + 1}}$$

Two Additional Bounds

We present two more bounds satisfied by $\binom{2n}{n}$. The first one was established by K. Chandrasekhar and the other by R. E Shafer of Berkeley, California:

$$\frac{2^{2n}}{2\sqrt{n}} < \binom{2n}{n} < \frac{2^{2n}}{\sqrt{2n}}$$

where $n \geq 2$, and

$$\frac{2^{2n}}{\sqrt{\pi}(n^2 + n/2 + 1/8)^{1/4}} < \binom{2n}{n} < \frac{2^{2n}}{\sqrt{(n + 1/4)\pi}}$$

where $n \geq 1$.

3

The Central Binomial Coefficient Revisited

In the preceding chapter, we saw a number of occurrences of the ubiquitous CBC and the quotient $\frac{1\cdot3\cdot5\cdots(2n-1)}{2\cdot4\cdot6\cdots(2n)}$. We study both further in this chapter.

The following are a few additional occurreces of the quotient $\dfrac{1\cdot3\cdot5\cdots(2n-1)}{2\cdot4\cdot6\cdots(2n)}$:

$$\sum_{n=1}^{\infty}\frac{1}{n}\left[\frac{1\cdot3\cdot5\cdots(2n-1)}{2\cdot4\cdot6\cdots(2n)}\right]=\log 4$$

$$\int_0^{\pi/2}\sin^{2n}x\,\mathrm{d}x=\frac{1\cdot3\cdot5\cdots(2n-1)}{2\cdot4\cdot6\,\cdots\,(2n)}\cdot\frac{\pi}{2}$$

$$\int_0^{\pi/2}\cos^{2n}x\,\mathrm{d}x=\frac{1\cdot3\cdot5\cdots(2n-1)}{2\cdot4\cdot6\,\cdots\,(2n)}\cdot\frac{\pi}{2}$$

$$\int_0^{\pi/2}\sin^{2n+1}x\,\mathrm{d}x=\frac{1\cdot3\cdot5\cdots(2n-1)}{2\cdot4\cdot6\,\cdots\,(2n)}\cdot\frac{1}{2n+1}$$

$$\int_0^{\pi/2}\cos^{2n+1}x\,\mathrm{d}x=\frac{1\cdot3\cdot5\cdots(2n-1)}{2\cdot4\cdot6\,\cdots\,(2n)}\cdot\frac{1}{2n+1}$$

$$\int_0^{\infty}\frac{\mathrm{d}x}{(x^2+a^2)^n}=\frac{1\cdot3\cdot5\cdots(2n-3)}{2\cdot4\cdot6\cdots(2n-2)}\cdot\frac{1}{a^{2n-1}}\cdot\frac{\pi}{2}$$

$$\int_0^{\infty}e^{-x^2}\,\mathrm{d}x=\frac{\pi}{2}\cdot\lim_{n\to\infty}\frac{1\cdot3\cdot5\cdots(2n-3)}{2\cdot4\cdot6\cdots(2n-2)}\cdot\sqrt{n}$$

$$\int_a^b (x-a)^n (b-x)^n \, dx = 2 \cdot \frac{2 \cdot 4 \cdot 6 \cdots (2n)}{3 \cdot 5 \cdot 7 \cdots (2n+1)} \cdot \left(\frac{b-a}{2}\right)^{2n+1}, \quad n > 1$$

$$\sqrt{\frac{2n(2n+1)}{4n+1}}\, \pi < \frac{2 \cdot 4 \cdot 6 \cdots (2n)}{1 \cdot 3 \cdot 5 \cdots (2n-1)} < \sqrt{\frac{(4n+3)(2n+1)}{4n+4}} \cdot \frac{\pi}{2}$$

$$\int_0^1 \frac{x^{2n+1}}{\sqrt{1-x^2}} \, dx = \frac{2 \cdot 4 \cdot 6 \cdots (2n)}{3 \cdot 5 \cdot 7 \cdots (2n+1)}$$

Wallis's Product

The quotient $\frac{1 \cdot 3 \cdot 5 \cdots (2n-1)}{2 \cdot 4 \cdot 6 \cdots (2n)}$ and hence the CBC occurs in Wallis's product for $\sqrt{\frac{\pi}{2}}$:

$$\sqrt{\frac{\pi}{2}} = \lim_{n \to \infty} \frac{2 \cdot 4 \cdot 6 \cdots (2n)}{1 \cdot 3 \cdot 5 \cdots (2n-1)}$$

To see this, we have

$$\frac{\sin x}{x} = \prod_{n=1}^{\infty} \left(1 - \frac{x^2}{n^2 \pi^2}\right)$$

where $x \neq 0$. This formula was conjectured by Euler. Letting $x = \pi/2$, this yields

$$\frac{\pi}{2} = \prod_{n=1}^{\infty} \frac{2n \cdot 2n}{(2n-1)(2n+1)}$$

$$= \frac{2 \cdot 2}{1 \cdot 3} \cdot \frac{4 \cdot 4}{3 \cdot 5} \cdot \frac{6 \cdot 6}{5 \cdot 7} \cdot \frac{8 \cdot 8}{7 \cdot 9} \cdot \frac{10 \cdot 10}{9 \cdot 11} \cdots$$

$$= \frac{2}{1} \cdot \frac{2}{3} \cdot \frac{4}{3} \cdot \frac{4}{5} \cdot \frac{6}{5} \cdot \frac{6}{7} \cdots \frac{2n}{2n-1} \cdot \frac{2n}{2n+1} \cdots$$

$$\frac{\pi}{2} = \lim_{n \to \infty} \left[\frac{2 \cdot 4 \cdot 6 \cdots (2n)}{1 \cdot 3 \cdot 5 \cdots (2n-1)}\right]^2 \cdot \frac{2n}{2n+1}$$

This is *Wallis's formula* for $\frac{\pi}{2}$. Since

$$\frac{1}{2^{2n}} \binom{2n}{n} = \frac{1 \cdot 3 \cdot 5 \cdots (2n-1)}{2 \cdot 4 \cdot 6 \cdots (2n)}$$

John Wallis (1616–1703) was born in Ashford, Kent, England, where his father was a priest. He attended Felstead School, where he studied Latin, Greek, and rudimentary logic, but no mathematics. When he was fifteen he became curious about the signs and symbols in his brother's arithmetic book, and with his brother's help, he mastered the subject in two weeks.

Around Christmas 1632, Wallis entered Queen's College, Cambridge, where he studied theology, anatomy, geography, physics, and mathematics. While at Queen's, he defended the theory of the circulation of the blood in a public debate. He graduated in 1637 and received his master's degree in 1640. The same year, he was ordained by the bishop of Winchester.

Wallis was appointed the Savilian professor of geometry at Oxford in 1649 by Lord Protector Oliver Cromwell (1599–1658); he held the chair until his death, more than fifty years later. In 1657, he was appointed keeper of the University Archives, creating more controversies.

His phenomenal computational skill in deciphering Royalist messages for the government and his appointment at Oxford paved the way for a fruitful mathematical career. His love of mathematics was cemented by his reading of William Ougtred's *Clavis Mathematicae* in 1647.

Wallis's mathematical contributions include *Conic Sections* (1655), *Arithmetica Infinitorum* (1656), *Algebra* (1685), and *Opera Mathematica* (published between 1693 and 1699). The most important of his works is *Arithmetica Infinitorum*, in which he introduced the infinity symbol ∞ and developed the infinite product for $\pi/2$ while computing $\int_0^1 (1 - x^2)^{1/2}\, dx$. This infinite product motivated his friend Lord W. V. Brouncker (1620–1684), the first president of the Royal Society, to discover the remarkable formula

(continued)

(*Continued*)

$$\frac{4}{\pi} = 1 + \cfrac{1^2}{2 + \cfrac{3^2}{2 + \cfrac{5^2}{2 + \cfrac{7^2}{2 + \cfrac{9^2}{2 + \cdots}}}}}$$

which led to the development of the theory of continued fractions. Besides his mathematical works, Wallis also wrote on theology, logic, and philosophy, and was the first to devise a system for teaching deaf-mutes. He helped found the Royal Society.

---

this formula can also be written as

$$\frac{\pi}{2} = \lim_{n \to \infty} \left[\frac{2^{2n}}{\binom{2n}{n}^2} \right]^2 \cdot \frac{2n}{2n + 1}$$

The next example, an application of Wallis's formula, was proposed in 1952 by H. P. Thielman of what was then Iowa State College. The solution is based on the one given two years later by R. H. Boyer of what was then Carnegie Institute of Technology, Pittsburgh.

Example 3.1 Let $\{a_n\}_{n=1}^{\infty}$ be a sequence of real numbers such that $a_{n+1} = \frac{1}{n^m a_n}$, where $m > 0$. Prove that $\{a_n\}$ is monotone nonincreasing if and only if $a_1 = (\pi/2)^{m/2}$.

Proof   It follows from the recurrence relation that

$$a_{2n+1} = \left[\frac{1 \cdot 3 \cdots (2n-1)}{2 \cdot 4 \cdots (2n)} \right]^m a_1 \quad \text{and} \quad \frac{1}{a_{2n}} = \left[\frac{3 \cdot 5 \cdots (2n-1)}{2 \cdot 4 \cdots (2n-2)} \right]^m a_1$$

$\{a_n\}$ is monotone nonincreasing if and only if $a_{2n+1} \leq a_{2n} \leq a_{2n-1}$ for every n, that is, if and only if

$$\left[\frac{1 \cdot 3 \cdots (2n-1)}{2 \cdot 4 \cdots (2n)} \right]^m a_1 \leq \left[\frac{2 \cdot 4 \cdots (2n-2)}{3 \cdot 5 \cdots (2n-1)} \right]^m \frac{1}{a_1} \leq \left[\frac{1 \cdot 3 \cdots (2n-3)}{2 \cdot 4 \cdots (2n-2)} \right]^m a_1$$

that is, if and only if

$$\frac{2 \cdot 4 \cdots (2n)}{1 \cdot 3 \cdots (2n-1)} \cdot \frac{2 \cdot 4 \cdots (2n-2)}{3 \cdot 5 \cdots (2n-1)} \geq a_1^{2/m} \geq \frac{2 \cdot 4 \cdots (2n-2)}{1 \cdot 3 \cdots (2n-3)} \cdot \frac{2 \cdot 4 \cdots (2n-2)}{3 \cdot 5 \cdots (2n-1)}.$$

By Wallis's formula,

$$\left[\frac{2\cdot4\cdots(2n-2)}{1\cdot3\cdots(2n-1)}\right]^2 \cdot \frac{2n}{2n-1} \geq \frac{\pi}{2} \geq \left[\frac{2\cdot4\cdots(2n-2)}{1\cdot3\cdots(2n-3)}\right]^2 \cdot \frac{2n-2}{2n-1}.$$

Therefore, taking limits as $n\to\infty$, it follows that $\{a_n\}$ is monotone non-increasing if and only if $a_1^{2/m}=\pi/2$, that is, if and only if $a_1=(\pi/2)^{m/2}$ ∎

Cesáro First Order Mean

The next example, another application of Wallis's formula, was proposed in 1957 by J. Barlaz of Rutgers University, New Brunswick, New Jersey. The solution is based on the one given the following year by R. Breusch of Amherst College, Massachusetts.

Example 3.2 Evaluate the Cesáro first order mean for the series $\sum_{n=2}^{\infty}(-1)^n \log n$.

Solution   Let $S_m = \sum_{n=2}^{m}(-1)^n \log n$. Then, by Wallis's formula,

$$S_{2n} = \frac{2\cdot4\cdots(2n)}{3\cdot5\cdots(2n-1)} \quad \text{and} \quad S_{2n+1} = \frac{2\cdot4\cdots(2n)}{3\cdot5\cdots(2n+1)}$$

$$S_{2n} + S_{2n+1} = \log\frac{2\cdot2}{1\cdot3}\cdot\frac{4\cdot4}{3\cdot5}\cdots\frac{(2n)\cdot(2n)}{(2n-1)\cdot(2n+1)}$$

$$= \log(\pi/2) + O(1/n)$$

$$\sum_{n=2}^{m} S_n = \frac{m}{2}\log(\pi/2) + O(\log m)$$

Thus the Cesáro first order mean is given by

$$\lim_{m\to\infty}\frac{1}{m}\sum_{n=2}^{m} S_n = \frac{1}{2}\log(\pi/2)$$ ∎

The following example is based on a problem proposed in 1938 by R. H. Bardell of the University of Wisconsin. The solution, given by T. C. Fry of Bell Telephone Laboratories, New York, employs calculus.

Example 3.3 Show that

$$\sum_{r=0}^{n} \frac{(-1)^r}{2r+1}\binom{n}{r} = \frac{n+1}{2n+1}2^{2n}C_n$$

Proof   Let

$$f(x) = \sum_{r=0}^{n} \frac{(-1)^r}{2r+1}\binom{n}{r}x^{2r+1}$$

So the desired sum is $f(1)$. We have:

$$f'(x) = \sum_{r=0}^{n} \binom{n}{r}(-x^2)^r$$

$$= (1 - x^2)^n$$

Then

$$f(1) = \int_0^1 (1 - x^2)^n \, dx$$

$$= \int_0^{\pi/2} \cos^{2n+1} \theta \, d\theta$$

$$= \frac{(2n)(2n-2)\cdots 2}{(2n+1)(2n-1)\cdots 3}$$

$$= \frac{2^{2n}}{(2n+1)\binom{2n}{n}}$$

$$= \frac{n+1}{2n+1}2^{2n}C_n \qquad\qquad \blacksquare$$

As another example, we would like to evaluate the sum

$$\sum_{n=1}^{m} \frac{n}{2^{2n}}\binom{2n}{n}$$

proposed in 1951 by L. C. Hsu of National Tsing-Hua University, Peiping, China. To find the sum, we rewrite it as a telescoping sum, using

$$v_n = \frac{n(n-1)}{3 \cdot 2^{2n-1}} \binom{2n}{n}$$

as M. R. Spiegel of Renesselaer Polytechnic Institute, Troy, New York, did in 1952:

$$v_{n+1} - v_n = \frac{(n+1)n}{3 \cdot 2^{2n+1}} \binom{2n+2}{n+1} - \frac{n(n-1)}{3 \cdot 2^{2n-1}} \binom{2n}{n}$$

$$= \frac{n(n+1)}{3 \cdot 2^{2n+1}} \cdot \frac{(2n+2)(2n+1)}{(n+1)(n+1)} \binom{2n}{n} - \frac{n(n-1)}{3 \cdot 2^{2n-1}} \binom{2n}{n}$$

$$= \frac{n}{2^{2n}} \binom{2n}{n}$$

Therefore,

$$\sum_{n=1}^{m} \frac{n}{2^{2n}} \binom{2n}{n} = \sum_{n=1}^{m} (v_{n+1} - v_n)$$

$$= v_{m+1} - v_1$$

$$= \frac{m(m+1)}{3 \cdot 2^{2m+1}} \binom{2m+2}{m+1} - 0$$

$$= \frac{m(m+1)}{3 \cdot 2^{2m+1}} \binom{2m+2}{m+1}$$

$$= \frac{m(m+1)(m+2)}{3 \cdot 2^{2m+1}} C_{m+1}$$

The next problem, proposed in 1987 by L. M. Christophe Jr. of Wilmimgton, Delaware, also shows that ubiquitousness of the CBC.

Let

$$a_n(x) = \frac{\prod_{i=1}^{n-1}(x+i) \prod_{i=2}^{n-1}(x+i)}{(n!)^2}$$

where x is a nonnegative real number and $n \geq 1$. Then

$$
\begin{aligned}
a_n(1/2) &= \frac{\displaystyle\prod_{i=1}^{n-1}(2i+1)\prod_{i=2}^{n-1}(2i+1)}{2^{2n-2}\,(n!)^2} \\[2ex]
&= \frac{[3 \cdot 5 \cdots (2n-1)][5 \cdot 7 \cdots (2n+1)]}{2^{2n-2}\,(n!)^2} \\[2ex]
&= \frac{[1 \cdot 3 \cdot 5 \cdots (2n-1)] \cdot [1 \cdot 3 \cdot 5 \cdots (2n-1)](2n+1)}{3 \cdot 2^{2n-2}\,(n!)^2} \\[2ex]
&= \frac{(2n)!(2n)!(2n+1)}{3 \cdot 2^n\, n!\, 2^n\, n!\, 2^{2n-2}\,(n!)^2} \\[2ex]
&= \frac{4(2n+1)}{3} \cdot \frac{[(2n)!]^2}{(n!)^4\, 2^{4n}} \\[2ex]
&= \frac{4(2n+1)}{3}\left[\frac{1}{2^{2n}}\binom{2n}{n}\right]^2
\end{aligned}
$$

Using *Stirling's asymptotic formula*,

$$
\sqrt{n} \approx \sqrt{2\pi n}\,(n/e)^n
$$

it follows that

$$
\frac{1}{2^{2n}}\binom{2n}{n} \approx \frac{1}{\sqrt{\pi n}}
$$

Consequently,

$$
\begin{aligned}
\lim_{n \to \infty} a_n(1/2) &= \lim_{n \to \infty} \frac{4(2n+1)}{3} \cdot \frac{1}{\pi n} \\[2ex]
&= \frac{8}{3\pi}
\end{aligned}
$$

as I. C. Bivens and B. G. Klien of Davidson College, North Carolina, showed in 1988.

Here is yet another occurrence of the same quotient $\frac{1 \cdot 3 \cdot 5 \cdots (2n-1)}{2 \cdot 4 \cdot 6 \cdots (2n)}$. In 1989, L. V. Hamme of Vrije Universiteit Brussel, Belgium, discovered that

$$
1 + \sum_{n=1}^{\infty} \frac{1}{n+1}\left[\frac{1 \cdot 3 \cdot 5 \cdots (2n-1)}{2 \cdot 4 \cdot 6 \cdots (2n)}\right]^2 = \frac{4}{\pi}
$$

To establish this formula, let

$$S_n = 4n \left[\frac{1 \cdot 3 \cdot 5 \cdots (2n-1)}{2 \cdot 4 \cdot 6 \cdots (2n)} \right]^2$$

Then

$$S_{n+1} - S_n = 4(n+1) \left[\frac{1 \cdot 3 \cdot 5 \cdots (2n+1)}{2 \cdot 4 \cdot 6 \cdots (2n+2)} \right]^2 - 4n \left[\frac{1 \cdot 3 \cdot 5 \cdots (2n-1)}{2 \cdot 4 \cdot 6 \cdots (2n)} \right]^2$$

$$= \left[\frac{1 \cdot 3 \cdot 5 \cdots (2n-1)}{2 \cdot 4 \cdot 6 \cdots (2n)} \right]^2 \left[\frac{4(n+1)(2n+1)^2}{4(n+1)^2} - 4n \right]$$

$$= \frac{1}{n+1} \left[\frac{1 \cdot 3 \cdot 5 \cdots (2n-1)}{2 \cdot 4 \cdot 6 \cdots (2n)} \right]^2$$

Therefore,

$$S_n = S_1 + \sum_{k=1}^{n} (S_k - S_{k-1})$$

$$= 1 + \sum_{k=1}^{n} \frac{1}{k+1} \left[\frac{1 \cdot 3 \cdot 5 \cdots (2k-1)}{2 \cdot 4 \cdot 6 \cdots (2k)} \right]^2$$

Thus

$$\lim_{n \to \infty} S_n = 1 + \sum_{k=1}^{n} \frac{1}{k+1} \left[\frac{1 \cdot 3 \cdot 5 \cdots (2k-1)}{2 \cdot 4 \cdot 6 \cdots (2k)} \right]^2$$

By Wallis's product, this yields

$$1 + \sum_{n=1}^{\infty} \frac{1}{n+1} \left[\frac{1 \cdot 3 \cdot 5 \cdots (2n-1)}{2 \cdot 4 \cdot 6 \cdots (2n)} \right]^2 - \frac{4}{\pi}$$

as desired.

Limit of a Sum of Binomial Probabilities

Interestingly, the quotient

$$a_n = \frac{1}{2^{2n}} \binom{2n}{n} = \frac{1 \cdot 3 \cdot 5 \cdots (2n-1)}{2 \cdot 4 \cdot 6 \cdots (2n)}$$

that we employed in the proof of the double inequality (2.10) reappears in the solution to the following example. It was proposed in 1984 by G. Chang of the University of Science and Technology of China, Hefei, Anhui. The solution presented here is based on the one given in 1986 by M. Chamberlain of the U.S. Naval Academy, Annapolis, Maryland.

Example 3.4 Let

$$I_n = n\binom{2n}{n} \int_0^{1/2} x^n (1-x)^{n-1}\, dx$$

Show that the sequence $\{I_n\}_{n=1}^{\infty}$ is monotonically increasing and find its limit.

Solution   Let

$$J_n = n\binom{2n}{n} \int_0^{1/2} (1-x)^n x^{n-1}\, dx$$

Then

$$J_n - I_n = n\binom{2n}{n} \int_0^{1/2} \left[(1-x)^n x^{n-1} - x^n(1-x)^{n-1}\right] dx$$

$$= \binom{2n}{n} \int_0^{1/2} \left\{\frac{d}{dx}\left[(1-x)^n x^n\right]\right\} dx$$

$$= \binom{2n}{n} \left[(1-x)^n x^n\right]_0^{1/2}$$

$$= \frac{1}{2^{2n}}\binom{2n}{n}$$

On the other hand,

$$J_n + I_n = n\binom{2n}{n}\int_0^{1/2}(1-x)^n x^{n-1}\, dx + n\binom{2n}{n}\int_0^{1/2} x^n(1-x)^{n-1}\, dx$$

$$= n\binom{2n}{n}\int_{1/2}^{1} t^n(1-t)^{n-1}\, dt + n\binom{2n}{n}\int_0^{1/2} x^n(1-x)^{n-1}\, dx$$

$$= n\binom{2n}{n}\int_0^{1} x^n(1-x)^{n-1}\, dx$$

Using integration by parts, this yields

$$J_n + I_n = n \binom{2n}{n} \frac{n!(n-1)!}{(2n)!}$$

$$= 1$$

Thus we have the following 2×2 linear system:

$$J_n + I_n = 1$$

$$J_n - I_n = \frac{1}{2^{2n}} \binom{2n}{n}$$

Solving this system, we get

$$I_n = \frac{1}{2}\left[1 - \frac{1}{2^{2n}} \binom{2n}{n} \right] \tag{3.1}$$

Let

$$a_n = \frac{1}{2^{2n}} \binom{2n}{n}$$

$$= \frac{1 \cdot 3 \cdot 5 \cdots (2n-1)}{2 \cdot 4 \cdot 6 \cdots (2n)}$$

So

$$\frac{a_{n+1}}{a_n} = \frac{2n+1}{2n+2}$$

$$< 1$$

So the sequence $\{a_n\}$ is strictly decreasing and hence the sequence $\{I_n\}$ is strictly increasing.

From the foregoing proof of double inequality (2.10), $a_n^2 < \dfrac{1}{2n}$; and hence $a_n < \dfrac{1}{\sqrt{2n}}$. So

$$\lim_{x \to \infty} a_n = 0$$

Thus, from equation (3.1),

$$\lim_{x \to \infty} I_n = \tfrac{1}{2}(1 - 0)$$

$$= \tfrac{1}{2}$$

$\blacksquare$

Interestingly, the expression a_n makes its appearance once again in the next example. It was proposed in 1990 by M. B. Handelsman of Erasmus High School, Brooklyn, New York. The solution given is due to R. J. Wagner of Muhlenberg College, Allentown, Pennsylvania.

Example 3.5 Suppose a fair coin is tossed $2n$ times, where $n \geq 1$. Let h and t denote the number of heads and the number of tails obtained, respectively. Let $d = |h - t|$. Find each.

1. The most probable value of d.
2. The expected value of d.

Solution   Notice that $h + t = 2n$, and both h and t are binomial random variables. The variable d has the probability distribution

$$p(d = 2k) = \begin{cases} \binom{2n}{n}\left(\tfrac{1}{2}\right)^{2n} & \text{if } k = 0 \\[2ex] 2\binom{2n}{n+k}\left(\tfrac{1}{2}\right)^{2n} & \text{if } 1 \leq k \leq n \end{cases}$$

1. Suppose $n = 1$. Then

$$p(0) = \binom{2}{1}\left(\frac{1}{2}\right)^2 = \frac{1}{2}$$

and

$$p(2) = 2\binom{2}{2}\left(\frac{1}{2}\right)^2 = \frac{1}{2}$$

So $d = 0$ and $d = 2$ are equally most probable in this case.

On the other hand, suppose $n > 1$. Then

$$p(0) = \binom{2n}{n}\left(\frac{1}{2}\right)^{2n}$$

$$< \frac{2n}{n+1}\binom{2n}{n}\left(\frac{1}{2}\right)^{2n}$$

$$= 2\binom{2n}{n+1}\left(\frac{1}{2}\right)^{2n}$$

$$= p(2)$$

Besides, since $\binom{2n}{n+1} > \binom{2n}{n+k}$, it follows that $p(2) > p(2k)$, where $2 \leq k \leq n$.

Thus, 2 is the most probable value of d.

2. The expected value $E(d)$ of d is given by

$$E(d) = \sum_{k=0}^{n} (2k) p(2k)$$

$$= 4\left(\frac{1}{2}\right)^{2n} \sum_{k=1}^{n} k \binom{2n}{n+k}$$

Let $S = \sum_{k=1}^{n} k \binom{2n}{n+k}$. Notice that

$$\binom{2n}{n+k+1} = \frac{n-k}{n+k+1} \binom{2n}{n+k}$$

That is,

$$(n+k+1)\binom{2n}{n+k+1} = (n-k)\binom{2n}{n+k}$$

Rearranging the terms, this yields

$$k\binom{2n}{n+k} + (k+1)\binom{2n}{n+k+1} = -n\left[\binom{2n}{n+k+1} - \binom{2n}{n+k}\right]$$

So

$$\sum_{k=1}^{n-1}\left[k\binom{2n}{n+k} + (k+1)\binom{2n}{n+k+1}\right] = -n\sum_{k=1}^{n-1}\left[\binom{2n}{n+k+1} - \binom{2n}{n+k}\right]$$

Noting that the sum on the RHS is telescopic, this yields

$$(S-n) + S - \binom{2n}{n+1} = -n\left[1 - \binom{2n}{n+1}\right]$$

Solving this equation, we get

$$S = \frac{n+1}{2}\binom{2n}{n+1}$$

Thus

$$E(d) = 4\left(\frac{1}{2}\right)^n \cdot \frac{n+1}{2}\binom{2n}{n+1}$$

$$= \frac{n}{2^{2n-1}}\binom{2n}{n}$$

◼

The next example was proposed as a problem in 1973 by A. V. Boyd of the University of Witwatersrand, Johannesburg, South Africa. The solution presented here is based on the one by V. Linis of the University of Ottawa, Canada.

Example 3.6 Find the sum of the series

$$\sum_{n=0}^{\infty}\binom{2n}{n}^{-1}(4x)^n$$

Solution   First, notice that the series converges for $|x| < 1$ and diverges otherwise. We have:

$$\binom{2n}{n}^{-1} = \frac{n!n!}{(2n)!}$$

$$= (2n+1)\int_0^1 (1-u)^n u^n \, du$$

So

$$\sum_{n=0}^{\infty}\binom{2n}{n}^{-1}(4x)^n = \sum_{n=0}^{\infty}(2n+1)\int_0^1 [4x(1-u)u]^n \, du$$

Let $y = 4x(1-u)u$. Interchanging the order of summation and integration, this yields

$$\sum_{n=0}^{\infty}\binom{2n}{n}^{-1}(4x)^n = \int_0^1 \left\{\sum_{n=0}^{\infty}(2n+1)[4x(1-u)u]^n\right\} du$$

$$= \int_0^1 \left[2\sum_{n=0}^{\infty} ny^n + \sum_{n=0}^{\infty} y^n\right] dy$$

$$= \int_0^1 \left[\frac{2y}{(1-y)^2} + \frac{1}{1-y}\right] dy$$

$$= \int_0^1 \frac{1+y}{(1-y)^2} \, dy$$

$$= \int_0^1 \frac{1 + 4(1 - u)ux}{[1 - 4(1 - u)ux]^2} \, du$$

$$= \begin{cases} (1 - x)^{-1} \left(1 - \sqrt{\frac{-x}{1-x}} \sinh^{-1} \sqrt{-x}\right) & \text{if } -1 < x \leq 0 \\[2ex] (1 - x)^{-1} \left(1 + \sqrt{\frac{x}{1-x}} \sin^{-1} \sqrt{x}\right) & \text{if } 0 \leq x < 1 \end{cases} \quad \blacksquare$$

This example has an interesting application to the theory of probability, as observed in 1974 by I. I. Kotlarski of Oklahoma State University. To this end, suppose the random variable X has the probability distribution given by the density function

$$g(u) = \begin{cases} \frac{1}{\pi\sqrt{1-u^2}} & \text{if } |u| < 1 \\[2ex] 0 & \text{otherwise} \end{cases}$$

Then the moments of X are given by

$$m_{2n} = \frac{1}{4^n} \binom{2n}{n}, \quad n \geq 0$$

where all odd moments are 0. Thus the given series is the generating function for the reciprocals of the even moments of X.

A classical theorem in number theory, discovered independently by English mathematician J. J. Sylvester and Belarusian mathematician I. Schur, states that any set of k consecutive integers, each of which is greater than k, contains an integer with a prime factor greater than k. Erdös stated this result as follows:

$$\text{If } n \geq 2k, \text{ then } \binom{n}{k} \text{ has a prime factor} > k.$$

In the late 1960s, he conjectured that if $n \geq 2k$, then $\binom{n}{k}$ has a prime factor $\leq n/2$.

In 1969, E. F. Ecklund Jr. of Western Washington State College established that if $n \geq 2k$, then $\binom{n}{k}$ has a prime factor $\leq \max\{n/k, n/2\}$, with the exception of $\binom{7}{3}$. In particular, $\binom{2n}{n}$ has a prime factor $\leq \max\{2, n/2\}$.

For example, $\binom{6}{3} = 20 = 2^2 \cdot 5$ has a prime factor $2 = \max\{2, 3/2\}$; and so does $\binom{8}{4} = 70$. Notice that $\binom{16}{8} = 12870 = 2 \cdot 3^2 \cdot 5 \cdot 11 \cdot 13$ has a prime factor $3 \leq \max\{2, 8/2\}$.

In fact, 2 is a factor of every binomial coefficient $\binom{2n}{n}$, where $n \geq 1$.

James Joseph Sylvester (1814–1897), a leading Victorian mathematician, was born in London, England. (He adopted his surname later in life, following his brother's lead.) At fourteen, he entered University College, London, and became a student of Augustus De Morgan. From 1829 to 1831, he studied at the Royal Institution in Liverpool and then to St. John's College, Cambridge. Because of his religion, he could not graduate, nor was he eligible for Smith's prize or a fellowship.

In 1838, Sylvester was appointed professor of natural philosophy (essentially physics) at the University of London, where he became a colleague of his former professor. The following year, he was elected to the Royal Society.

At twenty-seven, Sylvester left London to become professor of mathematics at the University of Virginia. After three months, due to an unfortunate classroom incident, Sylvester returned to London and became an actuary at a life insurance company. In 1846, he entered the Inner Temple to study law, and started practicing law four years later; nonetheless, he remained mathematically active. In 1852, Sylvester met another lawyer-mathematician, Arthur Cayley (1821–1825) at the courts of Lincoln's Inn; the pair often discussed mathematics as they walked around the courts, and they became lifelong friends. Together, they developed the theory of algebraic invariants. In 1851, Sylvester discovered the discriminant of a cubic equation and coined the term *discriminant*. Three years later, Sylvester applied for a lectureship in geometry at Gresham College, London, but was not accepted. He then applied for a professorship in mathematics at the Royal Military College, Woolwich; again, he failed. He was, however, appointed to the post the following year, after the death of the successful candidate, and remained there until the mandatory retirement age of fifty-five. He was elected to the French Academy of Sciences in 1863.

In 1877, Sylvester joined Johns Hopkins University, Baltimore, Maryland, and in the following year founded the *American Journal of Mathematics*, the first mathematics journal in the United States. He remained at Johns Hopkins for the next six years.

In 1883, at sixty-nine, Sylvester was appointed the Savilian professor of geometry at Oxford University, where he remained until 1894, when his failing eyesight forced him to retire. Three years later, Sylvester suffered a stroke and died in London.

Sylvester, along with Cayley, founded the theory of matrices. His avid interest in the classics and languages led him to introduce new mathematical terminology; for instance, he is credited with coining the term *matrix*.

Sylvester is also known for his contributions to number theory and combinatorics. For example, in 1888, he showed that every odd perfect number must have at least five distinct prime factors.

Issai Schur (1875–1941) was born in Mogilyov, Belarus. At thirteen, he went to Latvia and attended the Gymnasium in Libau (now Liepaja).

In 1894, Schur entered the University of Berlin to study mathematics and physics. There he was greatly influenced by German mathematician Georg Ferdinand Frobenius (1849–1917), who along with William Burnside (1852–1927) founded the theory of representations of groups. In 1901, Schur received his doctorate for his work on rational representations of the general linear group over the complex field. The functions he introduced in his dissertation are today called *S-functions*, in his honor.

Two years later, Schur was appointed a lecturer at the university. In 1911, he became a professor at the University of Bonn. Five years later, he returned to Berlin, where he founded his famous school of mathematicians and spent most of the rest of his life. He became a full professor at Berlin in 1919.

Besides his significant contributions to the representation theory of groups, he also worked in number theory, analysis, cohomology groups, Galois groups, and Laguerre and Hermite polynomials.

In 1922, Schur was elected to the Prussian Academy, but Nazi Germany made his life increasingly difficult and painful because he was Jewish. In 1935, the Nazis dismissed him from his chair at Berlin and forced him to resign from the academy three years later.

Broken in both mind and body, Schur left Germany in 1939 for Palestine. Without sufficient funds, he was forced to sell his beloved mathematics books to the Institute of Advanced Study at Princeton, New Jersey. Schur died two years later in Tel Aviv, on his 66th birthday.

In 1964, M. Faulkner of the University of Alberta, Edmonton, established that $\binom{2n}{n}$ has a prime factor $\geq \left\lceil \frac{7n}{5} \right\rceil$. For example, $\binom{6}{3}$ has a prime factor $5 \geq \left\lceil \frac{7 \cdot 3}{5} \right\rceil$, and $\binom{16}{8}$ has a prime factor $5 \geq \left\lceil \frac{7 \cdot 8}{5} \right\rceil$: $\binom{6}{3} = 20 = 5 \cdot 4$ and $\binom{16}{8} = 12870 = 5 \cdot 2574$.

In fact, because $\binom{2n}{n} = \frac{(2n)!}{n!n!}$, it contains a prime factor p for every p such that $n < p \leq 2n$; this is so since $p|(2n)!$, but $p \nmid n!$. For example, both 11 and 13 are prime factors of $\binom{16}{8}$.

Next, we present two results by Erdös involving the binomial coefficient $\binom{2n}{n}$. The first result gives an upperbound for $\binom{2n}{n}$.

Erdös and the CBC

In 1934, Erdös established that $\binom{2n}{n} < 4^{n-1}$. This can be confirmed using PMI. For example, $\binom{20}{10} = 184{,}756 < 4^9$.

Let $a_1 = \lceil n/2 \rceil, a_2 = \lceil n/2^2 \rceil, \ldots, a_k = \lceil n/2^k \rceil, \ldots$. Let m be the least positive integer such that $n/2^m \leq 1$; that is, $a_m = 1$. Erdös then showed that

$$\binom{2a_1}{a_1} \binom{2a_2}{a_2} \cdots \binom{2a_m}{a_m} < 4^n$$

For example, let $n = 13$. Then $a_1 = 6, a_2 = 4, a_3 = 2$, and $a_4 = 1$; so

$$\binom{2a_1}{a_1} \binom{2a_2}{a_2} \binom{2a_3}{a_3} \binom{2a_4}{a_4} = \binom{12}{6}\binom{8}{4}\binom{4}{2}\binom{2}{1}$$

$$= 924 \cdot 70 \cdot 6 \cdot 2$$

$$= 776{,}160$$

$$< 4^{13}$$

Next, we present an interesting result about the binomial coefficient $\binom{2n}{n}$, discovered in 1963 by Leo Moser of the University of Alberta, Canada.

Moser and the CBC

Moser, in his quest for solving the combinatorial equation $\binom{2n}{n} = \binom{2a}{a}\binom{2b}{b}$, noted

an interesting divisibility property involving $\binom{2n}{n}$: $\binom{2n}{n}$ | $\left(\begin{array}{c} 2\binom{2n}{n} - 2 \\ \binom{2n}{n} - 1 \end{array} \right)$ for

every $n \geq 0$.

For example, let $n = 3$. Then $\binom{2n}{n} = \binom{6}{3} = 20$ and $\left(\begin{array}{c} 2\binom{2n}{n} - 2 \\ \binom{2n}{n} - 1 \end{array} \right) =$

$\binom{2 \cdot 20 - 2}{20 - 1} = \binom{38}{19} = 35{,}345{,}263{,}800$. Clearly, $20 \mid 35345263800$, as expected.

Although $\frac{(2n)!}{n!(n+1)!}$ is always an integer, $G(n) = \frac{(2n)!}{(n+1)!(n+1)!}$ need not be an integer. For example, when $n = 2$, $\frac{(2n)!}{(n+1)!(n+1)!} = \frac{4!}{3!3!}$ is not an integer. In 1929, H. Balakram of Punjab, India, found that $G(n)$ is an integer for just eight values of $n \leq 100$, namely, 5, 14, 27, 41, 44, 65, 76, and 90. For example, $\frac{10!}{6!6!} = 7$.

Hoon Balakram (1876–1929) was born in Jullunder, Punjab, India. An exceptionally diligent student, he showed remarkable intelligence from early childhood. After receiving his master's degree from Punjab University, he joined St. John's College, Cambridge, England. In 1899, Balakram passed the Mathematical Tripos and was placed fourth in the list of wranglers; the same year, he joined the extremely competitive Indian Civil Service.

The pursuit of mathematics was his avocation; his favorites were the theory of relativity and the theory of numbers, on which he published a number of articles. He discovered all positive integers < 20,000,000 that can be expressed as the sum of two cubes in two different ways, following the well-known *Ramanujan number* 1729, the smallest such number. A month before his sudden and premature death, Balakram was appointed a High Court judge in Bombay. He was then actively working in number theory.

Balakram proved that $\frac{(2n)!}{(n+1)!(n+1)!}$ is an integer for infinitely many values of n. Consequently, $(n+1)^2 \mid \binom{2n}{n}$ for infinitely many positive integers n. Using his technique, it can be shown that $(n+1)^k \mid \binom{2n}{n}$ for infinitely many positive integers n, according to Erdös et al.

Balakram also showed:

- If n is an even perfect number, then $G(n-1)$ is an integer.
- $G(n-1)$ is an integer if and only if $\sum_{i=1}^{\infty} \left\lfloor \frac{2n-2}{p^i} \right\rfloor \geq 2 \sum_{i=1}^{\infty} \left\lfloor \frac{n}{p^i} \right\rfloor$.

- If $n = pq$, where p and q are primes and $\frac{3}{2}p < q < 2p$, then $G(n-1)$ is an integer.

Next, we present a few unsolved problems involving the CBC.

A Few Unsolved Problems

Notice that $\binom{2}{1} = 2$ and $\binom{4}{2} = 6$ are square-free, but $\binom{6}{3} = 20 = 2^2 \cdot 5$ is not.

In fact, it is not known if $\binom{2n}{n}$ is square-free if $n \geq 4$. However, it is known that $n = 23$ is the largest integer such that every binomial coefficient $\binom{n}{r}$ is square-free, where $0 \leq r \leq n$.

In 1975, Erdös, Graham, I. Z. Ruzsa of Eötvös Lornad University (Budapest, Hungary), and E. G. Strauss of the University of California at Los Angeles established that for any two primes p and q, there are infinitely many positive integers n such that $\binom{2n}{n}$ and pq are relatively prime; that is, $\left(\binom{2n}{n}, pq\right) = 1$. For three primes, however, nothing is known. For instance, it is not known if $\left(\binom{2n}{n}, 105\right) = 1$ for infinitely many integers n.

Let $g(n)$ denote the smallest odd prime factor of $\binom{2n}{n}$. It has been shown by a computer analysis that $g(n) \leq 11$ for $n < 3160$.

V. Vyssotsky of then Bell Telephone Laboratories, Murray Hill, New Jersey, pursued this further. He showed that the least odd prime factor of $\binom{2n}{n}$ is 13 when $n = 3160$ and at most 11 for $3160 < n < 10^{90}$. In 1979, Robert J. Kimble of the U.S. Naval Academy, Annapolis, Maryland, extended this result to 5.3×10^{100}.

It is worth noting that we encounter the CBC in integral calculus also. For example, using the substituition $x = a \tan \theta$,

$$\int_0^\infty \frac{dx}{(a^2 + x^2)^{n+1}} = \frac{1}{a^{2n+1}} \int_0^{\pi/2} \sin^{2n} \theta \, d\theta$$

$$= \frac{\pi}{2a^{2n+1}} \cdot \frac{1 \cdot 3 \cdot 5 \cdots (2n-1)}{2 \cdot 4 \cdot 6 \cdots (2n)}$$

$$= \frac{\pi}{(2a)^{2n+1}} \binom{2n}{n}$$

Next, we pursue the convergence of an infinite series involving the binomial coefficient $\binom{2n}{n}$.

Moore and the CBC

We saw earlier that

$$\binom{2n}{n} = \frac{1 \cdot 3 \cdot 5 \cdots (2n-1)}{2 \cdot 4 \cdot 6 \cdots 2n} 2^{2n}$$

where $n \geq 1$. In 1919, R. E. Moore of Lockheed Missiles and Space proposed a simple but delightful problem related to the CBC. His problem was to test the convergence of the series $\sum_{n=1}^{\infty} a_n^2$, where $a_n = \frac{1 \cdot 3 \cdot 5 \cdots (2n-1)}{2 \cdot 4 \cdot 6 \cdots 2n} 2^{2n}$. The solution presented in the next example follows the one given in the following year by O. Dunkel of Washington University.

Example 3.7 Test the convergence of the series $\sum_{n=1}^{\infty} a_n^2$, where $a_n = \frac{1 \cdot 3 \cdot 5 \cdots (2n-1)}{2 \cdot 4 \cdot 6 \cdots 2n} 2^{2n}$.

Solution   We have

$$4a_n^2 = \left(\frac{3}{4}\right)^2 \left(\frac{5}{6}\right)^2 \cdots \left(\frac{2n-3}{2n-2}\right)^2 \left(\frac{2n-1}{2n}\right)^2$$

Since

$$\left(\frac{2k-1}{2k}\right)^2 = \left(1 - \frac{1}{2k}\right)^2$$

$$= 1 - \frac{1}{k} + \frac{1}{4k^2}$$

$$> \frac{k-1}{k}$$

it follows that

$$4a_n^2 > \left(\frac{1}{2}\right) \cdot \left(\frac{2}{3}\right) \cdot \left(\frac{3}{4}\right) \cdots \frac{n-2}{n-1} \cdot \frac{n-1}{n}$$

$$= \frac{1}{n}$$

Thus, by the comparison test, $4 \sum_{n=1}^{\infty} a_n^2 = \sum_{n=1}^{\infty} 4a_n^2$ diverges, so the series $\sum_{n=1}^{\infty} a_n^2$ also diverges. In other words, $\sum_{n=1}^{\infty} \frac{1}{2^{4n}} \binom{2n}{n}^2$ is also a divergent series.   ∎

The next example, proposed in 1957 by A. E. Currier of the U.S. Naval Academy, Annapolis, Maryland, continues to reveal the omnipresence of the CBC. The solution is due to D. G. Cantor of the University of California at Los Angeles.

Example 3.8 Show that

$$\lim_{n\to\infty}\sum_{j=0}^{n}\frac{(-4)^j}{\binom{2n}{n}}\binom{2n-j}{n}=\frac{1}{3}$$

Proof   Let

$$f(x)=\frac{1}{\binom{2n}{n}}\sum_{j=0}^{n}\binom{2n-j}{n}x^j$$

Then by direct substitution,

$$2(x-1)f_n(x)=x-2+\frac{n}{2n-1}x^2f_{n-1}(x) \tag{3.2}$$

and by induction,

$$f_n(x)=\frac{x-2}{2x-2}\sum_{k=1}^{n}\prod_{i=k}^{n-1}\left[\frac{2i+2}{2i+1}\cdot\frac{x^2}{4(x-1)}\right]+\prod_{i=0}^{n-1}\left[\frac{2i+2}{2i+1}\cdot\frac{x^2}{4(x-1)}\right]$$

The sequence $\{f_n(x)\}$ converges when $\left|\frac{x^2}{4(x-1)}\right|<1$, that is, when $-2-2\sqrt{2}<x<-2+2\sqrt{2}$. Then $f(x)=\lim_{n\to\infty}f_n(x)$ exists; and it follows from (3.2) that

$$2(x-1)f(x)=x-2+\frac{x^2}{2}f(x)$$

$$f(x)=\frac{2}{2-x}$$

$$f(-4)=\frac{1}{3}$$

as required.   ∎

In 1936, F. Ayres Jr. of Dickinson College developed the following identities; we omit their proofs in the interest of brevity:

$$\sum_{r=0}^{n}(-1)^r2^{2r}\binom{n}{r}\binom{2n-2r}{n-r}=(-1)^n\binom{2n}{n}$$

$$\sum_{r=0}^{n-1}(-1)^r2^{2r}\binom{n}{r}\binom{2n-2r}{n+1-r}=(-1)^{n-1}\binom{2n}{n+1}$$

Two years later, E. Lehmer of Bethlehem, Pennsylvania, generalized them as follows:

$$\sum_{r=0}^{m}(-1)^r2^{2r}\binom{n}{r}\binom{2n-2r}{k-r}=(-1)^k\binom{2n}{k}$$

where $m=\min\{n,2n-k\}$.

Guessing Card Colors

The next example, proposed in 1988 by M. Andreoli of Miami-Dade Community College, Florida, is an interesting application of Lagrange's identity. The solution given is due to H. Steelman of Indiana University of Pennsylvania and contains the ubiquitous CBC.

Example 3.9 From a well-shuffled, standard deck of 52 cards, one card at a time is selected at random, turned face up, and is shown to a player. Before each card is turned, the player must guess its color. The player always chooses the color that constitutes the majority of the cards unturned, and guesses whimsically when the numbers of red and black cards are the same. Determine the expected number of correct guesses for this strategy.

Solution   Suppose the deck contains m red cards and n black cards. Let $E(m, n)$ denote the expected number of correct guesses. Since $E(m, n) = E(n, m)$, we assume that $m \geq n$. Then $E(m, n)$ can be defined recursively:

$$E(m, n) = \begin{cases} \frac{n}{m+n} E(m, n-1) + \frac{m}{m+n} [E(m-1, n) + 1] & \text{if } m > n > 0 \\ E(n, n-1) + \frac{1}{2} & \text{if } m = n > 0 \\ m & \text{if } n = 0 \end{cases}$$

We will now prove by induction on the sum $m + n$ that

$$E(m, n) = m + \frac{1}{\binom{m+n}{m}} \sum_{r=0}^{n-1} \binom{m+n}{r}$$

where $m \geq n$. The result is trivially true when $m = 0$ or $n = 0$. Assume that it is true for all sums $< m + n$. Then

$$E(m, n) = \frac{n}{m+n} E(m, n-1) + \frac{m}{m+n} [E(m-1, n) + 1]$$

$$= \frac{n}{m+n} \left[m + \frac{1}{\binom{m+n-1}{n-1}} \sum_{r=0}^{n-2} \binom{m+n-1}{r} \right]$$

$$+ \frac{m}{m+n} \left[m + \frac{1}{\binom{m+n-1}{n}} \sum_{r=0}^{n-1} \binom{m+n-1}{r} \right]$$

$$= \frac{mn}{m+n} \cdot \frac{1}{\binom{m+n}{n}} \sum_{r=0}^{n-1} \binom{m+n-1}{r-1}$$

$$+ \frac{m^2}{m+n} + \frac{1}{\binom{m+n}{n}} \sum_{r=0}^{n-1} \binom{m+n-1}{r}$$

$$= \frac{m(m+n)}{m+n} + \frac{1}{\binom{m+n}{n}} \sum_{r=0}^{n-1} \left[\binom{m+n-1}{r-1} + \binom{m+n-1}{r} \right]$$

$$= m + \frac{1}{\binom{m+n}{n}} \sum_{r=0}^{n-1} \binom{m+n}{r}$$

Thus by induction, the result is true for all $m \geq n \geq 0$.
In particular, let $m = n$. Then

$$E(n,n) = n + \frac{1}{\binom{2n}{n}} \sum_{r=0}^{n-1} \binom{2n}{r}$$

But

$$\sum_{r=0}^{2n} \binom{2n}{r} = 2^{2n}$$

That is,

$$\sum_{r=0}^{n-1} \binom{2n}{r} + \binom{2n}{n} + \sum_{r=n+1}^{2n} \binom{2n}{r} = 2^{2n}$$

$$2 \sum_{r=0}^{n-1} \binom{2n}{r} + \binom{2n}{n} = 2^{2n}$$

So

$$\sum_{r=0}^{n-1} \binom{2n}{r} = 2^{2n-1} - \frac{1}{2} \binom{2n}{n}$$

Therefore,

$$E(n,n) = n - \frac{1}{2} + \frac{2^{2n-1}}{\dbinom{2n}{n}}$$

Returning now to the original problem,

$$E(26,26) = 26 - \frac{1}{2} + \frac{2^{51}}{\dbinom{52}{26}}$$

$$\approx \frac{37244306009965475}{1239796332370266}$$

$$\approx 30.04 \qquad \blacksquare$$

Next, we present an example that shows that the ubiquitous CBC occurs even in probability.

CBC and Probability

In 1918, H. R. Howard of the University of St. Francis Xavier's College, Nova Scotia, Canada, proposed an interesting problem in discrete probability, the solution of which involves the CBC. The solution, given by him in the following year, requires a basic knowledge of discrete probability.

Example 3.10 A well-shuffled deck of $2(p+q)$ cards contains $2p$ honors. Suppose one-half of the pack is selected at random. Show that:

a. The probability of of obtaining half the honors is $\frac{[f(p,q)]^2}{f(2p,2q)}$, where $f(p,q)$
   denotes the number of sets of p cards that can be selected from $p+q$ cards.
b. If one honor is removed from the deck, the probability is not thereby affected.
c. Is this true for the probability of getting any other assigned number of honors?

Solution   Notice that $f(p,q) = \dbinom{p+q}{p}$.

(a) First, we need to find the number of ways we can get exactly p honors in $p+q$ cards. Clearly, we can choose p honors in $\dbinom{2p}{p}$ ways and the other q cards $\dbinom{2q}{q}$ ways. So the number of ways of selecting one-half the honors from the half pack is $\dbinom{2p}{p}\dbinom{2q}{q}$.

But the number of ways of selecting $p+q$ cards from the given full pack is $\binom{2p+2q}{p+q}$. Thus the desired probability is

$$\frac{\binom{2p}{p}\binom{2q}{q}}{\binom{2p+2q}{p+q}} = \frac{(2p)!(2q)!}{(p!)^2(q!)^2} \cdot \frac{[(p+q)!]^2}{(2p+2q)!} \tag{3.3}$$

$$= \left[\frac{(p+q)!}{p!q!}\right]^2 \Bigg/ \frac{(2p+2q)!}{(2p!)(2q!)}$$

$$= \frac{[f(p,q)]^2}{f(2p,2q)}$$

(b) Suppose one honor is removed from the given pack. Then the number of ways of selecting $p+q$ cards from the rest and obtaining exactly p honors is $\binom{2p-1}{p}\binom{2q}{q}$; so the desired probability is

$$\frac{\binom{2p-1}{p}\binom{2q}{q}}{\binom{2p+2q-1}{p+q}} \tag{3.4}$$

Dividing (3.3) by (3.4), we get

$$\frac{\binom{2p}{p}\binom{2q}{q}}{\binom{2p+2q}{p+q}} \cdot \frac{\binom{2p+2q-1}{p+q}}{\binom{2p-1}{p}\binom{2q}{q}} = \frac{\binom{2p}{p}\binom{2p+2q-1}{p+q}}{\binom{2p-1}{p}\binom{2p+2q}{p+q}}$$

$$= \frac{(2p)!}{p!p!} \cdot \frac{(2p+2q-1)!}{(p+q)!(p+q-1)!} \cdot \frac{p!(p-1)!}{(2p-1)!} \cdot \frac{(p+q)!(p+q)!}{(2p+2q)!}$$

$$= 1$$

Therefore, the two probabilites are equal. In other words, removing one honor from the pack does not affect the probability.

(c) Let x be the assigned number of honors. Then the probability of selecting x honors is given by

$$\frac{\binom{2p}{x}\binom{2q}{p+q-x}}{\binom{2p+2q}{p+q}} \tag{3.5}$$

Suppose one honor is removed from the pack. Then the corresponding probability is

$$\frac{\binom{2p-1}{x}\binom{2q}{p+q-x}}{\binom{2p+2q-1}{p+q}} \tag{3.6}$$

Dividing (3.5) by (3.6), these two probabilities can be equal if and only if

$$\frac{p}{2p-x} = 1; \text{ that is, if and only if } x = p.$$

Thus, removing a honor from the pack will not affect the probability if and only if the assigned number of honors is p. ∎

The following two identities* were developed in 2005 by Heinz-Jürgen Seiffert of Berlin, Germany:

$$\sum_{k=0}^{\lfloor n/2 \rfloor} 2^{n-2k} \frac{\binom{n}{2k}\binom{2k}{k}}{\binom{k+m}{m}} = \frac{\binom{2n+2m}{n+m}}{\binom{n+2m}{m}}$$

$$\sum_{k=0}^{\lfloor n/2 \rfloor} \frac{\binom{n}{2k}\binom{2k}{k}}{\binom{2k+m}{m}\binom{2k+2m+1}{m+1}} = \frac{2^n}{\binom{n+2m+1}{m+1}}$$

where n and m are nonnegative integers.

Fibonacci, Lucas, and the CBCs

Interestingly, the CBCs pop up in the study of Fibonacci numbers F_n and Lucas numbers L_n also, as the next example shows. It is based on a problem proposed in 2006 by Michel Bataille of Rouen, France.

We will employ the following facts in the proofs of the formulas in the example:

Lucas's Formula

$$F_n = \sum_{i=0}^{\lfloor (n-1)/2 \rfloor} \binom{n-i-1}{i}, \quad n \geq 1$$

This formula** was discovered in 1876 by French mathematician François-Edouard-Anatole Lucas (1842–1891).

Fermat's Little Theorem   Let p be a prime and a any integer such that $p \nmid a$. Then $a^{p-1} \equiv 1 \pmod{p}$.

This result[†] was discovered in 1640 by French mathematician Pierre de Fermat (1601–1665).

---

* For a proof, see *The American Mathematical Monthly*, **114** (May 2007), 457–458.
**See author's *Fibonacci and Lucas Numbers with Applications*, Wiley, New York, 2001.
[†] See author's *Elementary Number Theory with Applications*, 2nd ed., Academic Press, Boston, Massachusetts, 2007.

Example 3.11 Let p be a prime. Then:

1. $$\sum_{n=0}^{\lfloor (p-1)/2 \rfloor} (-1)^n \binom{2n}{n} \equiv F_p \pmod{p}$$

2. $$\sum_{n=0}^{\lfloor (p-1)/2 \rfloor} (-1)^n \binom{2n}{n} (n+1)^{p-2} \equiv L_{p-1} \pmod{p}$$

Solution

1) We have:

$$\binom{p-n-1}{n} = \frac{(p-n-1)\cdots(p-2n)}{n!}$$

$$= (-1)^n \frac{(2n-p)\cdots(n+1-p)}{n!}$$

$$\equiv (-1)^n \frac{(n+1)\cdots(2n)}{n!} \pmod{p}$$

$$\equiv (-1)^n \binom{2n}{n} \pmod{p} \tag{3.7}$$

Therefore, by Lucas's formula,

$$F_p = \sum_{n=0}^{\lfloor (p-1)/2 \rfloor} \binom{p-n-1}{n}$$

$$\equiv \sum_{n=0}^{\lfloor (p-1)/2 \rfloor} (-1)^n \binom{2n}{n} \pmod{p} \tag{3.8}$$

as desired.

2) By Lucas's formula,

$$F_{p-2} = \sum_{n=0}^{\lfloor (p-3)/2 \rfloor} \binom{p-n-3}{n}$$

$$= \sum_{n=1}^{\lfloor (p-1)/2 \rfloor} \binom{p-n-2}{n}$$

But

$$\binom{p-n-2}{n-1} = \frac{n}{p-n-1}\binom{p-n-1}{n}$$

$$\equiv (-1)^{n+1}\frac{n}{n+1}\binom{2n}{n} \pmod{p}, \quad \text{by equation (3.7)}$$

So

$$F_{p-2} \equiv \sum_{n=1}^{\lfloor(p-1)/2\rfloor} (-1)^{n+1}\frac{n}{n+1}\binom{2n}{n} \pmod{p}$$

$$\equiv \sum_{n=0}^{\lfloor(p-1)/2\rfloor} (-1)^{n+1}\frac{n}{n+1}\binom{2n}{n} \pmod{p} \tag{3.9}$$

It is well known* that $L_m = F_{m-1} + F_{m+1}$, so $L_{p-1} = F_{p-2} + F_p$. Thus, by adding congruences (3.8) and (3.9),

$$L_{p-1} \equiv \sum_{n=0}^{\lfloor(p-1)/2\rfloor} (-1)^n \left(1 - \frac{n}{n+1}\right)\binom{2n}{n} \pmod{p}$$

$$\equiv \sum_{n=0}^{\lfloor(p-1)/2\rfloor} (-1)^n \frac{1}{n+1}\binom{2n}{n} \pmod{p}$$

Since $0 \le n \le \frac{p-1}{2}$, $p \nmid (n+1)$. Therefore, by Fermat's little theorem, $(n+1)^{p-1} \equiv (n+1)^{-1} \pmod{p}$. Thus,

$$L_{p-1} \equiv \sum_{n=0}^{\lfloor(p-1)/2\rfloor} (-1)^n \binom{2n}{n}(n+1)^{p-2} \pmod{p} \tag{3.10}$$

again as desired. $\blacksquare$

Norton's Formula

In 1978, R. M. Norton of the College of Charleston, South Carolina, investigated the determinant $|A|$ of the matrix $A = (a_{ij})_{(n+1)\times(n+1)}$, where

$$a_{ij} = \begin{cases} 0 & \text{if } i+j \text{ is odd} \\ \binom{2k}{k} & \text{if } i+j = 2k \text{ is even} \end{cases}$$

* See author's *Fibonacci and Lucas Numbers with Applications*, Wiley, New York, 2001.

He showed that $|A| = 2^n$.

For example, when $n = 3$,

$$|A| = \begin{vmatrix} \binom{0}{0} & 0 & \binom{2}{1} & 0 \\ 0 & \binom{2}{1} & 0 & \binom{4}{2} \\ \binom{2}{0} & 0 & \binom{4}{2} & 0 \\ 0 & \binom{4}{2} & 0 & \binom{6}{3} \end{vmatrix}$$

$$= 8$$

Carlitz's Formula

In 1967, L. Carlitz of Duke University developed a combinatorial identity. Let

$$S_n = \sum_{\substack{i,j,k \geq 0 \\ i+j+k=n}} \binom{i+j}{i}\binom{j+k}{j}\binom{k+i}{k}$$

Then $S_n - S_{n-1} = \binom{2n}{n}$.

For example, $S_2 = 9$ and $S_3 = 29$, so $S_3 - S_2 = 20 = \binom{6}{3}$.

Since $S_0 = 1$, S_n can be defined recursively:

$$S_0 = 1$$

$$S_n = S_{n-1} + \binom{2n}{n}, \quad n \geq 1$$

It follows from the recurrence relation that

$$\sum_{r=1}^{n} S_r = \sum_{r=1}^{n} S_{r-1} + \sum_{r=1}^{n} \binom{2r}{r}$$

This implies that

$$S_n = S_0 + \sum_{r=1}^{n} \binom{2r}{r}$$

$$= \sum_{r=0}^{n} \binom{2r}{r}$$

Leonard Carlitz (1907–1999) was born in Philadelphia. After receiving his Ph.D. in mathematics from the University of Pennsylvania, he spent a year with Eric T. Bell at the California Institute of Technology as a National Research Council Scholar and the following year with Godfrey H. Hardy at Cambridge University as an International Research Fellow.

After returning from Cambridge, Carlitz joined the faculty at Duke University, Durham, North Carolina, where he remained until retirement in 1977. In 1964, he was named James B. Duke Professor of Mathematics, thus becoming the first member of the department to hold the distinguished professorship.

Carlitz published 770 research papers. He wrote a record number of them (forty-four) in 1953. A great majority of his forty-five doctoral students worked in finite fields, an area of abiding interest for him. According to David R. Hayes of the University of Massachusetts, a doctoral student of Carlitz, he "leaves a remarkable mathematical legacy, including work in number theory, finite field theory, combinatorics, special functions, and the arithmetic of polynomials over a finite field."

For example,

$$S_4 = \sum_{r=0}^{4} \binom{2r}{r}$$

$$= 1 + 2 + 6 + 20 + 70$$

$$= 99$$

Srinivasa Aiyangar Ramanujan (1887–1920), the greatest Indian mathematical genius, was born in Erode, near Madras. He was the son of a bookkeeper at a cloth store in Kumbakonam. After two years of elementary school and at the age of seven, he transferred to the high school there. At age ten, he placed first in the district

primary examination. In 1903, his passion for mathematics was sparked when he borrowed from a university student's copy of George Schoobridge Carr's *A Synopsis of Elementary Results in Pure and Applied Mathematics*. Without any formal training or outside help, Ramanujan established all 6000-plus theorems in the book, stated without proofs or any explanation, and kept their proofs in a notebook.

Graduating from high school in 1904, he entered the University of Madras on a scholarship. However, his excessive neglect of all subjects except mathematics caused him to lose the scholarship after a year, and Ramanujan dropped out of college. He returned to the university after some traveling through the countryside, but never graduated. At the university he pursued his passion, rediscovering previously known results and discovering new ones in hypergeometric series and elliptic functions.

His marriage in 1909 compelled him to earn a living. Three years later, he secured a low-paying clerk's job with the Madras Port Trust. He published his first article in 1911 on Bernoulli numbers in the *Journal of the Indian Mathematical Society* and two more the following year.

In 1913, Ramanujan began corresponding with eminent English mathematician Godfrey H. Hardy of Cambridge University. His first letter included more than 100 theorems, some without proofs. After examining them carefully, Hardy concluded that "they could only be written down by a mathematician of the highest class; they must be true because if they were not true, no one would have the imagination to invent them."

With the help of a scholarship arranged by Hardy, Ramanujan arrived in Cambridge in 1914. During his five-year stay, the two collaborated on a number of articles in the theory of partitions, analytic number theory, continued fractions, infinite series, and elliptic functions.

(*continued*)

In 1917, Ramanujan became seriously ill. He was incorrectly diagnosed with tuberculosis; however, it is now believed that he suffered from a vitamin deficiency caused by his strict vegetarianism.

When Ramanujan was sick in a nursing home, Hardy visited him. Hardy told him that the number of the cab he came in, 1729, was a "rather dull number"and that he hoped it wasn't a bad omen. "No, sir," Ramanujan responded. "It is a very interesting number. It is the smallest number expressible as the sum of two cubes in two different ways."

In 1918, Ramanujan became one of the youngest fellows of the Royal Society and a fellow of Trinity College.

Ramanujan returned to India in the following year. He pursued his mathematical passion even on his deathbed. His very short but extremely productive life came to an end when he was only thirty-two.

Summation Formulas and the CBC

There are numerous infinite series that contain the CBC. They were studied by S. Ramanujan, D. H. Lehmer, J. Riordan, Z. A. Melzak, and A. J. van der Poorten, among others. For instance, one of the fifteen infinite series expansions for $\frac{1}{\pi}$, discovered by Ramanujan, contains the CBC:

$$\frac{1}{\pi} = \sum_{n=0}^{\infty} \binom{2n}{n}^3 \frac{42n + 5}{2^{12n+4}}$$

It is also known* that

$$\sum_{n-1}^{\infty} \frac{1}{\binom{2n}{n}} = \frac{1}{3} + \frac{2\pi\sqrt{3}}{27}$$

and

$$\sum_{n=1}^{\infty} \frac{(-1)^{n-1}}{\binom{2n}{n}} = \frac{1}{5} + \frac{4\sqrt{5}}{25} \log \alpha$$

where $\alpha = \frac{1+\sqrt{5}}{2}$ denotes the well-known golden ratio[†]; both sums are transcendental numbers.

---

* See E. R. Hansen's *A Table of Series and Products.*
[†] See T. Koshy, *Fibonacci and Lucas Numbers with Applications,* for a discussion of the golden ratio.

Some additional summation formulas[‡] are listed below, where $|x| < 1$:

$$\sum_{r=0}^{n} (-1)^r \binom{2n}{r} = \frac{(-1)^n}{2}\binom{2n}{n}$$

$$\sum_{r=0}^{n} \binom{2r}{r}\binom{2n-2r}{n-r}\frac{1}{2r+1} = \frac{2^{4n}}{(2n+1)\binom{2n}{n}}$$

$$\sum_{n=1}^{\infty} \frac{\binom{2n}{n}}{(2n+1)16^n} = \frac{\pi}{3}$$

$$\sum_{n=1}^{\infty} \frac{n}{\binom{2n}{n}} = \frac{2}{27}(\sqrt{3}\pi + 9)$$

$$\sum_{n=1}^{\infty} \frac{n2^n}{\binom{2n}{n}} = \pi + 3$$

$$\sum_{n=1}^{\infty} \frac{3^n}{\binom{2n}{n}} = \frac{4\sqrt{3}\pi}{3} + 3$$

$$\sum_{n=1}^{\infty} \frac{3^n}{\binom{2n}{n}n^2} = \frac{2\pi^2}{9}$$

$$\sum_{n=1}^{\infty} \frac{(-1)^{n+1}}{\binom{2n}{n}n} = \frac{2\sqrt{5}}{5}\log\alpha$$

$$\sum_{n=1}^{\infty} \frac{(-1)^{n+1}}{\binom{2n}{n}n^2} = 2\log^2\alpha$$

[‡] I. J. Zucker, "On the Series $\sum_{k=1}^{\infty} \left(\frac{2k}{k}\right)^{-1} k^{-n}$ and Related Sums," *J. of Number Theory* 20 (1985), 92–102; J. M. Borwein and P. B. Borwein, *Pi and AGM*, Wiley, New York, 1987.

$$\sum_{r=0}^{n+1} r \binom{n+1}{r}^2 = (2n+1)\binom{2n}{n}$$

$$\sum_{r=0}^{n} (-1)^r \frac{\binom{n}{r}^2}{\binom{2n}{r}} = \frac{1}{\binom{2n}{n}}$$

$$\sum_{n=1}^{\infty} \frac{(-1)^n}{\binom{2n}{n} n^3} = 4 \int_0^{1/2} \log^2\left(y + \sqrt{(1+y^2)}\right) \frac{dy}{y}$$

$$\sum_{n=1}^{\infty} \frac{1}{\binom{2n}{n} n^4} = \frac{17\pi^6}{3240} \qquad \text{(Comtet's formula)}$$

$$\sum_{n=0}^{\infty} \binom{2n}{n} x^n = \frac{1}{\sqrt{1-4x}}, \quad |x| < 1/4$$

$$\sum_{n=0}^{\infty} \frac{1}{8^n} \binom{2n}{n} = \sqrt{2}$$

$$\sum_{n=0}^{\infty} \frac{(-1)^n}{8^n} \binom{2n}{n} = \frac{3\sqrt{2}}{3}$$

$$\sum_{n=0}^{\infty} \frac{1}{64^n} \binom{4n}{2n} = \frac{3\sqrt{2} + \sqrt{6}}{6}$$

$$\sum_{n=0}^{\infty} \frac{1}{10^n} \binom{2n}{n} = \frac{3\sqrt{5}}{3}$$

$$\sum_{n=0}^{\infty} \frac{1}{8^{4n}} \binom{8n}{4n} = \frac{15\sqrt{2} + 5\sqrt{6} + 6\sqrt{5}\sqrt{2+\sqrt{5}}}{60}$$

$$\sum_{n=1}^{\infty} \frac{1}{n} \binom{2n}{n} x^n = 2 \log\left(\frac{1 - \sqrt{1-4x}}{2x}\right), \quad |x| < 1/4$$

$$\sum_{n=1}^{\infty} \frac{(2x)^{2n}}{\binom{2n}{n}} = \frac{x^2}{1-x^2} + \frac{x \arcsin x}{(1-x^2)^{3/2}}$$

$$\sum_{n=1}^{\infty} \frac{(2x)^{2n}}{\binom{2n}{n} n} = \frac{2x \arcsin x}{\sqrt{1-x^2}}$$

$$\sum_{n=1}^{\infty} \frac{1}{\binom{2n}{n} n} = \frac{\sqrt{3}\pi}{9}$$

$$\sum_{n=1}^{\infty} \frac{(2x)^{2n}}{\binom{2n}{n} n^2} = 2(\arcsin x)^2 \qquad \text{(Euler's formula)}$$

$$\sum_{n=1}^{\infty} \frac{1}{\binom{2n}{n} n^2} = \frac{\pi^2}{18}$$

$$\sum_{n=1}^{\infty} \frac{\binom{2n}{n}}{(2n+1)} x^{2n} = \frac{1}{2x}(\arcsin 2x)$$

$$\sum_{n=0}^{\infty} \binom{2n}{n} \frac{1}{2^{2n}(2n+1)^2} = \frac{\pi}{2} \log 2$$

$$\sum_{n=0}^{\infty} \binom{2n}{n} \frac{1}{4^{2n}(2n+1)^3} = \frac{7\pi^3}{216}$$

$$\sum_{n=0}^{\infty} \binom{2n}{n} \frac{(-1)^n}{2^{3n}(2n+1)^2} = \frac{\pi^2}{10}$$

$$\sum_{n=0}^{\infty} \binom{2n}{n} \frac{x^n}{n+1} = \frac{1-\sqrt{1-4x}}{2x}, \quad |x| < 1/4$$

$$\sum_{n=0}^{\infty} \frac{1}{4^n(n+1)} \binom{2n}{n} = 2$$

$$\sum_{n=1}^{\infty} \binom{2n}{n} \frac{x^n}{n} = 2\log \frac{1-\sqrt{1-4x}}{2x}, \quad |x| < 1/4$$

$$\sum_{n=1}^{\infty} \frac{1}{n4^n} \binom{2n}{n} = \log 4$$

$$\sum_{n=1}^{\infty} \frac{(-1)^{n-1}}{n4^n} \binom{2n}{n} = \log \frac{1+\sqrt{2}}{2}$$

$$\sum_{n=1}^{\infty} \binom{2n}{n} \frac{x^{n+1}}{n(n+1)} = 2x\log \frac{1-\sqrt{1-4x}}{x} + \frac{\sqrt{1-4x}}{2}$$
$$+ (\log 4 - 1)x - 1/2, \quad |x| < 1/4$$

$$\sum_{n=1}^{\infty} \binom{2n}{n} nx^n = 2x(1-4x)^{-3/2}, \quad |x| < 1/4$$

$$\sum_{n=1}^{\infty} \binom{2n}{n} n^2 x^n = \frac{2x(2x+1)}{(1-4x)^{5/2}}, \quad |x| < 1/4$$

$$\sum_{n=0}^{\infty} \frac{(2x)^{2n}}{\binom{2n}{n}(2n+1)} = \frac{\arcsin x}{x\sqrt{1-x^2}} \qquad \text{(Pennisi, 1995)}$$

$$x^2 + \sum_{n=1}^{\infty} \frac{2^{2n+1}x^{2n+2}}{\binom{2n}{n}(2n+1)(2n+2)} = (\arcsin x)^2 \qquad \text{(Pennisi, 1995)}$$

$$\sum_{k=1}^{n-1} \frac{(-1)^{k-1}[(k-1)!]^2}{(n^2-1^2)\cdots(n^2-k^2)} = \frac{1}{n^2} - \frac{2(-1)^{n-1}}{\binom{2n}{n}n^2}$$

4

Binomial Coefficients Revisited

This chapter continues the investigation of binomial coefficients and develops a few of the myriad combinatorial identities. Many of them are applications of the binomial theorem.

We begin with the well-known *Lagrange's identity*, named after the great French mathematician Joseph L. Lagrange; it involves the ubiquitous CBC.

Theorem 4.1 (*Lagrange's identity*) Let $0 \leq r \leq n$. Then

$$\sum_{r=0}^{n} \binom{n}{r}^2 = \binom{2n}{n} \tag{4.1}$$

Proof   We shall expand $(1 + x)^{2n}$ in two different ways and then equate the coefficients of x^n from both sides. We have

$$(1 + x)^{2n} = (1 + x)^n (1 + x)^n$$

$$= \left[\binom{n}{r} x^{n-r}\right]\left[\binom{n}{r} x^r\right]$$

Equate the coefficients of x^n from both sides, we get:

$$\binom{2n}{r} = \sum_{r=0}^{n} \binom{n}{r}\binom{n}{r}$$

$$= \sum_{r=0}^{n} \binom{n}{r}^2$$

Joseph Louis Lagrange
(1736–1813), who ranks with Leonhard Euler as one of the greatest mathematicians of the eighteenth century, was the eldest of eleven children in a wealthy family in Turin, Italy. His father, an influential Cabinet official, became bankrupt due to unsuccessful financial speculations, which forced Lagrange to pursue a profession.

As a young man studying the classics at the College of Turin, his interest in  mathematics was kindled by an essay by astronmer Edmund Halley on the superiority of the analytical methods of calculus over geometry in the solution of optical problems. In 1754, he began corresponding with several outstanding mathematicians in Europe. The following year, Lagrange was appointed professor of mathematics at the Royal Artillery School in Turin. Three years later, he helped found a society that later became the Turin Academy of Sciences. While at Turin, Lagrange developed revolutionary results in the calculus of variations, mechanics, sound, and probability, winning the presitgious Grand Prix of the Paris Academy of Sciences in 1764 and 1766.

In 1766, when Euler left the Berlin Academy of Sciences, Frederick the Great wrote to Lagrange that "the greatest king in Europe" would like to have "the greatest mathematician of Europe" at his court. Accepting the invitation, Lagrange moved to Berlin to head the academy and remained there for twenty years. When Frederick died in 1786, Lagrange moved to Paris at the invitation of Louis XVI. Lagrange was appointed professor at the École Normale and then at the École Polytechnique, where he taught until 1799.

Lagrange made significant contributions to analysis, analytical mechanics, calculus, probability, and number theory and also helped set up the French metric system.

That is,

$$\sum_{r=0}^{n} \binom{n}{r}^2 = \binom{2n}{n}$$ ∎

For example, let $n = 4$. Then

$$\sum_{r=0}^{4} \binom{4}{r}^2 = \binom{4}{0}^2 + \binom{4}{1}^2 + \binom{4}{2}^2 + \binom{4}{3}^2 + \binom{4}{4}^2$$

$$= 1^2 + 4^2 + 6^2 + 4^2 + 1^2$$

$$= 70$$

$$= \binom{8}{4}$$

Vandermonde's Identity

Lagrange's identity is a special case of the well-known Vandermonde's identity, presented in the following theorem. Vandermonde's identity is named after the French mathematician and musician Alexandre-Théophile Vandermonde. We shall prove it, first using the binomial theorem and then using a simple combinatorial argument.

Alexandre-Théophile Vandermonde (1735–1796) was born in Paris. Because of his ill health as a child, his physician father guided him toward a career in music. Vandermonde's later interest in mathematics lasted for just two years. His mathematical output consists of four papers published during this brief period; they include significant and influential contributions to the theory of determinants, theory of equations, and the well-known knight's tour problem. In 1771, Vandermonde was elected to the French Academy of Sciences.

Later, Vandermonde returned to music and wrote several papers on harmony, as well as on experiments with cold and the manufacturing of steel. Ironically, mathematicians considered Vandermonde to be a musician, and musicians considered him to be mathematician.

Theorem 4.2 (*Vandermonde's identity*) Let $m, n, r \geq 0$, where $0 \leq r \leq m, n$. Then

$$\binom{m+n}{r} = \sum_{k=0}^{r} \binom{m}{k}\binom{n}{r-k}$$ (4.2)

Proof   Since $(1 + x)^{m+n} = (1 + x)^m (1 + x)^n$, using the binomial theorem, this yields

$$\sum_{r=0}^{m+n} \binom{m+n}{r} x^r = \left[\sum_{k=0}^{m} \binom{m}{k} x^k \right] \left[\sum_{k=0}^{n} \binom{n}{r-k} x^{r-k} \right]$$

Equating the coefficients of x^r from both sides, we get the desired result.   ∎

We now give a combinatorial proof of this identity.

A Combinatorial Argument

Suppose there are m men and n women, and we would like to select r of the $m + n$ people. This can be done in $\binom{m+n}{r}$ ways.

Another way of selecting r people from the $m + n$ people is by picking k people from the m men and the remaining $r - k$ people from the n women. By the multiplication principle, this can be accomplished in $\binom{m}{k}\binom{n}{r-k}$ ways. Since $0 \le k \le r$, by the addition principle, the total number of ways of selecting r people is given by $\sum_{k=0}^{r} \binom{m}{k}\binom{n}{r-k}$. Thus

$$\binom{m+n}{r} = \sum_{k=0}^{r} \binom{m}{k}\binom{n}{r-k}$$ ∎

For example, let $m = 7, n = 10$, and $r = 5$. Then

$$\sum_{k=0}^{5} \binom{7}{k}\binom{10}{5-k} = \binom{7}{0}\binom{10}{5} + \binom{7}{1}\binom{10}{4} + \binom{7}{2}\binom{10}{3}$$

$$+ \binom{7}{3}\binom{10}{2} + \binom{7}{4}\binom{10}{1} + \binom{7}{5}\binom{10}{0}$$

$$= 252 + 1470 + 2520 + 1575 + 350 + 21$$

$$= 6,188$$

$$= \binom{7+10}{5}$$

Notice that Lagrange's identity follows from Vandermonde's identity, by letting $m = n = r$ and $k = 0$. Also, by letting $m = n$, it yields the identity

$$\sum_{r=0}^{n-1} \binom{n}{r}\binom{n}{r+1} = \binom{2n}{n+1}$$

By letting $r = m - s$, Vandermonde's identity yields

$$\binom{m+n}{m-s} = \sum_{k=0}^{m-s} \binom{m}{m-s-k} \binom{n}{k}$$

That is,

$$\sum_{k=0}^{m} \binom{m}{s+k} \binom{n}{k} = \binom{m+n}{n+s} \tag{4.3}$$

where although we have changed the upper limit to m, $k \leq m - s$.

The binomial coefficient $\binom{m+n}{m} = \binom{m+n}{n}$ has a delightful geometric interpretation. It was originally given in 1938 in the Italian journal *Periodico di Matematiche* by C. Ciamberlini and A. Margengoni, and then reported in *Scripta Mathematica* in the same year by J. Ginsburg of Yeshiva University, New York.

Before presenting it, we state the following result from combinatorics but omit its proof for convenience.

Theorem 4.3 The number of permutations of n items of which n_1 items are of one type, n_2 are of a second type, $\cdots$, and n_k are of a kth type, is $\frac{n!}{n_1! n_2! \cdots n_k!}$. ■

For example, there are $\frac{9!}{2! 4!} = 7560$ permutations of the letters of the word REFERENCE. We now pursue the geometric interpretation of the binomial coefficient $\binom{m+n}{m}$.

Lattice Walking

Consider a rectangular grid of avenues and streets in a metropolis, as Figure 4.1 shows. The avenues run west to east, and the streets run south to north. Suppose we would like to find the number of paths from point O to point A. We can travel in the easterly or northerly direction only.

Two possible paths are shown in the figure. The heavy route can be represented by the eight-letter string EENNENEE; it means, travel two blocks east, two blocks north, one block east, one block north, and two blocks east. The second route, NEEEENEN, can be interpreted similarly.

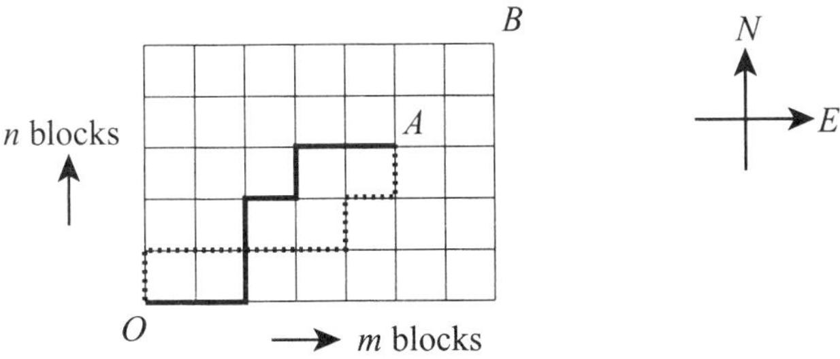

Figure 4.1 A Portion of a City Map

Thus every route from O to A can be represented a "word" of eight letters, of which five are alike (E's) and three are alike (N's). Therefore,

$$
\begin{pmatrix} \text{total number of paths} \\ \text{from O to A} \end{pmatrix} = \begin{pmatrix} \text{number of eight-letter words of which five} \\ \text{are alike and the other five are alike} \end{pmatrix}
$$

$$
= \frac{8!}{5!3!}
$$

$$
= \binom{8}{5}
$$

More generally, point B is m blocks east of O and n blocks north of O, as in Figure 4.1. Then there are $\binom{m+n}{m} = \binom{m+n}{n}$ paths from O to B, covering exactly $m + n$ blocks. In particular, if $m = n$, then there are $\binom{2n}{n}$ such paths from O to B.

Figure 4.1 can be employed to present a geometric interpretation of the identity in Corollary 2.1:

$$
\binom{n}{0} + \binom{n}{1} + \binom{n}{2} + \cdots + \binom{n}{n} = 2^n
$$

To this end, first we define a *lattice point* on the cartesian plane as a point with integral coordinates. For example, $(3, -5)$ is a lattice point, whereas $(3,2.8)$ is *not*.

To interpret the identity geometrically, consider the line segment $x + y = 4$, in Figure 4.2, where $0 \le x, y \le 4$. There are five points on this line segment: (4,0), (3,1), (2,2), (1,3), and (0,4). It is easy to verify that there are exactly 1, 4, 6, 4, and 1 paths from O to each of them. In other words, the number of paths to each lattice point on the line segment is given by $\binom{x+y}{r} = \binom{4}{r}$, where $0 \le x, y \le 4$.

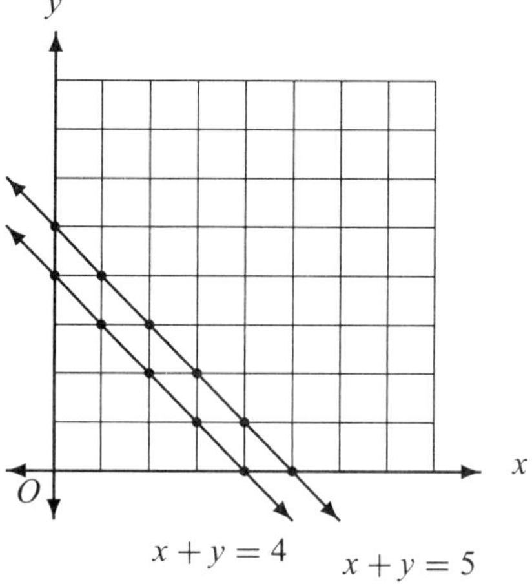

Figure 4.2 Lattice Points in the First Quadrant

More generally, the number of paths to each lattice point on the line segment $x + y = n$, where $0 \leq x, y \leq n$, is given by $\binom{n}{r}$, where $0 \leq r \leq n$. Thus the total number of paths from O to the lattice points on the line $x + y = n$ in the first quadrant is given by $\sum_{r=0}^{n} \binom{n}{r}$.

Now consider the lines $x + y = 4$ and $x + y = 5$ in the first quadrant. Notice that every path ending at a lattice point on the line segment $x + y = 4$ branches off to form two new paths on to two adjacent lattice points on the line segment $x + y = 5$. But there are $1 + 4 + 6 + 4 + 1 = 16$ paths from O leading to the various lattice points on the line segment $x + y = 4$, so there are $2 \cdot 16 = 32 = 1 + 5 + 10 + 10 + 5 + 1$ paths leading to the lattice points on the line segment $x + y = 5$, where $0 \leq x, y \leq 5$.

More generally, the number of paths from O to the various lattice points on the line segment $x + y = n$, where $0 \leq x, y \leq n$, equals twice the number of such paths to the various lattice points on the line segment $x + y = n - 1$, where $0 \leq x, y \leq n - 1$.

There are two paths from O to the lattice points on the line segment $x + y = 1$; four to $x + y = 2$; eight to $x + y = 3$; and so on. Inductively, there are 2^n paths from O to the various lattice points on the line segment $x + y = n$, where $0 \leq x, y \leq n$. Thus

$$\sum_{r=0}^{n} \binom{n}{r} = 2^n$$

as desired. ■

As a byproduct, it follows that the CBC $\binom{2n}{n}$ gives the number of paths from O to the middle lattice point (n, n) on the line segment $x + y = 2n$, where $0 \leq x, y \leq 2n$.

Another Combinatorial Formula

The geometric technique just illustrated can be applied to derive yet another delightful formula for $\binom{m+n}{m}$, using the lattice points in the first quadrant on the cartesian plane. It was developed in 1946 by H. D. Grossman of New York City.

To this end, consider the rectangle $OABC$ in Figure 4.3 and the line segment $x + y = n + i$, where $0 \leq x, y \leq n + i, m \geq n$, and $0 \leq i \leq m - n$. The line segment contains the lattice points $P_1(i, n), P_2(i + 1, n - 1), \cdots, P_{n+1}(i + n, 0)$.

From the earlier discussion, we know that the number of distinct paths from O to B is given by $\binom{m+n}{m} = \binom{m+n}{n}$. Every path from O to B passes through one and only one P_i. By the multiplication and addition formulas, the number of paths from O to B equals the sum of the products of the number of paths from O to P_i

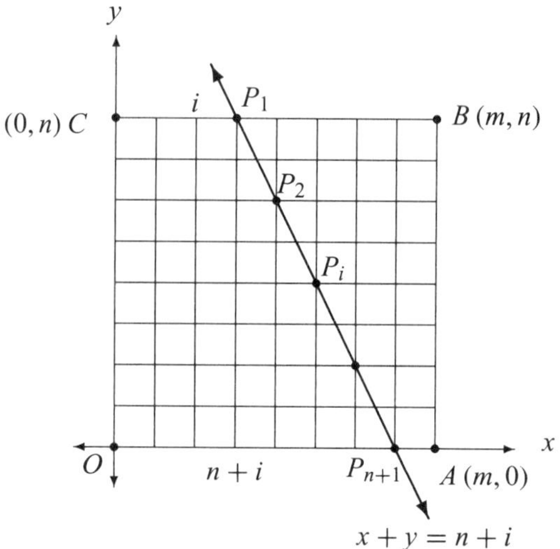

Figure 4.3

and the number of paths from P_i to B. Thus

$$\binom{m+n}{m} = \binom{n+i}{n}\binom{m-i}{0} + \binom{n+i}{n-1}\binom{m-i}{1} + \cdots + \binom{n+i}{0}\binom{m-i}{n} \qquad (4.4)$$

For example, let $m = 5, n = 3$, and $i = 2$. Then the RHS of formula (4.4) yields

$$\binom{5}{3}\binom{3}{0} + \binom{5}{2}\binom{3}{1} + \binom{5}{1}\binom{3}{2} + \binom{5}{0}\binom{3}{3} = 10 + 30 + 15 + 1 = 56$$

and the LHS yields $\binom{8}{5} = 56$. Thus the formula holds when $m = 5, n = 3$, and $i = 2$.

In particular, let $m = n$ and $i = 0$. Then formula (4.4) yields

$$\binom{2n}{n} = \binom{n}{0}^2 + \binom{n}{1}^2 + \cdots + \binom{n}{n}^2$$

which is Lagrange's identity.

By slightly altering this strategy, Grossman derived a different combinatorial identity:

$$\binom{m+r+1}{n} = \sum_{i=0}^{n}\binom{r+i}{i}\binom{m-i}{n-i} \qquad (4.5)$$

Three Interesting Special Cases

Formula (4.5) yields three interesting special cases. When $r = 0$, it yields the identity

$$\binom{m}{n} + \binom{m-1}{n-1} + \cdots + \binom{m-n}{0} = \binom{m+1}{n} \tag{4.6}$$

When $r = 1$, it gives

$$\binom{m}{n} + 2\binom{m-1}{n-1} + 3\binom{m-2}{n-2} \cdots + (n+1)\binom{m-n}{0} = \binom{m+2}{n} \tag{4.7}$$

and when $m = n = r$, formula (4.5) yields

$$\binom{n}{0} + \binom{n}{1} + \cdots + \binom{2n}{n} = \binom{2n+1}{n} \tag{4.8}$$

The following example expands $(1 + x)^{2n}$ in two different ways and derives a new combinatorial identity. In the procees, we encounter the CBC again.

Example 4.1 Prove that

$$\sum_{r=0}^{\lfloor n/2 \rfloor} \binom{n}{r}\binom{n-r}{r} 2^{n-2r} = \binom{2n}{n}$$

Proof   We have

$$(1+x)^{2n} = [1 + x(2+x)]^n$$

$$= \sum_{k=0}^{n} \binom{n}{k} x^k (2+x)^k$$

$$= \sum_{k=0}^{n} \binom{n}{k} x^k \left[\sum_{r=0}^{k} \binom{k}{r} 2^{k-r} x^r \right]$$

$$= \sum_{k=0}^{n} \binom{n}{k} \left[\sum_{r=0}^{k} \binom{k}{r} 2^{k-r} x^{k+r} \right]$$

Equating the coefficents of x^n from both sides, we get:

$$\binom{2n}{n} = \sum_{r=\lfloor n/2 \rfloor}^{0} \binom{n}{r}\binom{n-r}{r} 2^{n-2r}$$

$$= \sum_{r=0}^{\lfloor n/2 \rfloor} \binom{n}{r}\binom{n-r}{r} 2^{n-2r}$$

This yields the desired result. ∎

The next example develops an identity using Corollary 2.2 and calculus, and identifies another occurrence of Catalan numbers C_n.

Example 4.2 Prove that

$$\sum_{r=1}^{n} r \binom{n}{r}^2 = n(2n-1)!C_{n-1}$$

Proof   By Corollary 2.2, we have

$$(1+x)^n = \binom{n}{0} + \binom{n}{0}x + \binom{n}{2}x^2 + \cdots + \binom{n}{n}x^n$$

Differentiating both sides with respect to x, we get:

$$n(1+x)^{n-1} = \binom{n}{1} + 2\binom{n}{2}x + 3\binom{n}{3}x^2 + \cdots + n\binom{n}{n}x^{n-1}$$

Then

$$nx(1+x)^{n-1} = \binom{n}{1}x + 2\binom{n}{2}x^2 + 3\binom{n}{3}x^3 + \cdots + n\binom{n}{n}x^n$$

Now replace x with $\frac{1}{x}$:

$$\frac{n}{x}\left(1+\frac{1}{x}\right)^{n-1} = \frac{1}{x}\binom{n}{1} + \frac{2}{x^2}\binom{n}{2} + \frac{3}{x^3}\binom{n}{3} + \cdots + \frac{n}{x^n}\binom{n}{n}$$

$$n(1+x)^{n-1} = n\binom{n}{n} + (n-1)\binom{n}{n-1}x + \cdots + 2\binom{n}{2}x^{n-2} + \binom{n}{1}x^{n-1}$$

Multiply both sides by $(1 + x)^n = \sum_{r=0}^{n} \binom{n}{r} x^r$:

$$n(1 + x)^{2n-1} = \left[\sum_{r=1}^{n} r \binom{n}{r} x^{n-r} \right] \left[\sum_{r=0}^{n} r \binom{n}{r} x^r \right]$$

Now equate the coefficients of x^n from both sides:

$$\sum_{r=1}^{n} r \binom{n}{r}^2 = n \binom{2n-1}{n}$$

$$= n(2n - 1)C_{n-1} \qquad \blacksquare$$

For example,

$$\sum_{r=1}^{5} r \binom{5}{r}^2 = 1\binom{5}{1}^2 + 2\binom{5}{2}^2 + 3\binom{5}{3}^2 + 4\binom{5}{4}^2 + 5\binom{5}{5}^2$$

$$= 25 + 200 + 300 + 100 + 5$$

$$= 5 \cdot 9 \cdot 14$$

$$= 5(2 \cdot 5 - 1)C_4$$

The next example provides a delightful confluence of central binomial coefficients, calculus, and generating functions.

Example 4.3 Prove that

$$\sum_{r=0}^{n} \binom{2r}{r} \binom{2n - 2r}{n - r} = 4^n \tag{4.9}$$

Proof   First, we construct a generating function for the sequence a_n, where $a_n = \binom{2n}{n}$ and $n \geq 0$. To this end, notice that

$$a_{n+1} = \frac{4n + 2}{n + 1} \binom{2n}{n}$$

$$= \frac{4n + 2}{n + 1} a_n$$

Thus, a_n satisfies the recurrence relation

$$(n + 1)a_{n+1} = (4n + 2)a_n.$$

Let $f(x) = \sum_{n=0}^{\infty} a_n x^n$. Then this recurrence relation yields

$$\sum_{n=0}^{\infty}(n + 1)a_{n+1}x^n = 4\sum_{n=0}^{\infty} na_n x^n + 2\sum_{n=0}^{\infty} a_n x^n$$

That is,

$$f'(x) = 4xf'(x) + 2f(x)$$

where $f'(x)$ denotes the derivative of $f(x)$ with respect to x. Then

$$\frac{f'(x)}{f(x)} = \frac{2}{1 - 4x}$$

Integrating this with respect to x, we get

$$\ln f(x) = -\tfrac{1}{2}\ln(1 - 4x) + C$$

where C is a constant. Because $f(0) = a_0 = 1$, this implies $C = 0$. Thus $\ln f(x) = -\tfrac{1}{2}\ln(1 - 4x)$; that is,

$$f(x) = \frac{1}{\sqrt{1 - 4x}}$$

In other words,

$$\frac{1}{\sqrt{1 - 4x}} = \sum_{n=0}^{\infty}\binom{2n}{n}x^n$$

Squaring both sides,

$$\frac{1}{1 - 4x} = \left[\sum_{r=0}^{\infty}\binom{2r}{r}x^r\right]\left[\sum_{n=0}^{\infty}\binom{2n}{n}x^n\right]$$

That is,

$$\sum_{n=0}^{\infty} 4^n x^n = \sum_{n=0}^{\infty}\left[\sum_{r=0}^{n}\binom{2r}{r}\binom{2n - 2r}{n - r}\right]x^n$$

Equating the coefficients of x^n from both sides, we get the desired result:

$$\sum_{r=0}^{n} \binom{2r}{r} \binom{2n-2r}{n-r} = 4^n$$

For example, let $n = 3$. Then

$$\sum_{r=0}^{3} \binom{2r}{r} \binom{6-2r}{3-r} = \binom{0}{0} \binom{6}{3} + \binom{2}{1} \binom{4}{2} + \binom{4}{2} \binom{2}{1} + \binom{6}{0} \binom{0}{0}$$

$$= 20 + 12 + 12 + 20$$

$$= 4^3$$

The Ubiquitous CBC Revisited

Next we pursue a finite multiple summation formula, proposed in 1956 by J. Winter and R. C. Kao of the Rand Corporation, Santa Monica, California. The proof by Carlitz employs identity

$$\binom{r}{r} + \binom{r+1}{r} + \binom{r+2}{r} + \cdots + \binom{n}{r} = \binom{n+1}{r} \tag{4.10}$$

Example 4.4 Establish the identity

$$\sum_{a_n=0}^{n} \sum_{a_{n-1}=0}^{a_n} \sum_{a_{n-2}=0}^{a_{n-1}} \cdots \sum_{a_1=0}^{a_2} 1 = \binom{2n}{n}$$

Proof By identity (4.10), we have

$$\sum_{a_1=0}^{a_2} \binom{a_1}{k} = \binom{a_2+1}{k+1}$$

$$\sum_{a_2=0}^{a_3} \binom{a_2+1}{k+1} = \binom{a_3+1}{k+2}$$

$$\vdots$$

$$\sum_{a_n=0}^{a_{n+1}} \binom{a_n+n-1}{k+n-1} = \binom{a_{n+1}+n}{k+n}$$

Thus

$$\sum_{a_n=0}^{a_{n+1}} \sum_{a_{n-1}=0}^{a_n} \sum_{a_{n-2}=0}^{a_{n-1}} \cdots \sum_{a_1=0}^{a_2} \binom{a_1}{k} = \binom{a_{n+1}+n}{k+n}$$

In particular, let $a_{n+1} = n$ and $k = 0$. This yields the desired result. ∎

The following example, proposed in 1971 by S. Heller of Brookhaven National Laboratory, Upton, New York, deals with summations with ordered indices. The combinatorial argument is by M. Shimshoni of Weizman Institute of Science, Rehovot, Israel.

Example 4.5 Show that

$$\sum_{i_m=1}^{n-m+1} \sum_{i_{m-1}=i_m+1}^{n-m+2} \cdots \sum_{i_3=i_4+1}^{n-2} \sum_{i_2=i_3+1}^{n-1} \sum_{i_1=i_2+1}^{n} 1 = \binom{n}{m}$$

Proof   Notice that $n \geq i_1 > i_2 > \cdots > i_m \geq 1$; so any combination of the i's satisfying these inequalities will appear exactly once. Therefore, the desired sum is the number of m-element subsets of the set $\{1, 2, 3, \ldots, n\}$, namely, $\binom{n}{m}$.

In particular, let $n = 2m$. Then the identity yields

$$\sum_{i_m=1}^{m+1} \sum_{i_{m-1}=i_m+1}^{m+2} \cdots \sum_{i_3=i_4+1}^{2m-2} \sum_{i_2=i_3+1}^{2m-1} \sum_{i_1=i_2+1}^{2m} 1 = \binom{2m}{m}$$ ∎

F. G. Schmitt Jr. of Berkeley, California, gave an alternate proof of this using identity (4.10) and induction.

5

Catalan Numbers

To date, nearly 400 articles and problems have appeared on Catalan numbers. In 1965, William G. Brown of the University of British Columbia, Canada, presented a historical background with a list of sixty references. Five years later, Jo Anne Simpson Growney investigated finitely generated free groupoids for her doctoral dissertation at the University of Oklahoma, and found occurrences of those numbers in such algebraic structures. The dissertation also includes a brief historical introduction to Catalan numbers. In 1976, Henry W. Gould of West Virginia University, Morgantown, published an extensive bibliography on Catalan numbers, containing 470 listings. Richard P. Stanley of Massachusetts Institute of Technology has listed over seventy occurrences of Catalan numbers in his *Enumerative Combinatorics*, vol. 2, and another seventy on his Web site Catalan Addendum. In this chapter, we present a historical introduction to this delightful class of numbers.

In the preceding chapters, we encountered a number of occurrences of the CBC $\binom{2n}{n}$ and Catalan numbers. Belgian mathematician Eugene C. Catalan "discovered" Catalan numbers in 1838, while studying well-formed sequences of parentheses. Although they are named after Catalan, they were not first discovered by him. Around 1751, Euler found them while studying the triangulations of convex polygons. However, according to a 1988 article by Chinese mathematician J. J. Luo, Chinese mathematician Antu Ming discovered them about 1730 through his geometric models. Ming's work was published in Chinese, so it was not known in the West.

Antu Ming (1692?–1763?), according to Luo, was a Zhengxianbai tribesman of Inner Mongolia and a famous scientist during the Qing dynasty. His childhood mathematical education, carefully directed by the emperor, specialized in astronomy and mathematics. After mastering the scientific knowledge of the period, Ming became a mandarin, a high-ranking government official, at the national astronomical center. In 1759, he became director of the center. His work included problem solving in astronomy, meteorology, geography, surveying, and mathematics.

Around 1730, he started a book titled *Efficient Methods for the Precise Values of Circular Functions*, which clearly demonstrates his understanding of Catalan numbers. Although the book was completed by Ming's students by 1774, it was not published until 1839.

Leonhard Euler (1707–1783) was born in Basel, Switzerland. His father, a mathematician and Calvinist pastor, wanted him also to become a pastor. Although Euler had different ideas, he followed his father's wishes and studied Hebrew and theology at the University of Basel. His hard work at the university and remarkable ability brought him to the attention of well-known mathematician Johann Bernoulli (1667–1748). Realizing the young Euler's talents, Bernoulli persuaded the father to change his mind, and Euler pursued his studies in mathematics.

At the age of nineteen, Euler published his first paper. Although his article failed to win the prestigious Paris Prize in 1727, he won it twelve times in later years.

Euler was undoubtedly the most prolific mathematician, making significant contributions to every branch of mathematics. With his phenomenal memory, he had every formula at his fingertips. A genius, he could work anywhere and under any conditions. Euler belongs to a class by himself.

Eugene Charles Catalan (1814–1894) was born in Bruges, Belgium. He studied at École Polytechnique, Paris, and received his doctor of science in 1841. After resigning his position with the Department of Bridges and Highways, he became professor of mathematics at Collège de Chalons-sur-Marne and then at Collège Charlemagne. Catalan went on to teach at Lycée Saint Louis and in 1865 became professor of analysis at the University of Liège in Belgium. Besides authoring *Élements de Geometriè* and *Notions d'astronomie*, published in 1843 and 1860, respectively, he published numerous articles on multiple integrals, the theory of surfaces, mathematical analysis, calculus of probability, and geometry. He did extensive research on spherical harmonics, analysis of differential equations, transformation of variables in multiple integrals, continued fractions, series, and infinite products.

> **Richard Peter Stanley** (1944–) received his B.S. in 1966 from California Institute of Technology, and Ph.D. in 1971 from Harvard University under the supervision of Gian-Carlo Rota. After spending the next two years as a Miller Research Fellow at the University of California, Berkeley, he joined the faculty at the Massachusetts Institute of Technology. The Norman Levinson Professor of Applied Mathematics since 2000, he is a leading expert in combinatorics and its applications. His two-volume *Enumerative Combinatorics* is "a far-reaching and authoritative treatise on enumerative methods that is widely regarded for its elegant exposition, usefulness as a reference, and creative exercises." He is the author of *Combinatorics and Commutative Algebra* and over 100 articles. A member of the National Academy of Sciences since 1995 and a fellow of the American Academy of Arts and Sciences since 1998, Stanley won the SIAM George Pólya Prize in Applied Combinatorics in 1975, the Leroy P. Steele Prize in 2001 for mathematical exposition, and the Rolf Schock Prize in 2003. Stanley received an honorary doctor of mathematics from the University of Waterloo in 2007.

As we saw in Chapter 1, the *Catalan numbers C_n* are often defined by

$$C_n = \frac{(2n)!}{(n+1)!n!} \tag{5.1}$$

$$= \frac{1}{n+1}\binom{2n}{n}, \quad n \geq 0 \tag{5.2}$$

In Corollary 1.2, we found that $n+1 \mid \binom{2n}{n}$; so every Catalan number is an integer. The various Catalan numbers are

$$1, 1, 2, 5, 14, 42, 132, 429, 1430, 4862, 16796, \ldots.$$

This is the integer sequence numbered 577 in N. J. A. Sloane's *A Handbook of Integer Sequences*.

There is a second way of defining C_n:

$$\binom{2n}{n} - \binom{2n}{n-1} = \frac{(2n)!}{(n!)^2} - \frac{(2n)!}{(n-1)!(n+1)!}$$

$$= \frac{(2n)!}{n!(n+1)!}$$

$$= C_n$$

This also confirms that every Catalan number is an integer.

There are several additional characterizations of C_n, as we shall find here and in the following chapters.

We begin our discussion with Euler's triangulation problem, which he proposed in 1751 to Prussian mathematician Christian Goldbach (1690–1764). Over the centuries, a number of outstanding mathematicians, besides Goldbach and Catalan, have worked on the polygonal disection problem. They include Hungarian Johnann Andreas von Segner (1704–1777); Germans Johann Friedrich Pfaff (1765–1825) and J. A. Grunert (1797–1872); Swiss Nicolaï Ivanovich Fuss (1755–1826); Frenchmen Gabriel Lamé (1795–1870), B. O. Rodrigues (1794–1851), Jean Marie Constant Duhamel (1797–1872), Jacques Phillippe Marie Binet (1786–1865), Joseph Louiville (1809–1882), and François-Edouard-Anatole Lucas (1842–1891); Englishman Arthur Cayley (1821–1895); and Johann L. Tellkampf.

Example 5.1 Find the number of ways T_n (the interior of) a convex n-gon* can be divided into triangles by drawing nonintersecting diagonals, where $n \geq 3$.

Solution   There is exactly one way of triangulating a triangle, two different ways triangulating a square, five different ways of triangulating a pentagon, and fourteen different ways of triangulating a hexagon, as shown in Figure 5.1. Thus, we have the Catalan numbers 1, 2, 5, and 14.

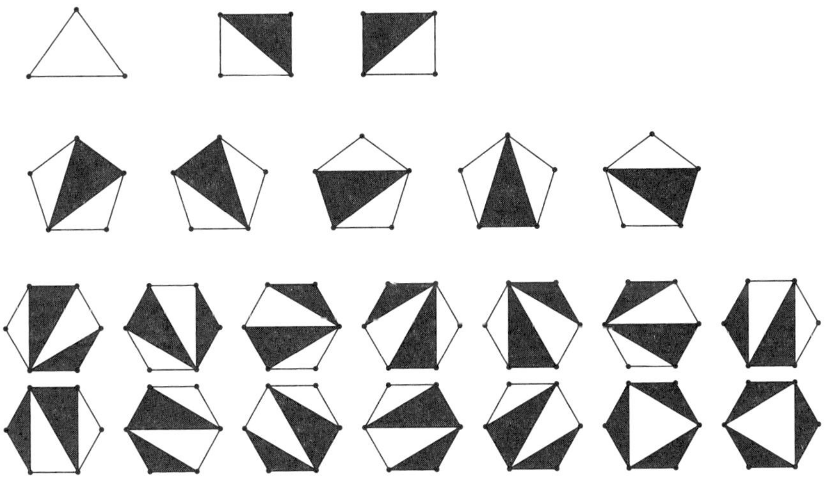

Figure 5.1 Triangulations of an *n*-gon, where $3 \leq n \leq 6$

* A convex *n*-gon is a polygon with *n* sides such that every diagonal lies in its interior.

Euler, using an inductive argument that he described as "quite laborious," established that

$$T_n = \frac{2 \cdot 6 \cdot 10 \cdots (4n - 10)}{(n - 1)!}, \quad n \geq 3$$

Although Euler's formula, published in 1761, makes sense only for $n \geq 3$, we can extend it to include the cases $n = 0, 1$, and 2. To this end, let $k = n - 3$. Then

$$T_{k+3} = \frac{2 \cdot 6 \cdot 10 \cdots (4k + 2)}{(k + 2)!}, \quad k \geq 0$$

Then $T_3 = 1, T_4 = 2$, and $T_5 = 5$. These are the Catalan numbers C_1, C_2, and C_3, respectively, shifted by two spaces to the right. So we define $C_n = T_{k+2}$. Thus

$$C_n = \frac{2 \cdot 6 \cdot 10 \cdots (4n - 2)}{(n + 1)!}, \quad n \geq 1$$

This can be rewritten as

$$C_n = \frac{4n - 2}{n + 1} \cdot \frac{2 \cdot 6 \cdot 10 \cdots (4n - 6)}{n!}$$

$$= \frac{4n - 2}{n + 1} C_{n-1}$$

When $n = 1$, this yields $C_1 = C_0$. But $C_1 = 1$. So we can define $C_0 = 1$.

Consequently, C_n can be defined recursively as follows. Euler published it in 1761 in the St. Petersburg Proceedings, *Novi Commentarii Academiae Scientarum Imperialis Petropolitanae* (New Proceedings of the Imperial Academy of Sciences of Petropolis).

A Recursive Definition of C_n

$$C_0 = 1$$

$$C_n = \frac{4n - 2}{n + 1} C_{n-1}, \quad n \geq 1 \tag{5.3}$$

For example,

$$C_4 = \frac{4 \cdot 4 - 2}{4 + 1} C_3$$

$$= \frac{14}{5} \cdot 5$$

$$= 14$$

Five years after Euler published his formula, S. Kotelnikow wrote an article on Euler's triangulation decomposition problem in the same journal; its title, interestingly enough, includes the expression

$$\frac{4 \cdot 6 \cdot 10 \cdot 14 \cdot 18 \cdot 22 \cdots (4n - 10)}{(n - 1)!}$$

(It contains an extra factor of 2 in the numerator.) It was the third article to appear on the subject, after Euler and Segner published theirs.

Urban's Conjecture

It is interesting to note that in 1941 H. Urban conjectured the recurrence relation (5.3) by noticing a pattern. First, he computed C_3, C_4, and C_5 using Segner's recursive formula (5.6), and then he made the following observation:

$$\frac{C_1}{C_0} = \frac{1}{1} = \frac{2}{2} = \frac{4 \cdot 1 - 2}{1 + 1}$$

$$\frac{C_2}{C_1} = \frac{2}{1} = \frac{6}{3} = \frac{4 \cdot 2 - 2}{2 + 1}$$

$$\frac{C_3}{C_2} = \frac{5}{2} = \frac{10}{4} = \frac{4 \cdot 3 - 2}{3 + 1}$$

$$\frac{C_4}{C_3} = \frac{14}{5} = \frac{14}{5} = \frac{4 \cdot 4 - 2}{4 + 1}$$

$$\frac{C_5}{C_4} = \frac{42}{14} = \frac{18}{6} = \frac{4 \cdot 5 - 2}{5 + 1}$$

Noticing a clear pattern, Urban inferred that $C_n = \dfrac{4n - 2}{n + 1} C_{n-1}$, where $n \geq 1$.

An Explicit Formula for C_n

The recursive formula (5.3) can be employed to derive the explicit formula (5.1) and hence formula (5.2):

$$C_n = \frac{4n - 2}{n + 1} C_{n-1}$$

$$= \frac{(4n - 2)(4n - 6)}{(n + 1)n} C_{n-2}$$

$$= \frac{(4n-2)(4n-6)(4n-10)}{(n+1)n(n-1)}C_{n-3}$$

$$\vdots$$

$$= \frac{(4n-2)(4n-6)(4n-10)\cdots 6\cdot 2}{(n+1)n\cdots 3\cdot 2}C_0$$

$$= \frac{(2n-1)(2n-3)(2n-5)\cdots 3\cdot 1}{(n+1)!}\cdot 2^n$$

$$= \frac{(2n)!2^n}{(n+1)!2^n n!}$$

$$= \frac{(2n)!}{(n+1)!n!}$$

$$= \frac{1}{n+1}\binom{2n}{n}$$

An Approximate Value of C_n

An approximate value of C_n can be found using *Stirling's approximation for factorials*, namely, $n! \approx (n/e)^n \sqrt{2\pi n}$:

$$\binom{2n}{n} = \frac{(2n)!}{(n!)^2}$$

$$\approx \frac{(2n/e)^{2n}\sqrt{2\pi \cdot 2n}}{(n/e)^{2n}\cdot 2\pi n}$$

$$= \frac{2^{2n}}{\sqrt{n\pi}}$$

So

$$C_n \approx \frac{2^{2n}}{(n+1)\sqrt{n\pi}}$$

$$\approx \frac{2^{2n}}{n\sqrt{n\pi}}$$

For example, when $n = 5$:

$$\frac{2^{2n}}{n\sqrt{n\pi}} = \frac{2^{10}}{5\sqrt{5\pi}}$$

$$\approx 52$$

whereas $C_5 = 42$. When $n = 10$:

$$\frac{2^{2n}}{n\sqrt{n\pi}} = \frac{2^{20}}{10\sqrt{10\pi}}$$

$$\approx 18,708$$

whereas $C_{10} = 16,796$.

It follows from equation (5.3) that

$$\lim_{n \to \infty} \frac{C_{n+1}}{C_n} = 4$$

Thus, when n is sufficiently large, $C_{n+1} \approx 4C_n$.

Decimal Digits in C_{10^n}

The digits in C_{10^n} manifest an interesting behavior. The number of digits in C_{10^n} are 1, 5, 57, 598, 6015, 60199, 602051, 6020590, ..., where $n \geq 0$. This sequence converges to the number formed by the digits on the right side of the decimal point in $\log 4 = 0.60205999132 \ldots$.

A Related Problem

Let e_n and o_n denote the number of dissections of a convex n-gon by nonintersecting digonals into an even and odd number of regions, respectively. Then $e_n - o_n = (-1)^n$. This result* was discovered in 2005 by Emeric Deutsch of Polytechnic University, Brooklyn, New York.

Next, we present an interesting relationship between Catalan numbers and the gamma function.

Catalan Numbers and the Gamma Function

The *gamma function* Γ, an extension of the factorial function, was introduced by French mathematician Adrien-Marie Legendre (1752–1833). It is defined by

$$\Gamma(z) = \int_0^\infty t^{z-1} e^{-t} \, dt$$

where z is a complex variable with a positive real part. Using integration parts, we can show that

$$\Gamma(z) = (z-1)\Gamma(z-1) \tag{5.4}$$

---

* For a proof, see *American Mathematical Monthly* **114** (May 2007), 456–457.

When z is a positive integer, this yields $\Gamma(n) = (n-1)!$; so $\Gamma(n+2) = (n+1)!$.

It follows from *Euler's reflection formula*

$$\Gamma(1-z)\Gamma(z) = \frac{\pi}{\sin \pi z}$$

that $\Gamma(1/2) = \sqrt{\pi}$.

It follows from (5.4) that

$$\Gamma\left(n + \frac{1}{2}\right) = \left(n - \frac{1}{2}\right)\Gamma\left(n - \frac{1}{2}\right)$$

$$= \left(n - \frac{1}{2}\right)\left(n - \frac{3}{2}\right)\Gamma\left(n - \frac{3}{2}\right)$$

$$\vdots$$

$$= \left(n - \frac{1}{2}\right)\left(n - \frac{3}{2}\right)\cdots\left(n - \frac{2n-1}{2}\right)\Gamma\left(\frac{1}{2}\right)$$

$$= \frac{1 \cdot 3 \cdot 5 \ldots (2n-1)}{2^n}\sqrt{\pi}$$

$$= \frac{(2n)!}{2^n(2 \cdot 4 \cdot 6 \ldots n)}\sqrt{\pi}$$

$$= \frac{(2n)!}{4^n\,n!}\sqrt{\pi}$$

So

$$\frac{\Gamma\left(n + \frac{1}{2}\right)}{\Gamma(n+2)} = \frac{(2n)!}{4^n\,n!(n+1)!}\sqrt{\pi}$$

$$= \frac{C_n}{4^n}\sqrt{\pi}$$

Thus

$$C_n = \frac{4^n\Gamma\left(n + \frac{1}{2}\right)}{\sqrt{\pi}\,\Gamma(n+2)} \tag{5.5}$$

A Reappearance of Euler's Triangulation Problem

Interestingly, the triangulation problem reappeared in 1930, proposed by O. Dunkel of Washington University. The solution given in the following year by W. A. Bristol of the University of Pennsylvania, yields Table 5.1. (We have omitted the details for brevity.) The row sums yield the various Catalan numbers $T_n = C_{n-2}$, where $n \geq 3$.

Table 5.1

n		Row Sums
3	1	1
4	$1 + 1$	2
5	$2 + 2 + 1$	5
6	$5 + 5 + 3 + 1$	14
7	$14 + 14 + 9 + 4 + 1$	42
8	$42 + 42 + 28 + 14 + 5 + 1$	132

$$\uparrow$$
$$C_{n-2}$$

W. R. Church of the University of Pennsylvania made a fascinating observation of the sums in the middle column and cleverly used his observation to develop an explicit formula for T_n. To see this, we delete the first term in each sum, reverse the order of the summands in each resulting sum, and renumber the row; see Table 5.2. Interestingly, this is exactly the same as the array we shall encounter in our study of the paths of a rook in Example 8.3, with the main diagonal deleted.

r \ c	1	2	3	4	5
1	1				
2	1	2			
3	1	3	5		
4	1	4	9	14	
5	1	5	14	28	42

$$\nwarrow$$
$$T_n$$

Let $T(r, c)$ denote the rth element in column c, where $r, c \geq 1$. It can be defined recursively:

$$T(r, 1) = 1$$

$$T(r, c) = \sum_{i=c-1}^{r} T(i, c - 1), \quad c \geq 2$$

Notice that $T(r,c)$ is also given by

$$T(r,c) = \sum_{j=1}^{c} T(r-1,j)$$

For example,

$$T(5,3) = 14$$

$$= 2+3+4+5 \qquad \text{(see column 2)}$$

$$= 1+4+9 \qquad \text{(see row 4)}$$

Using the identity

$$\sum_{i=1}^{n} \prod_{j=i}^{i+s-1} j = \frac{1}{s+1} \prod_{t=n}^{n+s} t$$

it can be shown by induction that

$$T(r,c) = \frac{r-c+2}{r+1}\binom{r+c-1}{r}$$

In particular,

$$T_{n+2} = T(n,n)$$

$$= \frac{2}{n+1}\binom{2n-1}{n}$$

$$= \frac{1}{n+1}\binom{2n}{n}$$

$$= C_n$$

where $n \geq 1$.

Segner's Recursive Formula

Euler had conveyed the seven Catalan numbers 1, 2, 5, 14, 42, 132, and 429 to Hungarian mathematician Johann von Segner. In 1761, Segner, using the addition and multiplication principles, published a second order recurrence formula for T_n, where $n \geq 3$. He presented it to the St. Petersburg Academy.

To develop Segner's formula, consider an n-gon with vertices $A_1, A_2, \ldots, A_n$, where $n \geq 3$. See Figure 5.2. Let $1 < k < n$. Then $\triangle A_1 A_k A_n$ partitions the n-gon into two smaller polygons. If $3 \leq k \leq n-2$, we get a k-gon on its right side with vertices $A_1, A_2, \ldots, A_k$ and $(n-k+1)$-gon on the left with vertices $A_k, A_{k+1}, \ldots, A_n$. On the other hand, if $k = 2$, we get no polygon on the right side, but an $(n-1)$-gon on its left side; the case $k = n-1$ yields a similar result.

Johann Andreas von Segner (1704–1777), a mathematician, physicist, and physician, was born in Pressburg, Hungary (now Bratislava, Slovakia). The son of a merchant, he studied at the gymnasiums of Pressburg and Debrecen. He graduated from the University of Jena with an M.D. in 1730. While there he also studied mathematics and physics and developed a lifelong fascination for both. In 1728, Segner published an article on Descartes' Rule of Signs. He wrote a second one in 1758.

After practicing medicine at Pressburg and Debrecen for a brief period, Segner returned to Jena in 1732 as an assistant professor and became associate professor of mathematics in the following year.

(continued)

(*Continued*)

In 1735 Segner was appointed professor of mathematics and physics at the newly founded University of Göttingen, Germany. Twenty years later, he moved to the University of Halle, Germany, and remained there until his death.

A voluminous writer in both mathematics and physics, Segner ironically made no discoveries in medicine. In 1743, he developed a naval barometer. Seven years later, he invented a simple reaction hydraulic turbine that bears his name. Later Euler added his own improvements to Segner's hydraulic wheel. Segner spent considerable time constructing and perfecting scientific devices, including slide rules, clocks, and telescopes. In 1751, he introduced the concept of surface tension of liquids and studied the theory of the spinning top.

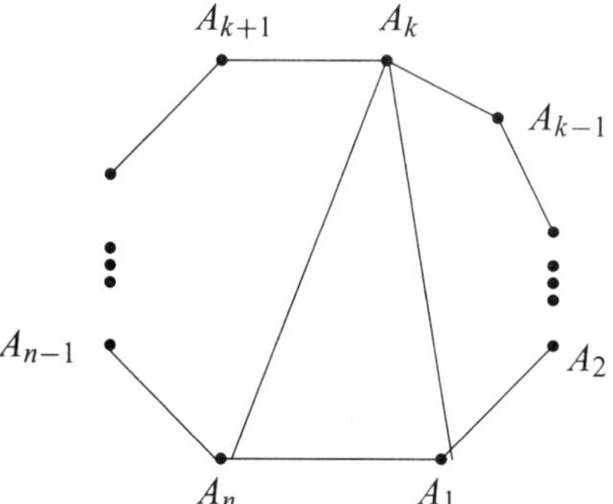

Figure 5.2 Polygonal Dissection

The k-gon can be triangulated in T_k ways and the $(n-k+1)$-gon in T_{n-k+1} ways. By the multiplication principle, there are $T_k T_{n-k+1}$ different triangulations of the n-gon that contain $\triangle A_1 A_k A_n$. Since $2 \leq k < n$, by the addition principle, the total number of triangulations of the n-gon is given by

$$T_n = \sum_{k=2}^{n-1} T_k T_{n-k+1}$$

$$= T_2 T_{n-1} + T_3 T_{n-2} + \cdots + T_{n-1} T_2, \qquad n \geq 3$$

where we have defined $T_2 = 1$.

Since $C_n = T_{n+2}$, this yields *Segner's recurrence relation* for C_n:

$$C_n = C_0 C_{n-1} + C_1 C_{n-2} + \cdots + C_{n-1} C_0 \tag{5.6}$$

$$= (C_0, C_1, \ldots, C_{n-1}) \cdot (C_{n-1}, C_{n-2}, \ldots, C_0) \tag{5.7}$$

where the dot indicates the *dot product* of the two vectors.

For example,

$$C_5 = (C_0, C_1, C_2, C_3, C_4) \cdot (C_4, C_3, C_2, C_1, C_0)$$

$$= (1, 1, 2, 5, 14) \cdot (14, 5, 2, 1, 1)$$

$$= 1 \cdot 14 + 1 \cdot 5 + 2 \cdot 2 + 5 \cdot 1 + 14 \cdot 1$$

$$= 42$$

Yet Another Recursive Formula

Using a technique quite similar to the one employed for deriving Segner's formula, we can derive yet another recursive formula for C_n. To this end, first we prove the following theorem.

Theorem 5.1 Let T_n denote the number of triangulations of an n-gon, where $n \geq 4$. Then

$$(2n - 6)T_n = n \sum_{k=3}^{n-1} T_k T_{n-k+2} \tag{5.8}$$

Proof   Consider an n-gon with vertices $A_1, A_2, \ldots, A_n$ (see Figure 5.3). Consider an arbitrary diagonal $\overline{A_1 A_k}$ leaving A_1, where $2 < k < n$. It partitions the n-gon into the k-gon $A_1 A_2 \ldots A_k$ on its right side and the $(n - k + 2)$-gon $A_1 A_k A_{k+1} \ldots A_n$ on its left side.

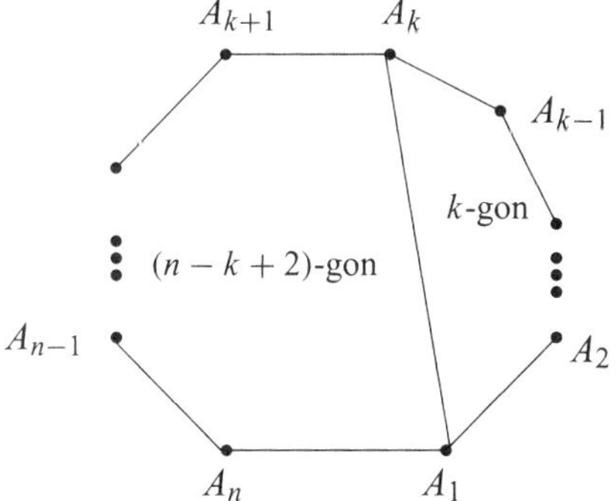

Figure 5.3 Polygonal Dissection

The k-gon can be triangulated in T_k different ways and the $(n - k + 2)$-gon in T_{n-k+2} different ways. By the multiplication principle, the n-gon can be triangulated in $T_k T_{n-k+2}$ different ways using $\overline{A_1 A_k}$ as a diagonal. Since $3 \le k \le n - 1$, by the addition principle, the total number of triangulations, using a diagonal at A_1, is given by

$$\sum_{k=3}^{n-1} T_k T_{n-k+2}$$

Because A_1 can be any one of the n vertices, the sum $n \sum_{k=3}^{n-1} T_k T_{n-k+2}$ denotes the total number of triangulations using every possible diagonal. But in this sum each diagonal $\overline{A_i A_j}$ is counted twice, once as a diagonal at A_i and again as a diagonal at A_j. Thus $\frac{n}{2} \sum_{k=3}^{n-1} T_k T_{n-k+2}$ denotes the total number of triangulations, counting every diagonal exactly once.

Since the n-gon has $n - 3$ diagonals at each vertex and every triangulation uses $n - 3$ diagonals, this sum counts the number T_n of triangulations exactly $n - 3$ times. Thus

$$(n - 3) T_n = \frac{n}{2} \sum_{k=3}^{n-1} T_k T_{n-k+2}$$

This yields the desired result. ∎

The next recursive formula for C_n follows from this theorem, as the following corollary shows.

Corollary 5.1 Let $n \ge 2$. Then

$$(2n - 2) C_n = (n + 2) \sum_{k=1}^{n-1} C_k C_{n-k} \tag{5.9}$$

Proof   This follows by letting $C_n = T_{n+2}$ in formula (5.8). ∎

For example, when $n = 5$, recursive formula (5.9) yields:

$$4 C_5 = 7 \sum_{k=1}^{4} C_k C_{5-k}$$
$$= 7(C_1 C_4 + C_2 C_3 + C_3 C_2 + C_4 C_1)$$
$$= 7(1 \cdot 14 + 2 \cdot 5 + 5 \cdot 2 + 14 \cdot 1)$$
$$C_5 = 42$$

as expected.

As a byproduct, Segner's formula, coupled with formula (5.9) can be used to derive the explicit formula (5.2) for C_n, as the next corollary demonstrates.

Corollary 5.2 Let $n \geq 0$. Then

$$C_n = \frac{1}{n+1}\binom{2n}{n}$$

Proof   By Segner's formula,

$$C_{n+1} = C_0 C_n + (C_1 C_{n-1} + \cdots + C_{n-1}C_1) + C_n C_0$$

Since $C_0 = 1$, this can be rewritten as

$$C_{n+1} - 2C_n = C_1 C_{n-1} + \cdots + C_{n-1}C_1$$

Using formula (5.9), this becomes

$$C_{n+1} - 2C_n = \frac{2n-2}{n+2}C_n$$

$$C_{n+1} = \left(2 + \frac{2n-2}{n+2}\right)C_n$$

$$= \frac{4n+2}{n+2}C_n$$

That is,

$$C_n = \frac{4n-2}{n+1}C_{n-1}$$

This is the same as the recursive formula (5.3). As before, it yields the desired explicit formula:

$$C_n = \frac{1}{n+1}\binom{2n}{n}$$

$\blacksquare$

A Derivation of the Explicit Formula from Segner's Recursive Formula

Using generating functions, Segner's recurrence relation can be employed to derive the explicit formula (5.2). To this end, consider the generating function

$$C(x) = C_0 + C_1 x + C_2 x^2 + \cdots + C_n x^n + \cdots \tag{5.10}$$

Then

$$[C(x)]^2 = C_0^2 + (C_0C_1 + C_1C_0)x + (C_0C_2 + C_1C_1 + C_2C_0)x^2 + \cdots$$
$$+ (C_0C_n + C_1C_{n-1} + \cdots + C_nC_0)x^n + \cdots$$
$$= C_1 + C_2x + C_3x^2 + \cdots + C_{n+1}x^n + \cdots$$
$$= \frac{C(x) - C_0}{x}$$

Then

$$x[C(x)]^2 - C(x) + 1 = 0$$

Solving for $C(x)$,

$$C(x) = \frac{1 \pm \sqrt{1 - 4x}}{2x}$$

Because C_n is always a positive integer, we choose the minus sign. So

$$C(x) = \frac{1 - \sqrt{1 - 4x}}{2x}$$

But from Chapter 2,

$$\sqrt{1 - 4x} = 1 - 2 \sum_{n=1}^{\infty} \frac{1}{n} \binom{2n - 2}{n - 1} x^n$$

$$\frac{1 - \sqrt{1 - 4x}}{2x} = \sum_{n=1}^{\infty} \frac{1}{n} \binom{2n - 2}{n - 1} x^{n-1}$$

That is,

$$C(x) = \sum_{n=0}^{\infty} \frac{1}{n + 1} \binom{2n}{n} x^n \tag{5.11}$$

Equating the coefficients of x^n from (5.10) and (5.11), we get the desired explicit formula:

$$C_n = \frac{1}{n + 1} \binom{2n}{n}$$

On the other hand, suppose that $C_n = \frac{1}{n+1} \binom{2n}{n}$. We would like to establish Segner's recursive formula. The proof is based on one given in 1979 by R. B. Nelsen of Lewis and Clark College, Portland, Oregon.

Let

$$C(x) = \sum_{n=0}^{\infty} C_n x^n$$

Then

$$[C(x)]^2 = \sum_{n=0}^{\infty} \left(\sum_{i=0}^{n} C_i C_{n-i} \right) x^n$$

We shall show that $[C(x)]^2 = \sum_{n=0}^{\infty} C_{n+1} x^n$.

Let $F(x) = \frac{\mathrm{d}}{\mathrm{d}x}[xC(x)]$. Then

$$F(x) = \sum_{n=0}^{\infty} \binom{2n}{n} x^n$$

$$= \frac{1}{\sqrt{1 - 4x}}$$

Therefore,

$$xC(x) = \int_0^x F(t)\,\mathrm{d}t$$

$$= \int_0^x \frac{\mathrm{d}t}{\sqrt{1 - 4t}}$$

$$= \tfrac{1}{2}(1 - \sqrt{1 - 4x})$$

Hence

$$[C(x)]^2 = \frac{1 - 2\sqrt{1 - 4x} + (1 - 4x)}{4x^2}$$

$$= \frac{1 - \sqrt{1 - 4x}}{2x^2} - \frac{1}{x}$$

$$= \frac{C(x) - 1}{x}$$

$$= \frac{1}{x} \sum_{n=0}^{\infty} C_n x^n$$

$$= \sum_{n=0}^{\infty} C_{n+1} x^n$$

Thus

$$C_{n+1} = \sum_{i=0}^{n} C_i C_{n-i}$$

That is,

$$C_n = \sum_{i=0}^{n-1} C_i C_{n-i-1}$$

as desired.

An Alternate Method

We now derive the generating function $C(x)$ for Catalan numbers using integral calculus and the generating function $A(x)$ for the CBCs $A_n = \binom{2n}{n}$ that we derived in Chapter 2. This method was used by W. G. Brown and V. E. Hoggatt Jr. of San Jose State College, California, in 1978. To this end, let $A(x) = \sum_{n=0}^{\infty} \binom{2n}{n} x^n$.

From Chapter 2,

$$\frac{1}{\sqrt{1-4x}} = \sum_{n=0}^{\infty} \binom{2n}{n} x^n$$

$$= \sum_{n=0}^{\infty} A_n x^n$$

Integrating the power series with respect to x, term by term,

$$\int \frac{dx}{\sqrt{1-4x}} = \sum_{n=0}^{\infty} \frac{1}{n+1} A_n x^{n+1} + C^*$$

$$-\frac{1}{2}\sqrt{1-4x} = \sum_{n=0}^{\infty} C_n x^{n+1} + C^*$$

$$-\frac{1}{2}\sqrt{1-4x} = xC(x) + C^*$$

where C^* is a constant. When $x = 0$, this yields $C^* = -\frac{1}{2}$. Thus

$$-\frac{1}{2}\sqrt{1-4x} = xC(x) - \frac{1}{2}$$

$$C(x) = \frac{1 - \sqrt{1-4x}}{2x}$$

as obtained earlier.

Shapiro's Identity

The generating functions

$$E(x) = \frac{C(x) + C(-x)}{2} = \sum_{n \geq 0} C_{2n} x^{2n}$$

and

$$O(x) = \frac{C(x) - C(-x)}{2} = \sum_{n \geq 0} C_{2n+1} x^{2n+1}$$

can be used to develop the following identity, discovered in 2002 by L. Shapiro of Howard University, Washington, DC:

$$\sum_{k=0}^{n} C_{2k} C_{2n-2k} = 4^n C_n \tag{5.12}$$

This can be established as follows:

$$[E(x)]^2 = \left(\sum_{m \geq 0} C_{2m} x^{2m} \right) \left(\sum_{n \geq 0} C_{2n} x^{2n} \right)$$

$$= \sum_{n \geq 0} \left(\sum_{k=0}^{n} C_{2k} C_{2n-2k} \right) x^{2n} \tag{5.13}$$

Since $C(x) = \sum_{n \geq 0} C_n x^n$,

$$C(4x^2) = \sum_{n > 0} 4^n C_n x^{2n} \tag{5.14}$$

But

$$[E(x)]^2 = C(4x^2) = 1 + 4x^2 + 32x^4 + 320x^6 + 3584x^8 + 43008x^{10} + \cdots .$$

Thus, the identity follows by equating the coefficients of x^{2n} in equations (5.13) and (5.14).

For example,

$$\sum_{k=0}^{3} C_{2k} C_{6-2k} = C_0 C_6 + C_2 C_4 + C_4 C_2 + C_6 C_0$$

$$= 1 \cdot 132 + 2 \cdot 14 + 14 \cdot 2 + 132 \cdot 1$$

$$= 320$$

$$= 4^3 \cdot 5$$

$$= 4^3 C_3$$

Shapiro's identity reminds us of Lagrange's identity (see Theorem 4.1); see Figures 5.4 and 5.5.

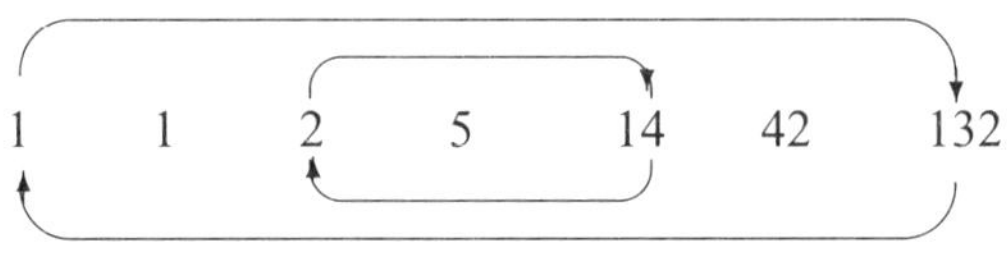

$$1 \cdot 132 + 2 \cdot 14 + 14 \cdot 2 + 132 \cdot 1 = 4^3 \cdot 5$$

Figure 5.4

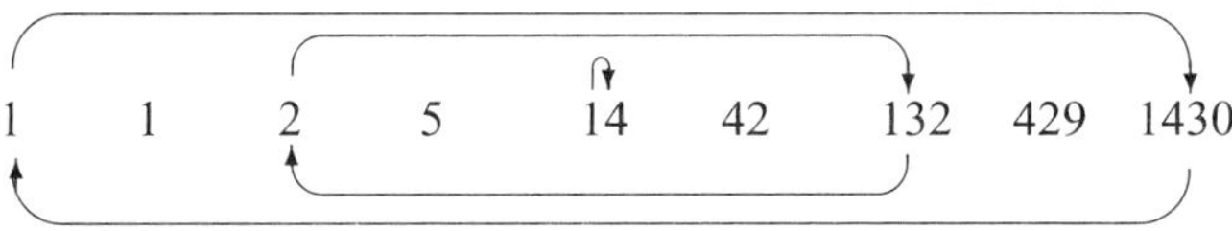

$$1 \cdot 1430 + 2 \cdot 132 + 14 \cdot 14 + 132 \cdot 2 + 1430 \cdot 1 = 4^4 \cdot 14$$

Figure 5.5

Next we present an interesting problem proposed in 1917 by A. A. Bennett of the University of Texas, Austin. The solution, given in 1918 by C. F. Gummer of Queen's University, Kingston, Ontario, Canada, is a bit complicated, so we omit most of it; the solution ties the answer with the generating function $\frac{1-\sqrt{1-4x}}{2x}$.

Example 5.2 Let $a_1, a_2, a_3, \ldots$ be positive integers such that

$$\begin{vmatrix} a_0 & a_1 & a_2 & \cdots & & \cdot \\ a_1 & a_2 & \cdot & \cdots & & \cdot \\ a_2 & \cdot & \cdot & \cdots & & \cdot \\ & & \vdots & & & \\ \cdot & \cdot & \cdot & \cdots & a_{n-1} \end{vmatrix} = 1 = \begin{vmatrix} a_1 & a_2 & a_3 & \cdots & & \cdot \\ a_2 & a_3 & \cdot & \cdots & & \cdot \\ a_3 & \cdot & \cdot & \cdots & & \cdot \\ & & \vdots & & & \\ \cdot & \cdot & \cdot & \cdots & a_n \end{vmatrix}$$

for every positive integer n. Evaluate a_n.

Solution   It follows from the definition that $a_0 = 1 = a_1$ and $a_2 = 2$. Gummer showed that a_n satisfies the recurrence relation

$$a_n = a_{n-1}a_0 + a_{n-2}a_1 + \cdots + a_0 a_{n-1} \tag{5.15}$$

where $n \geq 1$.

Let

$$u = a_0 + a_1 t + a_2 t^2 + \cdots + a_n t^n + \cdots$$

Then it follows from equation (5.15) that a_n, the coefficient of t^n in u, must be equal to the coefficient of t^{n-1} in u^2 for every $n \geq 1$. Therefore,

$$\frac{u-1}{t} = u^2$$

$$tu^2 - u + 1 = 0$$

$$u = \frac{1 \pm \sqrt{1 - 4t}}{2t}$$

We choose the minus sign to make u a power series with positive coefficients. Thus

$$u = \frac{1 - \sqrt{1 - 4t}}{2t}$$

Hence

$$a_n = \text{coefficient of } t^n \text{ in } \frac{1 - \sqrt{1 - 4t}}{2t}$$

$$= C_n$$

Note that this follows directly from the recurrence relation (5.15).   ∎

Superfactorials and Catalan Numbers

The next example, proposed in 1979 by M. Sholander and E. B. Leach of Case Western University, Cleveland, Ohio, presents an interesting confluence of super-factorials and Catalan numbers. Let $n!!$ denote the *superfactorial* $\prod_{i=1}^{n} i!$, where $0!! = 1$. For example, $5!! = 1!2!3!4!5! = 34{,}560$. The proof presented here is based on one given independently by H. L. Abbott of the University of Alberta and N. Franceschine of Sebastopol, California; it uses both recursion and the principle of induction.

Example 5.3 Let $A_n = (2n-1)!!/[(n-1)!!]^4$, where n is a positive integer. Prove that $(2n-1)!|A_n$. [Notice that A_n is the reciprocal of the determinant of the *Hilbert matrix* $H_n = (h_{ij})_{n \times n}$, where $h_{ij} = (i+j-1)^{-1}$.]

Proof   Let $B_n = \dfrac{A_n}{(2n-1)!}$. Then

$$\frac{B_{n+1}}{n+1} = \frac{1}{n+1} \cdot \frac{A_n}{(2n+1)!}$$

$$= \frac{1}{n+1} \cdot \frac{(2n+1)!!}{(n!!)^4(2n+1)!}$$

$$= \frac{1}{n+1} \cdot \frac{(2n-1)!!}{[(n-1)!!]^4} \cdot \frac{(2n)!(2n+1)!}{(n!)^4(2n+1)!}$$

$$= \frac{1}{n+1} \cdot A_n \cdot \frac{(2n)!}{(n!)^4}$$

$$= \frac{1}{n+1} \cdot \frac{A_n}{(2n-1)!} \cdot \frac{(2n)!(2n+1)!}{(n!)^4}$$

$$= \frac{B_n}{n} \cdot \frac{1}{n+1} \cdot \frac{(2n)!}{(n!)^2} \cdot \frac{(2n-1)!n}{n!n!}$$

$$= \frac{B_n}{n} \cdot C_n \cdot \binom{2n-1}{n}$$

Since A_1 and hence $B_1/1$ are integers, it follows by induction that B_n/n is an integer for every positive integer n; that is, $n \mid B_n$. Thus, $(2n-1)! \mid A_n$ for every integer $n \geq 1$. ∎

We now verify the recurrence relation

$$\frac{B_{n+1}}{n+1} = \frac{B_n}{n} \cdot C_n \cdot \binom{2n-1}{n}$$

for $n = 3$. We have:

$$A_3 = \frac{5!!}{(2!!)^4}$$

$$= \frac{1!2!3!4!5!}{16}$$

$$= 2,160$$

$$B_3 = \frac{A_3}{5!}$$

$$= 18$$

$$A_4 = \frac{7!!}{(3!!)^4}$$

$$= \frac{1!2!3!4!5!6!7!}{(2!3!)^4}$$

$$= 6,048,000$$

$$B_4 = \frac{A_4}{7!}$$

$$= 1,200$$

$$\mathrm{RHS} = \frac{B_3}{3} \cdot C_3 \cdot \binom{5}{3}$$

$$= 6 \cdot 5 \cdot 10$$

$$= 300$$

$$= \frac{B_4}{4}$$

$$= \mathrm{LHS}$$

Notice that $5! \mid A_3$ and $7! \mid A_4$.

Additional Results on B_n

In 1980, R. E. Shafer showed that B_n is also divisible by $1 \cdot 3 \cdot 5 \cdots (2n-3)$ if $n \geq 3$; O. G. Ruehr showed that $n! \mid B_n$; and T. Hermann found that $B_n = \prod_{i=1}^{n-1} \binom{n+1}{n}\binom{n}{i}$.

Lagrange Meets Catalan

The next example, proposed in 1972 by E. T. Ordman of the University of Kentucky, is a fascinating application of Lagrange's identity. Once again, it shows that Catalan numbers can occur in mysterious places. The solution is based on one given in 1973 by R. Gibbs and H. Stocker of Fort Lewis College, Durango, Colorado.

Example 5.4 Let $n \geq 0, m \geq 1$, and

$$S_m(n) = \sum_{r=0}^{\lfloor n/2 \rfloor} \left[\binom{n}{r} - \binom{n}{r-1} \right]^m$$

Evaluate $S_2(n)$.

Solution   Let

$$T(n) = \sum_{r=0}^{n+1} \left[\binom{n}{r} - \binom{n}{r-1} \right]^2$$

By virtue of the symmetry of the binomial coefficients, it follows that $T(n) = 2S_2(n)$. Since

$$\binom{n}{r-1} + \binom{n}{r} = \binom{n+1}{r}$$

it follows that

$$\binom{n+1}{r}^2 - \binom{n}{r}^2 - \binom{n}{r-1}^2 = 2\binom{n}{r-1}\binom{n}{r}$$

Therefore,

$$T_n = \sum_{r=0}^{n+1} \left[\binom{n}{r}^2 - 2\binom{n}{r}\binom{n}{r-1} + \binom{n}{r-1}^2 \right]$$

$$= 2\sum_{r=0}^{n+1} \binom{n}{r}^2 + 2\sum_{r=0}^{n+1} \binom{n}{r-1}^2 - \sum_{r=0}^{n+1} \binom{n+1}{r}^2$$

$$= 2\sum_{r=0}^{n} \binom{n}{r}^2 + 2\sum_{r=0}^{n} \binom{n}{r}^2 - \sum_{r=0}^{n+1} \binom{n+1}{r}^2$$

$$= 2\binom{2n}{n} + 2\binom{2n}{n} - \binom{2n+2}{n+1}$$

$$= 4\binom{2n}{n} - 2\left[\binom{2n}{n-1} + \binom{2n}{n} \right]$$

$$= 2\binom{2n}{n} - 2\binom{2n}{n-1}$$

So

$$S_2(n) = \binom{2n}{n} - \binom{2n}{n-1}$$

$$= \frac{1}{n+1}\binom{2n}{n}$$

$$= C_n$$

We shall use the property that $C_n = \binom{2n}{n} - \binom{2n}{n-1}$ in the next chapter.   ∎

The next example, proposed in 1930 by B. C. Wong of Berkeley, California, presents yet another occurrence of the Catalan numbers C_n. It appeared in the 1965 William Lowell Putnam Mathematical Competition. The solution is based on one given in the same year by E. T. Lehmer of Brown University, Providence, Rhode Island; it employes Lagrange's and Vandermonde's identities.

Example 5.5 Prove that

$$\sum_{r=0}^{\lfloor (n-1)/2 \rfloor} \left[\binom{n}{r} \frac{n-2r}{n} \right]^2 = C_{n-1}$$

Proof Let

$$f(n) = \sum_{r=0}^{\lfloor (n-1)/2 \rfloor} \left[\binom{n}{r} \frac{n-2r}{n} \right]^2 \tag{5.16}$$

Changing r to $n-j$ in (5.16) and adding the result to (5.16) yield:

$$f(n) = \frac{1}{2} \sum_{r=0}^{n} \left[\binom{n}{r} \frac{n-2r}{n} \right]^2$$

$$= \frac{1}{2} \sum_{r=0}^{n} \binom{n}{r}^2 - 2 \sum_{r=0}^{n} \binom{n}{r}^2 \frac{r}{n} + 2 \sum_{r=0}^{n} \binom{n}{r}^2 \frac{r^2}{n^2}$$

$$= \frac{1}{2} \sum_{r=0}^{n} \binom{n}{r}^2 - 2 \sum_{r=0}^{n} \binom{n}{r}\binom{n-1}{r-1} + 2 \sum_{r=0}^{n} \binom{n-1}{r-1}^2$$

$$= \frac{1}{2} \sum_{r=0}^{n} \binom{n}{r}^2 - 2 \sum_{r=1}^{n} \binom{n}{r}\binom{n-1}{r-1} + 2 \sum_{r=1}^{n} \binom{n-1}{r-1}^2$$

$$= \frac{1}{2} \sum_{r=0}^{n} \binom{n}{r}^2 - 2 \sum_{r=0}^{n-1} \binom{n}{r+1}\binom{n-1}{r} + 2 \sum_{r=0}^{n-1} \binom{n-1}{r}^2$$

$$= \frac{1}{2} \binom{2n}{n} - 2 \binom{2n-1}{n} + 2 \binom{2n-2}{n-1}$$

$$= \binom{2n-2}{n-1} \left[\frac{2n-1}{n} - 2 \frac{2n-1}{n} + 2 \right]$$

$$= \frac{1}{n} \binom{2n-2}{n-1}$$

$$= C_{n-1} \qquad\qquad\blacksquare$$

For example, let $n = 6$. Then:

$$\sum_{r=0}^{2}\left[\binom{6}{r}\frac{6-2r}{6}\right]^2 = \binom{6}{0}^2 + \left[\binom{6}{1}\frac{4}{6}\right]^2 + \left[\binom{6}{2}\frac{2}{6}\right]^2$$

$$= 1 + 16 + 25$$

$$= 42$$

$$= C_5$$

Catalan Numbers and $\binom{m+n}{m}$

In 1917, A. A. Bennett derived two combinatorial identities involving Catalan numbers and the binomial coefficient $\binom{m+n}{m}$. Their proofs are a bit long and complicated, so we omit them.

Let $k_i(m,n) = \sum_{j}\binom{m-i+j}{i-j}\binom{n-j}{j}$, where $m + j \geq i$ and $n, i \geq j$. Then

$$\sum_{i}(-1)^i k_i(m,n)C_{m+n-i} = \binom{m+n}{m} \tag{5.17}$$

For example, let $m = 3$ and $n = 2$. Then

$$k_i(3,2) = \sum_{j}\binom{3-i+j}{i-j}\binom{2-j}{j}$$

where $i \leq 5$ and $0 \leq i \leq 5$. We then have:

$$k_0(3,2) = \binom{3}{0}\binom{2}{0} \qquad\qquad = 1$$

$$k_1(3,2) = \sum_{j}\binom{2+j}{1-j}\binom{2-j}{j} = 3$$

$$k_2(3,2) = \sum_{j}\binom{1+j}{2-j}\binom{2-j}{j} = 2$$

$$k_3(3,2) = \sum_{j}\binom{j}{3-j}\binom{2-j}{j} = 0$$

$$k_4(3,2) = \sum_{j}\binom{j-1}{4-j}\binom{2-j}{j} = 0$$

$$k_5(3,2) = \sum_{j}\binom{j-2}{5-j}\binom{2-j}{j} = 0$$

So

$$\sum_i k_i(3,2)C_{5-i} = k_0(3,2)C_5 - k_1(3,2)C_4 + k_2(3,2)C_3$$

$$-k_3(3,2)C_2 + k_4(3,2)C_1 - k_5(3,2)C_0$$

$$= 1 \cdot 42 - 3 \cdot 14 + 2 \cdot 5 - 0 \cdot 2 + 0 \cdot 1 - 0 \cdot 1$$

$$= 10$$

$$= \binom{3+2}{3}$$

as expected.

Formula (5.17) can be considered a recursive formula for Catalan numbers. Since $k_0(m,n) = \binom{m}{0}\binom{n}{0} = 1$, formula (5.17) can be rewritten as follows:

$$C_{m+n} = \binom{m+n}{m} - \sum_{i \geq 1}(-1)^i k_i(m,n)C_{m+n-i} \tag{5.18}$$

The following is the second formula by Bennett:

$$\sum_i (-1)^i \binom{m-i}{i}C_{m-n-i} = 0 \tag{5.19}$$

where $n \leq \lfloor m/2 \rfloor$.

For example, let $m = 7$ and $n = 3$. Then:

$$\sum_{i \geq 0}(-1)^i \binom{7-i}{i}C_{4-i} = \binom{7}{0}C_4 - \binom{6}{1}C_3 + \binom{5}{2}C_2 - \binom{4}{3}C_1$$

$$= 1 \cdot 14 - 6 \cdot 5 + 10 \cdot 2 - 4 \cdot 1$$

$$= 0$$

Next we present an interesting application of Catalan numbers. The application was proposed as a problem in 1918 by P. Franklin of the College of the City of New York.

Suppose we arrange in a row n letters of one kind (say, A), and $n-1$ letters of a second kind (say, B) such that moving from left to right, at any instant, the number of letters of the first kind exceeds that of the second kind. Find the number $f(n)$ of such arrangements.

For example, when $n = 2$, there are $\frac{3!}{2!} = 3$ permutations of three letters, of which two are alike. They are AAB, ABA, and BAA, of which only one permuation meets the criterion: AAB. So $f(2) = 1$.

When $n = 3$, there are $\frac{5!}{3!2!} = 10$ permutations of five letters, of which three are alike (A's) and two are alike (B's). Only two of them have the desired property: AAABB and AABAB. So $f(3) = 2$.

When $n = 4$ (4 A's and 3 B's), there are five such permuations: AAAABBB, AAABABB, AAABBAB, AABAABB, and AABABAB. Thus $f(5) = 5$.

More generally, in 1919 Gummer showed that using recursion,

$$f(n) = \frac{1}{n}\binom{2n-2}{n-1} = C_{n-1}$$

In fact, Gummer solved a generalized version of Franklin's problem. To see it, let $f(r,s)$ denote the number of permutations of r letters, some of kind A and the rest of kind B, such that the number of A's at any instant is more than that of B's, and there are s more A's than B's. Clearly, $f(r,s) = 0$ when $r - s$ is odd and $s < 1$.

Such a permutation will end in an A in $f(r-1, s-1)$ cases, where $r > 1$; it will end in a B in $f(r-1, s+1)$ cases, where $s \neq 0$. So, by the addition principle,

$$f(r,s) = f(r-1, s-1) + f(r-1, s+1)$$

To remove the restriction $s \neq 0$, we define

$$g(r,s) = \begin{cases} f(r,s) & \text{if } s \geq 0 \\ -f(r,-s) & \text{otherwise} \end{cases}$$

Then

$$g(r,s) = g(r-1, s-1) + g(r-1, s+1)$$

where $r > 1$.

Applying recursion successively, this yields:

$$\begin{aligned}
g(r,s) &= g(r-1, s-1) + g(r-1, s+1) \\
&= g(r-2, s-2) + 2g(r-2, s) + g(r-2, s+2) \\
&= g(r-3, s-3) + 3g(r-3, s-1) + 3g(r-3, s+1) + g(r-3, s+3) \\
&\;\;\vdots \\
&= g(1, s-r+1) + \binom{r-1}{1}g(1, s-r+3) + \binom{r-1}{2}g(1, s-r+5) + \cdots \\
&= \sum_{j \geq 0} \binom{r-1}{j} g(1, 1-(r-s-2j))
\end{aligned}$$

Now $g(1, s) = 0$, except when $g(1, 1) = 1$, and $g(1, -1) = -1$. Thus, since $r - s$ is even,

$$g(r, s) = \binom{r - 1}{(r - s - 2)/2} g(1, -1) + \binom{r - 1}{(r - s)/2} g(1, 1)$$

$$= -\binom{r - 1}{(r - s - 2)/2} + \binom{r - 1}{(r - s)/2}$$

$$= -\binom{r - 1}{(r - s)/2} \cdot \frac{r - s}{r + s} + \binom{r - 1}{(r - s)/2}$$

$$= \binom{r - 1}{(r - s)/2} \left(1 - \frac{r - s}{r + s} \right)$$

$$= \frac{2s}{r + s} \binom{r - 1}{(r - s)/2}$$

In particular, let $r = 2n - 1$ and $s = 1$. Then

$$f(r, s) = g(r, s)$$

$$= \frac{1}{n} \binom{2n - 2}{n - 1}$$

$$= C_{n-1}$$

as desired.

Next, we pursue Catalan's parenthesization problem. To this end, first notice that the expression $((ab)(cd))$ is correctly parenthesized, but $((ab)($ and $)(ab)cd)$ are not.

Example 5.6 Find the number of different ways of parenthesizing $n + 1$ elements with n pairs of left and right parentheses; that is, find the number P_n of well-formed sequences of left and right parentheses that can be formed with n pairs, where $n \geq 0$.

Solution   Table 5.3 lists the various possible correctly parenthesized expressions for $0 \leq n \leq 3$ and the corresponding values of P_n, where λ denotes the *null word* consisting of no characters.

Let S denote the set of well-formed sequences with n pairs of parentheses. Then S can be defined recursively, where the terminal clause is omitted for convenience:

- $\lambda \in S$.
- If $x, y \in S$, then $(x), xy \in S$, where xy denotes the concatenation of the symbols x and y.

Table 5.3 Well-Formed Sequences with n Pairs

n	Correctly Parenthesized Expressions					P_n
0	λ					1
1	()					1
2	()()	(())				2
3	()()()	(())()	()()()	()(())	((()))	5
4	()()()() (((()))() (())(()) ()((())) (())()() ()(())() ()()(()) (())()() ()((()) (()()()) ((()))() (()(())) (((())() (((()))))					14

$\uparrow$
Catalan Numbers

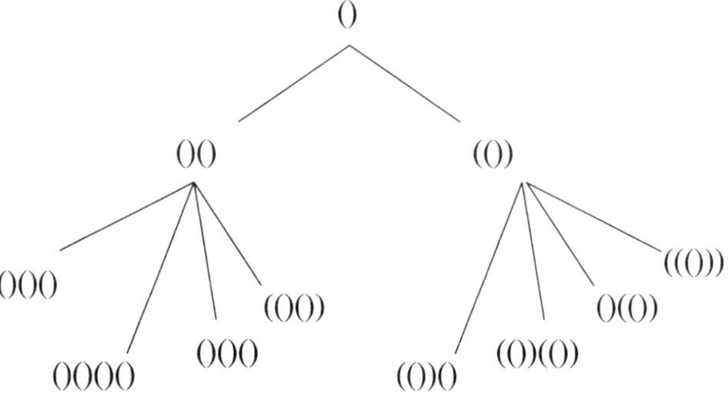

Figure 5.6 Well-Formed Sequences

Accordingly, the various elements of S can be generated using a tree diagram, as in Figure 5.6, where we have omitted λ for convenience. Notice that $x\lambda = x = \lambda x$ for every $x \in S$, and that the tree contains duplicates.

We shall now develop a recursive formula for P_n. First, notice that $P_0 = 1 = P_1$.

Suppose $n \geq 2$. Let $0 \leq i \leq n - 1$. The first i pairs can be correctly grouped in P_i ways and the remaining $n - 1 - i = n - i - 1$ pairs in P_{n-i-1} ways. Using the multiplication principle, these two events can take place together in $P_i P_{n-i-1}$ different ways. Because this is true for each value of i, by the addition principle,

$$P_n = \sum_{i=0}^{n-1} P_i P_{n-i-1}$$

Thus P_n satisfies the same recurrence relation and the same initial condition as C_n. Consequently, $P_n = C_n$. ∎

It is interesting to notice the similarity between the argument in Segner's development of the recurrence relation and this argument; they are basically the same.

An Interesting Observation

Since C_n denotes the number of ways of parenthesizing products with $n + 1$ symbols, the corresponding generating function can be written as follows:

$$P(x) = C_0x + C_1x^2 + C_2x^3 + C_3x^4 + \cdots$$

$$= x + x^2 + 2x^3 + 5x^4 + \cdots$$

$$= x + x^2 + [x^3 + x^3] + [x^4 + x^4 + x^4 + x^4 + x^4] + \cdots$$

$$= x + [(xx)] + [((xx)x) + (x(xx))]$$

$$+[(((xx)x)x) + ((x(xx))x) + ((xx)(xx)) + (x((xx)x)) + (x(x(xx)))] + \cdots$$

where each bracketed term shows the various possible ways of parenthesizing the product.

Forder's Bijection Algorithm

Let T denote the set of triangulations of an n-gon and P the set of correctly parenthesized sequences with n pairs of parentheses. It follows from Examples 5.1 and 5.6 that $|T| = |P| = C_n$, where $|S|$ denotes the cardinality of the set S. Since the sets T and P are finite and have the same cardinality, there must exist a bijection between the two sets.

Interestingly, in 1961 H. G. Forder of the University of Auckland, New Zealand, developed such an algorithm. To illustrate it, consider the triangulated hexagon in Figure 5.7. Label its sides, except the base, with the letters (operands) a through e. Each diagonal spanning adjacent side is labeled with the parenthesized concatenation of the labels of the sides. For example, the diagonal abutting the sides a and b is labeled with (ab). Each remaining diagonal is labeled the same way, using the labels of the other two sides of the corresponding triangle. Continue this procedure until the base is labeled. The dissection in Figure 5.7 yields the correctly parenthesized expression $(((ab)c)(de))$ with exactly four pairs of parentheses; see Figure 5.8.

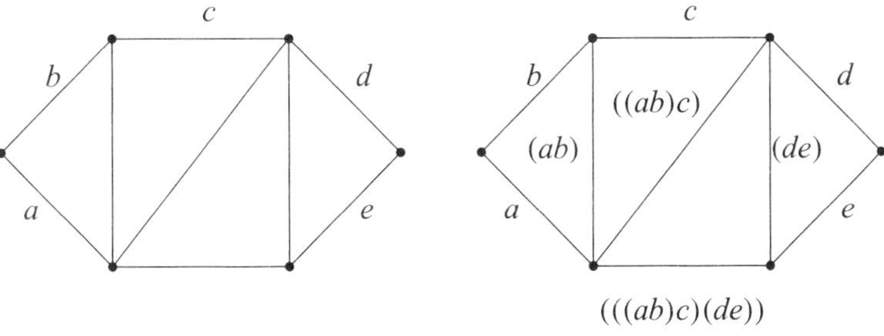

Figure 5.7   Figure 5.8

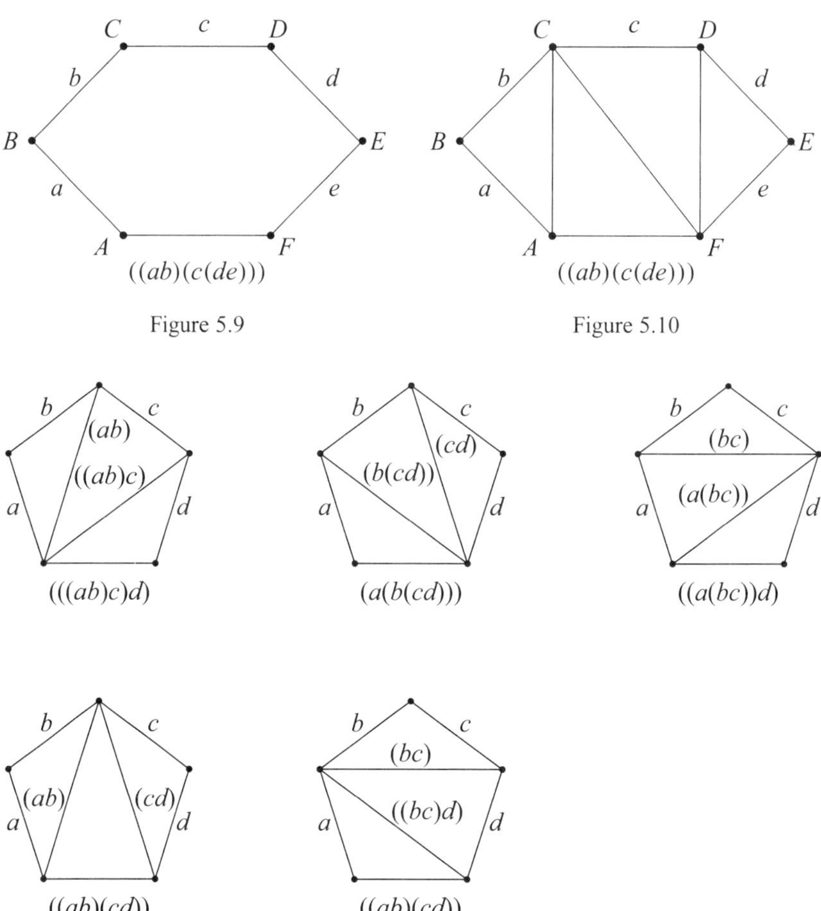

Figure 5.9

Figure 5.10

Figure 5.11 Triangulations of a Pentagon and Corresponding Expressions

This process is reversible. As an example, consider the correctly parenthesized expression $((ab)(c(de)))$ and the hexagon $ABCDEF$ in Figure 5.9, where the base is labeled with $((ab)(c(de)))$ and the remaining sides with the variables a through e. Corresponding to each valid subexpression (xy), except the original expression, draw the diagonal by joining the nonrepeating endpoints of the sides labeled x and y. For instance, corresponding to (ab), join the vertices A and C; corresponding to (de), join D and F; and corresponding to $(c(de))$, join C and F. Thus, we have accounted for each valid subexpression, terminating the procedure. By now the hexagon is fully triangulated; see Figure 5.10.

In general, every triangulation of an $(n + 2)$-gon yields a unique correctly parenthesized expression consisting of $n + 1$ operands and n binary operators and vice versa.

For the curious-minded, Figure 5.11 shows the various triangulations of a pentagon and the corresponding correctly parenthesized expressions.

We shall return later to the triangulations of n-gons and the corresponding parenthesized expressions.

The identity in the next example was proposed as a problem in the *American Mathematical Monthly* by Garrett Birkhoff of Harvard University, in 1934. It is the same as Segner's recurrence formula. The solution presented is due to E. D. Raineville, an engineer with the U.S. Bureau of Reclamation, Denver, Colorado. It uses the generating function $\sqrt{1 - 4x}$ of Catalan numbers we pursued earlier.

Example 5.7 Let $f_n = \frac{1}{2n-1}\binom{2n-1}{n}$. Establish the recurrence relation

$$f_n = \sum_{i=1}^{n-1} f_i f_{n-i} \tag{5.20}$$

Deduce that f_n is an integer.

Solution Let

$$f(x) = \sum_{n=1}^{\infty} f_n x^n \tag{5.21}$$

$$= \sum_{n=1}^{\infty} \frac{1}{2n-1}\binom{2n-1}{n} x^n$$

$$= \sum_{n=1}^{\infty} \frac{(2n-2)!}{n!(n-1)!} x^n$$

$$= \sum_{n=1}^{\infty} \frac{1}{n}\binom{2n-2}{n-1} x^n \tag{5.22}$$

$$= \frac{1}{2} - \frac{1}{2}\sqrt{1 - 4x}$$

$$[f(x)]^2 + x = f(x)$$

But from equation (5.21),

$$[f(x)]^2 = \sum_{n=2}^{\infty} \left(\sum_{i=1}^{n-1} f_i f_{n-i} \right) x^n$$

Therefore,

$$x + \sum_{n=2}^{\infty} \left(\sum_{i=1}^{n-1} f_i f_{n-i} \right) x^n = \sum_{n=1}^{\infty} f_n x^n$$

It follows from this identity that $f_1 = 1$ and $f_n = \sum_{i=1}^{n-1} f_i f_{n-i}$. Since $f_1 = 1 = f_2$, it follows by induction from this recurrence relation that f_n is an integer. ∎

A Few Observations

Notice that $f_n = \dfrac{(2n-2)!}{n!(n-1)!} = C_{n-1}$. Also,

$$f_n = \binom{2n-1}{n} - 2\binom{2n-2}{n}$$

as H. Gupta noted, and

$$f_n = \binom{2n-2}{n-1} - \binom{2n-2}{n}$$

as E. P. Starke observed, independently, both in 1935. It follows from both formulas also that f_n is indeed an integer for every $n \geq 1$.

M. Ward of the California Institute of Technology proved that f_n is an integer in yet another way. He observed that

$$f_n = \frac{1}{n}\binom{2n-2}{n-1} = \frac{1}{2n-1}\binom{2n-1}{n}$$

Therefore,

$$(2n-1)\binom{2n-2}{n-1} = n\binom{2n-1}{n}$$

Since $(2n-1, n) = 1$, this implies that $n \mid \binom{2n-2}{n-1}$; so $\frac{1}{n}\binom{2n-2}{n-1} = f_n$ is an integer.

Ward's proof of the recurrence relation involves the hypergeometric series and the gamma function, so we omit it.

In 1964, J. Petersen of Saskatchewan Power Corporation, Regina, Saskatchewan, Canada, proposed the problem in Example 5.7 in the reverse order: Let $f_n = \sum_{i=1}^{n-1} f_i f_{n-i}$, where $f_1 = 1$. Then $f_n = C_{n-1}$. The proof, as in the example, hinges on the generating function $\sqrt{1-4x}$.

The following example, proposed by R. Breusch of Amherst College, Massachusetts, in 1964 deals with a combinatorial sum in which each summand contains the ubiquitous CBC $\binom{2m}{m}$ as a factor; as will be seen shortly, the other factor is a Catalan number. The solution given is due to M. T. L. Bizley of London, England, and uses the power series expansion of $(1-4x)^{-1/2}$.

Example 5.8 Let $n \geq 0$ and A_i a positive integer such that

$$\sum_{i=0}^{n} \binom{2n-2i}{n-i} A_i = \binom{2n+1}{n}$$

Find A_i.

Solution   Let

$$f(x) = A_0 + A_1 x + A_2 x^2 + \cdots + A_i x^i + \cdots$$

Notice that $\binom{2n-2i}{n-i} A_i$ is the coefficient of x^{n-i} in the expansion of $(1-4x)^{-1/2} f(x)$.

But this is given to be $\binom{2n+1}{n}$, which is the coefficient of x^n in $[(1-4x)^{-1/2} - 1]/2x$. Because this is true for every $n \geq 0$,

$$(1-4x)^{-1/2} f(x) = [(1-4x)^{-1/2} - 1]/2x$$

$$f(x) = \frac{1 - (1-4x)^{1/2}}{2x}$$

Consequently,

$$A_i = \frac{1}{i+1} \binom{2i}{i}$$

$$= C_i \qquad\qquad \blacksquare$$

An Alternate Proof

M. T. Bird of San Jose State College gave an alternate proof using the identity

$$\sum_{i=j}^{n} \frac{1}{i+1}\binom{2n-2i}{n-i}\binom{2i}{i} = \frac{n-j+1}{2(n+1)}\binom{2j}{j}\binom{2n+2-2j}{n+1-j} \tag{5.23}$$

where $0 \leq j \leq n$. When $j = 0$, identity (5.23) yields

$$\sum_{i=0}^{n} \frac{1}{i+1}\binom{2n-2i}{n-i}\binom{2i}{i} = \frac{1}{2}\binom{2n+2}{n+1} \tag{5.24}$$

$$= \binom{2n+1}{n}$$

It follows from this that $A_i = \frac{1}{i+1}\binom{2i}{i} = C_i$.

Following the problem by Breusch, V. K. Rohatgi of the University of Alberta, Edmonton, Canada, generalized identity (5.23) to the following:

$$v\sum_{i=0}^{n} \frac{1}{ik+i+v}\binom{ik+i+v}{i}\binom{(n-i)(k+1)}{n-i} = \binom{nk+v+n}{n} \tag{5.25}$$

where $v, k \geq 1$. When $k = 1 = v$, this reduces to identity (5.23).

The next example, proposed in 1947 by C. D. Olds of San Jose State College, is closely related to the foregoing example. We shall give a few steps showing the close link between the two examples.

Example 5.9 Find a formula for $U_n + U_1 U_{n-1} + U_2 U_{n-2} + \cdots + U_{n-1} U_1 + U_n$, where $U_n = \dfrac{2 \cdot 6 \cdot 10 \cdots (4n-2)}{1 \cdot 2 \cdot 3 \cdots (n+1)}$.

Solution   Notice that

$$U_n = \frac{2 \cdot 6 \cdot 10 \cdots (4n-2)}{1 \cdot 2 \cdot 3 \cdots (n+1)}$$

$$= \frac{2^n \cdot 1 \cdot 3 \cdot 5 \cdot \cdots (2n-1)}{(n+1)!}$$

$$= \frac{2^n (2n)!}{(n+1)![2 \cdot 4 \cdot 6 \cdots (2n)]}$$

$$= \frac{2^n (2n)!}{2^n (n+1)! n!}$$

$$= \frac{(2n)!}{(n+1)! n!}$$

$$= C_n$$

From the preceding discussion, $U_n = C_n = f_{n+1}$, where $U_0 = f_0 = 1$. But, from equation (5.20),

$$f_{n+2} = \sum_{i=1}^{n+1} f_i f_{n+2-i}$$

That is,

$$U_{n+1} = \sum_{i=1}^{n+1} U_{i-1} U_{n+1-i} = \sum_{i=0}^{n} U_i U_{n-i}$$

as desired. ∎

Interestingly enough, the same problem resurfaced in a slightly different form in 1964, this time proposed by J. Petersen.

The following example, proposed in 1968 by L. Carlitz and R. A. Scoville of Duke University, Durham, North Carolina, presents yet another occurrence of Catalan numbers. The solution, given by M. G. Beumer of Technological University, Delft, The Netherlands, closely resembles the one in Example 5.7.

Example 5.10 Find the number of n-tuples $(a_1, a_2, \ldots, a_n)$ such that $1 \leq a_1 \leq a_2 \leq \cdots \leq a_n$, where each $a_i \leq i$ and $1 \leq i \leq n$.

Solution   Let A_n denote the number of such sequences. Setting $A_0 = 1$, we have the recurrence relation

$$A_n = \sum_{i=0}^{n-1} A_i A_{n-i-1}$$

Let $A(x) = \sum_{n=0}^{\infty} A_n x^n$. Then, as in Example 5.7,

$$A(x) = 1 + x[A(x)]^2$$

$$= \frac{1 - \sqrt{1-4x}}{2x}$$

$$= \sum_{n=0}^{\infty} C_n x^n$$

Thus $A_n = C_n$, the nth Catalan number. ∎

For example, there are $C_3 = 5$ 3-tuples (a_1, a_2, a_3) such that $1 \leq a_1 \leq a_2 \leq a_3$, where each $a_i \leq i$. They are $(1,1,1)$, $(1,1,2)$, $(1,1,3)$, $(1,2,2)$, $(1,2,3)$.

Heuer's Alternate Proof

We now present an alternate solution by G. A. Heuer of Concordia College, Montreal, Canada.

Let $f(n)$ denote the desired answer, and $f_i(n)$ the number of these n-tuples with $a_i = i$, where $1 \leq i \leq n$. Then $f_i(n + 1) = \sum_{j=1}^{i} f_j(n)$ and

$$f(n) = \sum_{i=1}^{n} f_i(n)$$

$$= f_n(n + 1)$$

$$= f_{n+1}(n + 1)$$

Clearly, $f_1(n) = 1$ for all n and we have the recurrence relation

$$f_{i+1}(n + 1) - f_i(n + 1) = \sum_{j=1}^{i+1} f_j(n) - \sum_{j=1}^{i} f_j(n)$$

$$= f_{i+1}(n)$$

That is,

$$f_{i+1}(n + 1) - f_{i+1}(n) = f_i(n + 1)$$

These conditions completely determine the value of $f_i(n)$.

It can be verified that

$$f_i(n) = \binom{n + i - 3}{n - 2} - \binom{n + i - 3}{n}$$

satisfies the recurrence relation and the initial condition. Thus

$$f(n) = f_{n+1}(n + 1)$$

$$= \binom{2n - 1}{n - 1} - \binom{2n - 1}{n + 1}$$

$$= \frac{(2n)!}{(n + 1)!n!}$$

$$= C_n$$

as desired.

The next example is equally interesting. Proposed in 1978 by E. T. H. Wang of Wilfred Laurier University, the problem asks for a recurrence relation satisfied by C_n. The proof given by R. B. Nelsen of Lewis and Clark College, has some resemblance to the one in Example 5.7 and uses generating functions and the calculus.

Example 5.11 Let $C_n = \frac{1}{n+1}\binom{2n}{n}$, where $n \geq 0$. Prove that $\sum_{i=0}^{n} C_i C_{n-i} = C_{n+1}$.

Proof Let $f(x) = \sum_{i=0}^{\infty} C_n x^n$. Then

$$f(x)^2 = \sum_{n=0}^{\infty}\left(\sum_{i=0}^{n} C_i C_{n-i}\right) x^n \tag{5.26}$$

Let

$$g(x) = \frac{\mathrm{d}}{\mathrm{d}x}\left(xf(x)\right)$$

$$= \sum_{n=0}^{\infty}\binom{2n}{n} x^n$$

where $g(0) = 1$. Since

$$(n+1)\binom{2n+2}{n+1} = (4n+2)\binom{2n}{n}$$

it follows that

$$\sum_{n=0}^{\infty}(n+1)\binom{2n+2}{n+1} x^n = 4x \sum_{n=1}^{\infty} n\binom{2n}{n} x^{n-1} + 2 \sum_{n=0}^{\infty} n\binom{2n}{n} x^n$$

That is,

$$g'(x) = 4xg'(x) + 2g(x)$$

$$\frac{g'(x)}{g(x)} = \frac{1}{1-4x}$$

Because $g(0) = 1$, this implies that $g(x) = (1-4x)^{-1/2}$. So

$$xf(x) = \int_0^x g(t)\,\mathrm{d}t$$

$$= \tfrac{1}{2}(1 - \sqrt{1-4x})$$

$$[f(x)]^2 = \frac{1 - 2\sqrt{1-4x} + (1-4x)}{4x^2}$$

$$= \frac{f(x) - 1}{x}$$

$$= \frac{1}{x} \sum_{n=1}^{\infty} C_n x^n$$

$$= \sum_{n=0}^{\infty} C_{n+1} x^n \tag{5.27}$$

It follows from equations (5.26) and (5.27) that $\sum_{i=0}^{n} C_i C_{n-i} = C_{n+1}$, as required. ∎

Another Summation Formula

The following summation formula also involves the nearly ubiquitous quotient $\dfrac{2 \cdot 4 \cdot 6 \cdots (2n)}{1 \cdot 3 \cdot 5 \cdots (2n-1)}$ we have enountered a number of times by now:

$$1 + \frac{1}{3} \cdot \frac{1-n}{1+n} + \frac{1}{5} \cdot \frac{(1-n)(2-n)}{(1+n)(2+n)} + \cdots = \frac{1}{4n} \left[\frac{2 \cdot 4 \cdot 6 \cdots (2n)}{1 \cdot 3 \cdot 5 \cdots (2n-1)} \right]^2 \tag{5.28}$$

This problem was proposed in 1953 by H. F. Sandham of Trinity College, Ireland. In fact, he generalized this formula to

$$\sum_{r=0}^{\infty} \frac{(1-n)_r (1-x)_r}{(1+n)_r (1+x)_r} = \frac{1}{2} \cdot \frac{n!(\frac{1}{2}+x)_{n-1}}{(\frac{1}{2})_n (1+x)_{n-1}}$$

where $(a)_r = a(a+1) \cdots (a+r-1)$ and $(a)_0 = 1$. We skip its proof for convenience.

Suppose we let $x = 1/2$ in this formula. Then

$$\sum_{r=0}^{\infty} \frac{(1-n)_r (1/2)_r}{(1+n)_r (3/2)_r} = \frac{1}{2} \cdot \frac{n!(1)_{n-1}}{(\frac{1}{2})_n (3/2)_{n-1}} \tag{5.29}$$

Notice that

$$(1/2)_r = \frac{1}{2}\left(\frac{1}{2}+1\right)\left(\frac{1}{2}+2\right)\cdots\left(\frac{1}{2}+r-1\right)$$

$$= \frac{1\cdot 3\cdot 5\cdots(2r-1)}{2^r}$$

$$(3/2)_r = \frac{3}{2}\left(\frac{3}{2}+1\right)\left(\frac{3}{2}+2\right)\cdots\left(\frac{3}{2}+r-1\right)$$

$$= \frac{3\cdot 5\cdot 7\cdots(2r+1)}{2^r}$$

$$(1)_{n-1} = 1(1+1)(1+2)\cdots[1+(n-1)+1]$$

$$= (n-1)!$$

Equation (5.29) becomes

$$\sum_{r=0}^{\infty}\frac{(1-n)_r}{(1+n)_r}\cdot\frac{1\cdot 3\cdot 5\cdots(2r-1)}{3\cdot 5\cdots(2r+1)} = \frac{1}{2}\cdot\frac{n!2^n(n-1)!2^{n-1}}{1\cdot 3\cdot 5\cdots(2n-1)\ \ 3\cdot 5\cdots(2n-1)}$$

$$\sum_{r=0}^{\infty}\frac{(1-n)_r}{(1+n)_r}\cdot\frac{1}{2r+1} = \frac{1}{4n}\left[\frac{2\cdot 4\cdot 6\cdots(2n)}{1\cdot 3\cdot 5\cdots(2n-1)}\right]^2$$

$$1+\frac{1}{3}\cdot\frac{1-n}{1+n}+\frac{1}{5}\cdot\frac{(1-n)(2-n)}{(1+n)(2+n)}+\cdots = \frac{2^{4n-2}}{n(n+1)^2 C_n^2} \tag{5.30}$$

Catalan Numbers and Nonassociative Algebras

Here we present an interesting application of Catalan numbers to nonassociative algebras, proposed in 1960 by H. Goheen of Oregon State College. The solution presented here is based on one by J. R. Brown of the University of Massachusetts and uses both Pascal's identity and the principle of induction.

Example 5.12 Consider an algebra of n elements that satisfies the commutative property but not the associative property. Find the number of distinct sums that can be formed.

Solution   Let S_n denote the number of distinct sums that can be formed. Such a sum must be composed of two subsums, composed of k and $n-k$ elements, where $1 \le k \le n-1$. Because the algebra is commutative, we then have:

$$S_n = \frac{1}{2}\sum_{k=1}^{n-1}\binom{n}{k}S_k S_{n-k}$$

Notice that $S_1 = 1$. Using Pascal's identity, we have:

$$S_{n+1} = \frac{1}{2} \sum_{k=1}^{n} \binom{n+1}{k} S_k S_{n+1-k}$$

$$= \frac{1}{2} S_n + \frac{1}{2} \sum_{k=1}^{n-1} \binom{n}{k} S_k S_{n+1-k} + \frac{1}{2} \sum_{k=2}^{n} \binom{n+1}{k} S_k S_{n+1-k} + \frac{1}{2} S_n$$

$$= \frac{1}{2} \sum_{k=1}^{n-1} \binom{n}{k} (S_k S_{n+1-k} + S_{k+1} S_{n-k}) + S_n$$

Suppose $S_{k+1} = (2k-1)S_k$ for $k < n$. Then

$$S_{n+1} = \frac{1}{2} \sum_{k=1}^{n-1} \binom{n}{k} [(2n-2k-1) + (2k-1)] S_k S_{n-k} + S_n$$

$$= (2n-2)S_n + S_n$$

$$= (2n-1)S_n$$

Since $S_2 = 1 = (2 \cdot 1 - 1)S_1$, this result is true for all $n \geq 2$.
Using iteration, it follows that

$$S_n = 1 \cdot 3 \cdot 5 \cdots (2n-3)$$

$$= \frac{(2n-3)!}{2 \cdot 4 \cdot 6 \cdots (2n-4)}$$

$$= \frac{(2n-3)!}{2^{n-2}} (n-2)!$$

$$= \frac{(2n-2)!}{2^{n-1}} (n-1)!$$

$$= \frac{1}{n} \binom{2n-2}{n-1} \frac{n!}{2^{n-1}}$$

$$= \frac{n! C_{n-1}}{2^{n-1}}$$

Catalan Numbers Again

The identity in the following example was developed in 1938 by R. H. Bardell of the University of Wisconsin. We established this identity in Example 3.3. We now reestablish it using Corollary 2.2 and integral calculus.

Example 5.13 Prove that

$$\sum_{r=0}^{n} \frac{(-1)^r}{2r+1} \binom{n}{r} = \frac{2^{2n+1}}{(n+1)(n+2)C_{n+1}}$$

Proof   We have:

$$(1 - x^2)^n = \sum_{r=0}^{n} (-1)^r \binom{n}{r} x^{2r}$$

Therefore,

$$\int_0^1 (1 - x^2)^n \, dx = \sum_{r=0}^{n} (-1)^r \binom{n}{r} \int_0^1 x^{2r} \, dx$$

$$\int_0^{\pi/2} \cos^{2n+1} \theta \, d\theta = \sum_{r=0}^{n} (-1)^r \binom{n}{r} \frac{1}{2r+1}$$

$$\frac{2 \cdot 4 \cdot 6 \cdots (2n)}{1 \cdot 3 \cdot 5 \cdots (2n+1)} = \sum_{r=0}^{n} \frac{(-1)^r}{2r+1} \binom{n}{r}$$

That is,

$$\sum_{r=0}^{n} \frac{(-1)^r}{2r+1} \binom{n}{r} - \frac{n! 2^n}{1 \cdot 3 \cdot 5 \cdots (2n+1)}$$

$$= \frac{2^n n! [2 \cdot 4 \cdot 6 \cdots (2n+2)]}{(2n+2)!}$$

$$= \frac{2^{2n+1} n! (n+1)!}{(2n+2)!}$$

$$= \frac{2^{2n+1} (n+1)!(n+1)!(n+2)}{(n+1)(n+2)(2n+2)!}$$

$$= \frac{2^{2n+1}}{(n+1)(n+2)C_{n+1}}$$

as desired.   ∎

6

The Ubiquity of Catalan Numbers I

Like Fibonacci and Lucas numbers,* Catalan numbers are also ubiquitous. "They have the same delightful propensity for popping up unexpectedly, particularly in combinatorial problems," as Martin Gardner wrote in 1976 in *Scientific American*. "Indeed," he continues, "the Catalan sequence is probably the most frequently encountered sequence that is still obscure enough to cause mathematicians lacking access to N. J. A. Sloane's *A Handbook of Integer Sequences* to expend inordinate amounts of energy re-discovering formulas that were worked out long ago." We now turn to a number of instances that demonstrate that Catalan numbers pop up in numerous and quite unexpected places, as we saw in the preceding chapter. We will revisit a number of these examples in the next chapter.

The most extensive study of the various occurrences of Catalan numbers and bijections among the various Catalan structures was undertaken by Michael J. Kuchinski while he was a graduate student at West Virginia University. His 1977 master's thesis, *Catalan Structures and Correspondences*, makes many references to Gould's extensive bibliography. It lists 31 occurrences of Catalan numbers and 158 bijections among the various Catalan structures, out of a possible $\binom{31}{2} = 465$ correspondences. Like Growney, Kuchinski included a brief history of Catalan numbers in his work.

The following example was originally studied in 1877 by French geometer Gohierre de Longchamps (1842–1906). It resurfaced in 1918 when Frank Irwin of the University of California proposed it as a problem. It reminds us of the different ways of interpreting the nonassociative product $a_1 a_2 \ldots a_{n+1}$ with n pairs of parentheses, which we investigated in Example 5.6.

---

* See T. Koshy, *Fibonacci and Lucas Numbers with Applications*, 2001.

Example 6.1 Find the number of different ways the complex fraction in Figure 6.1 can be interpreted in such a way that the resulting fraction would be valid.

$$\cfrac{a_1}{\cfrac{a_2}{\cfrac{a_3}{\begin{array}{c}\vdots\\ a_{n+1}\end{array}}}}$$

Figure 6.1

Solution   Table 6.1 lists the possible interpretations for $0 \le n \le 3$, where we have used parentheses to indicate the groupings.

Table 6.1

a_1	$\dfrac{a_1}{a_2}$	$\dfrac{\left(\dfrac{a_1}{a_2}\right)}{a_3}$	$\dfrac{a_1}{\left(\dfrac{a_2}{a_3}\right)}$
$n = 0$	$n = 1$		$n = 2$

$$\dfrac{\left(\dfrac{a_1}{a_2}\right)}{\left(\dfrac{a_3}{a_4}\right)} \qquad \dfrac{a_1}{\left(\dfrac{\left(\dfrac{a_2}{a_3}\right)}{a_4}\right)} \qquad \dfrac{\left(\dfrac{\left(\dfrac{a_1}{a_2}\right)}{a_3}\right)}{a_4} \qquad \dfrac{\left(\dfrac{a_1}{\left(\dfrac{a_2}{a_3}\right)}\right)}{a_4} \qquad \dfrac{a_1}{\left(\dfrac{a_2}{\left(\dfrac{a_3}{a_4}\right)}\right)}$$

$$n = 3$$

Clearly, there is a bijection between the set of possible interpretations of the complex fraction and the set of different ways of parenthesizing the product $a_1 a_2 \ldots a_{n+1}$. To see this, we first write the fraction horizontally: $a_1 | a_2 | \ldots | a_{n+1}$. Then we change it into a multiplication problem: $a_1 a_2 \ldots a_{n+1}$. Clearly, this process is reversible.

For example, the fraction

$$\cfrac{\left(\dfrac{a_1}{a_2}\right)}{\left(\dfrac{a_3}{a_4}\right)}$$

corresponds to the product $((a_1 a_2)(a_3 a_4))$ and the fraction

$$\cfrac{a_1}{\left(\cfrac{\left(\cfrac{a_2}{a_3}\right)}{a_4}\right)}$$

corresponds to the product $(a_1((a_2 a_3)a_4))$.

From Catalan's parenthesization problem, there are C_n different ways of parenthesizing the product $a_1 a_2 \ldots a_{n+1}$ with n pairs of parentheses. So, there are C_n different ways of interpreting the complex fraction in Figure 6.1.

> **Walther Franz Anton von Dyck** (1856–1934) was born in Munich, Germany. He studied mathematics at the Munich Polytechnikum under Alexander Brill and Felix Klien. In 1879, he received his doctorate and became an assistant to Klien. He joined the faculty at the University of Leipzig in 1882. Two years later, he was appointed professor at the Munich Polytechnikum, where he served as director, rector, and second chairman.
>
> Dyck made significant contributions to group theory, function theory, topology, and algebraic characteristic theory. He is credited with the modern definition of a mathematical group. The *Dyck language* in the theory of formal languages is named after him. Dyck was also the editor of Kepler's works.

Example 6.2 Find the number of mountain ranges that can be drawn with n upstrokes and n downstrokes. In other words, find the number of different paths we can choose from the origin to the lattice point $(2n, 0)$ on the xy-plane subject to the following conditions:

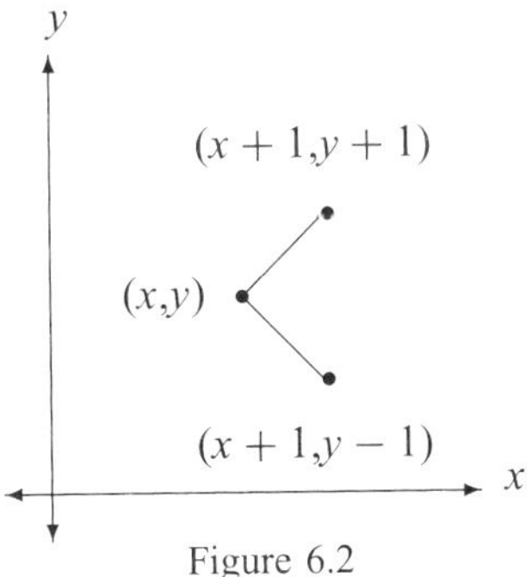

Figure 6.2

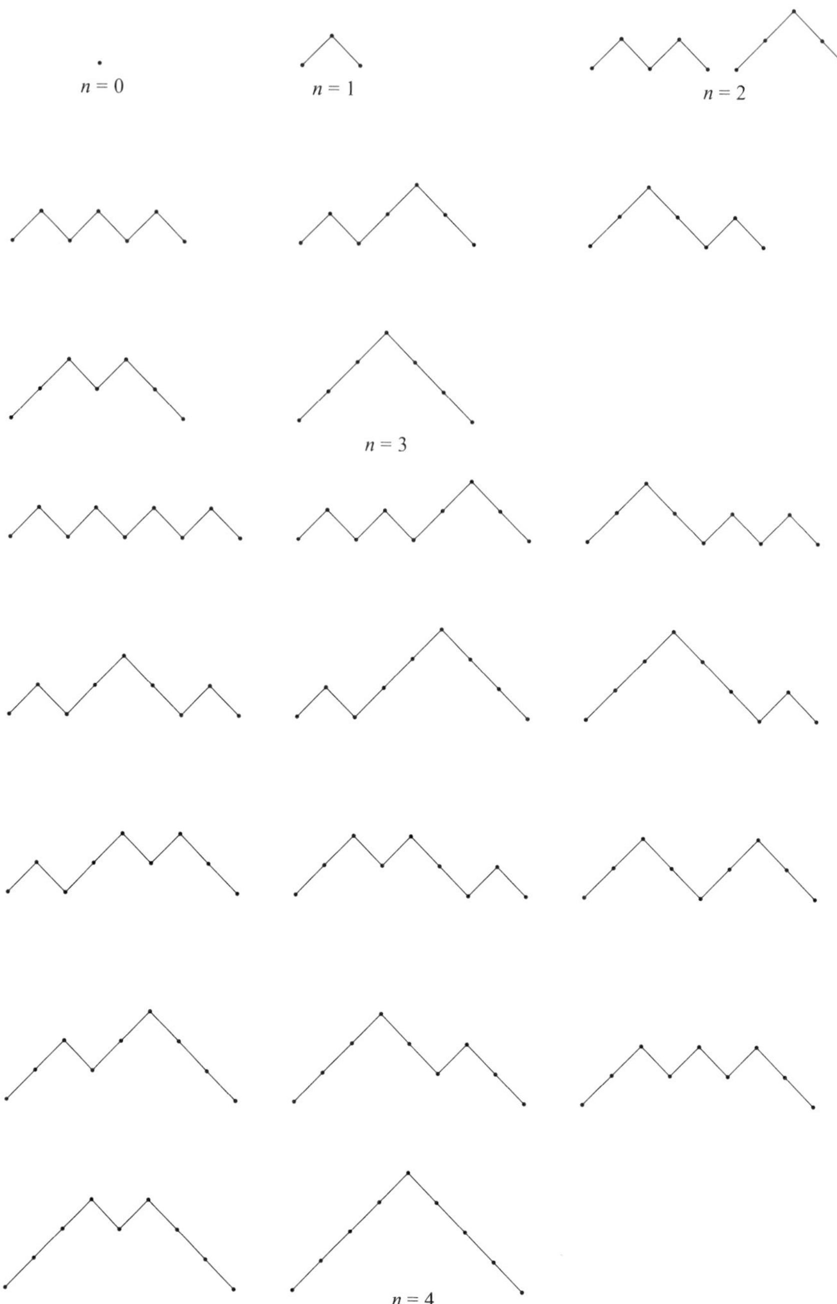

Figure 6.3 Possible Mountain Ranges for $0 \leq n \leq 4$

- We can touch the x-axis but cannot cross it.
- From the point (x, y), we can climb up to the point $(x + 1, y + 1)$ or climb down to the point $(x + 1, y - 1)$. See Figure 6.2.

Such mountain ranges are *lattice paths* with steps $(1,1)$ and $(1, -1)$ that never fall below the x-axis; they are called *Dyck paths* after Walther von Dyck.

Solution   Figure 6.3 shows the various possible mountain ranges for $0 \leq n \leq 4$. Notice that the answer in each case is a Catalan number.

Mountain Ranges and Correctly Parenthesized Sequences

Interestingly, the problem of mountain ranges in Example 6.2 and the problem of correctly parenthesized expressions are essentially the same. In other words, there is a bijection between the two. This can be exhibited by replacing an upstroke with a left parenthesis and a downstroke with a right parenthesis, as in Figure 6.4. Clearly, this procedure is reversible.

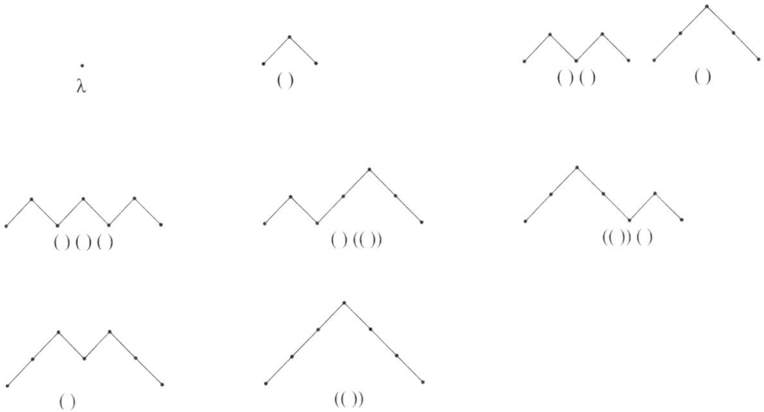

Figure 6.4 Mountain Ranges and Corresponding Parenthesized Sequences

The next two examples are closely related to the mountain range problem.

Example 6.3 Find the number of different Dyck paths (that is, mountain ranges) from the origin to the lattice point[†] $(2n + 2, 0)$ on the cartesian plane such that:

- From the lattice point (x, y), we can travel to $(x + 1, y + 1)$ or $(x + 1, y - 1)$, never crossing the x-axis.
- The length of the largest downward path ending on the x-axis is odd.

[†] A *lattice point* is a point on the cartesian plane with integral coordinates.

Solution Figure 6.5 shows the possible paths for $0 \leq n \leq 3$. In each case, the answer is a Catalan number.

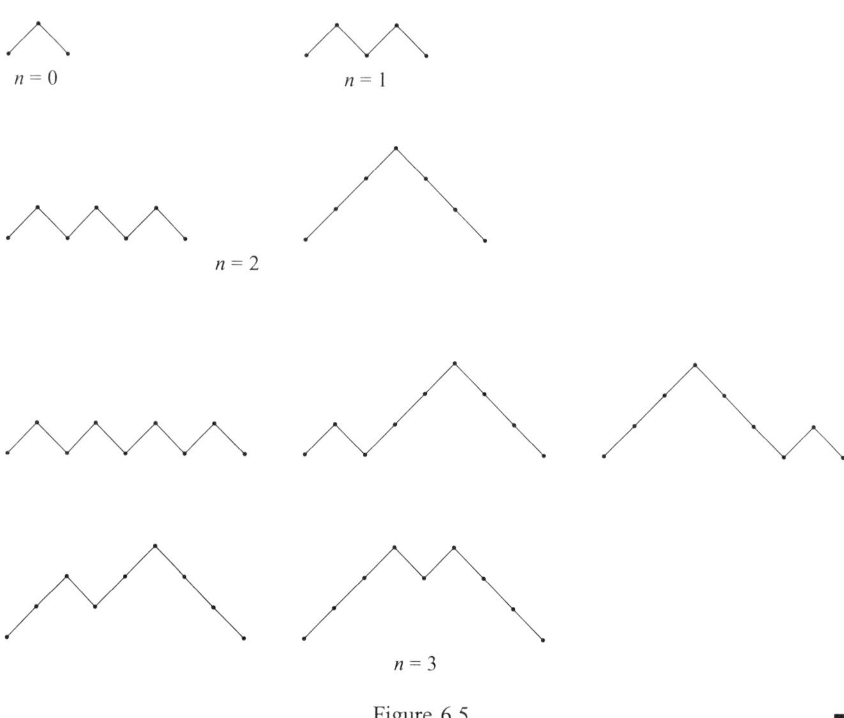

Figure 6.5

Example 6.4 Find the number of different Dyck paths (that is, mountain ranges) from the origin to the lattice point $(2n + 2, 0)$ on the cartesian plane, such that from the lattice point (x, y), we can travel to $(x + 1, y + 1)$ or $(x + 1, y - 1)$, but never cross the x-axis and have no peaks at height two.

Solution Figure 6.6 shows the possible paths for $0 \leq n \leq 3$. It appears that the answer once again is a Catalan number, which is in fact the case.

Example 6.5 Find the number of Dyck paths (mountain ranges) of length $2n + 2$ whose first downstep (downstroke) is followed by another downstep (downstroke).

Solution Figure 6.7 shows the possible mountain ranges, where $0 \leq n \leq 4$.

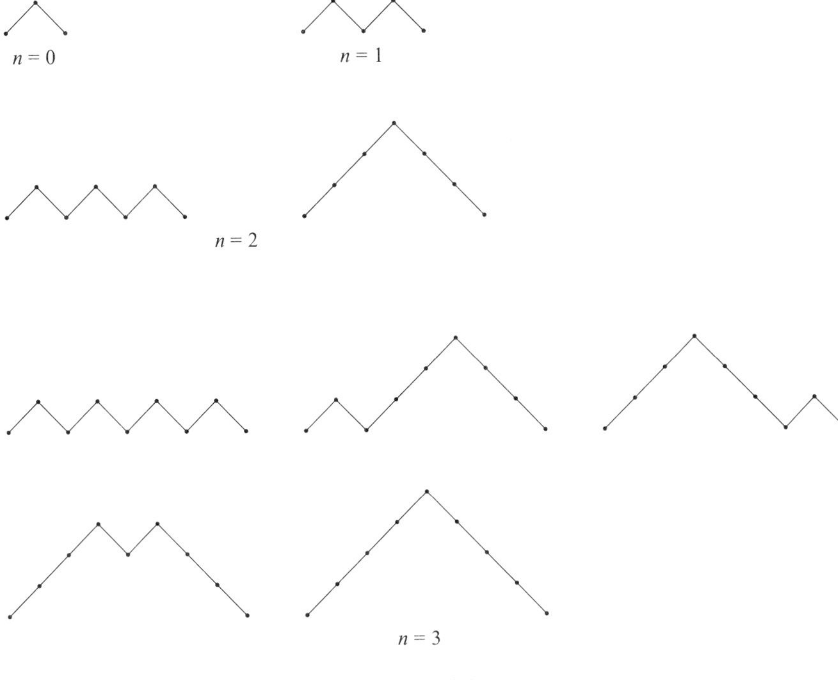

Figure 6.6

The next example, developed by Emeric Deutsch of Polytechnic University, Brooklyn, New York, is also related to Dyck paths (or mountain ranges).

Example 6.6 Find the number of (incomplete) Dyck paths that have $n - 1$ upsteps (upstrokes).

Solution   Figure 6.8 shows the various possibilities for $0 \leq n \leq 4$.   ■

Example 6.2 Revisited

It is obvious from the preceding discussion that there is a bijection between the set of (incomplete) Dyck paths in Example 6.6 and the set of Dyck paths (or mountain ranges) in Example 6.2. This can be exhibited directly as follows: At the end of each incomplete path in Example 6.6, add an upstep and then enough downsteps to reach the horizontal axis.

For example, consider the paths in Figure 6.9 from Example 6.6. Adding an upstep at the end of each path and then enough downsteps yields the valid Dyck paths in Figure 6.10 (see Figure 6.3 also).

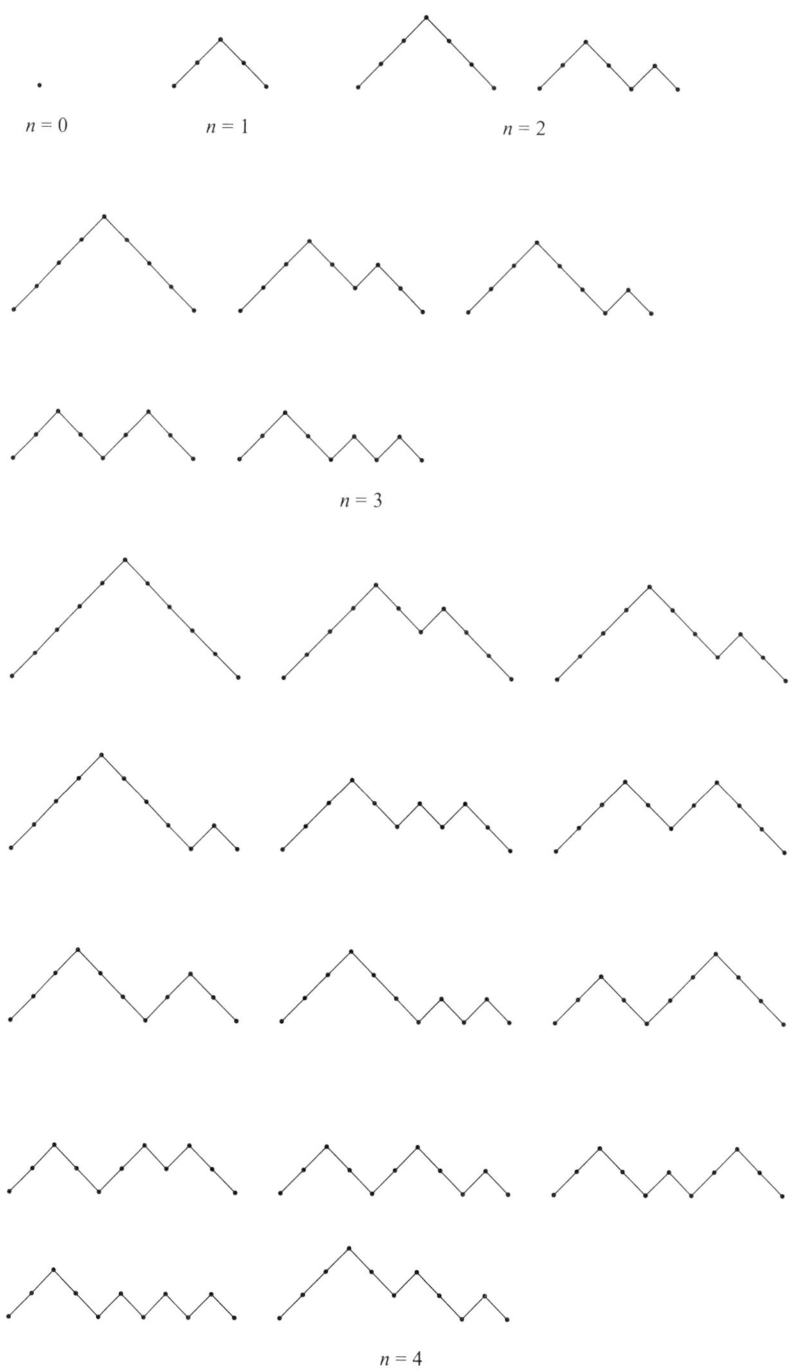

Figure 6.7

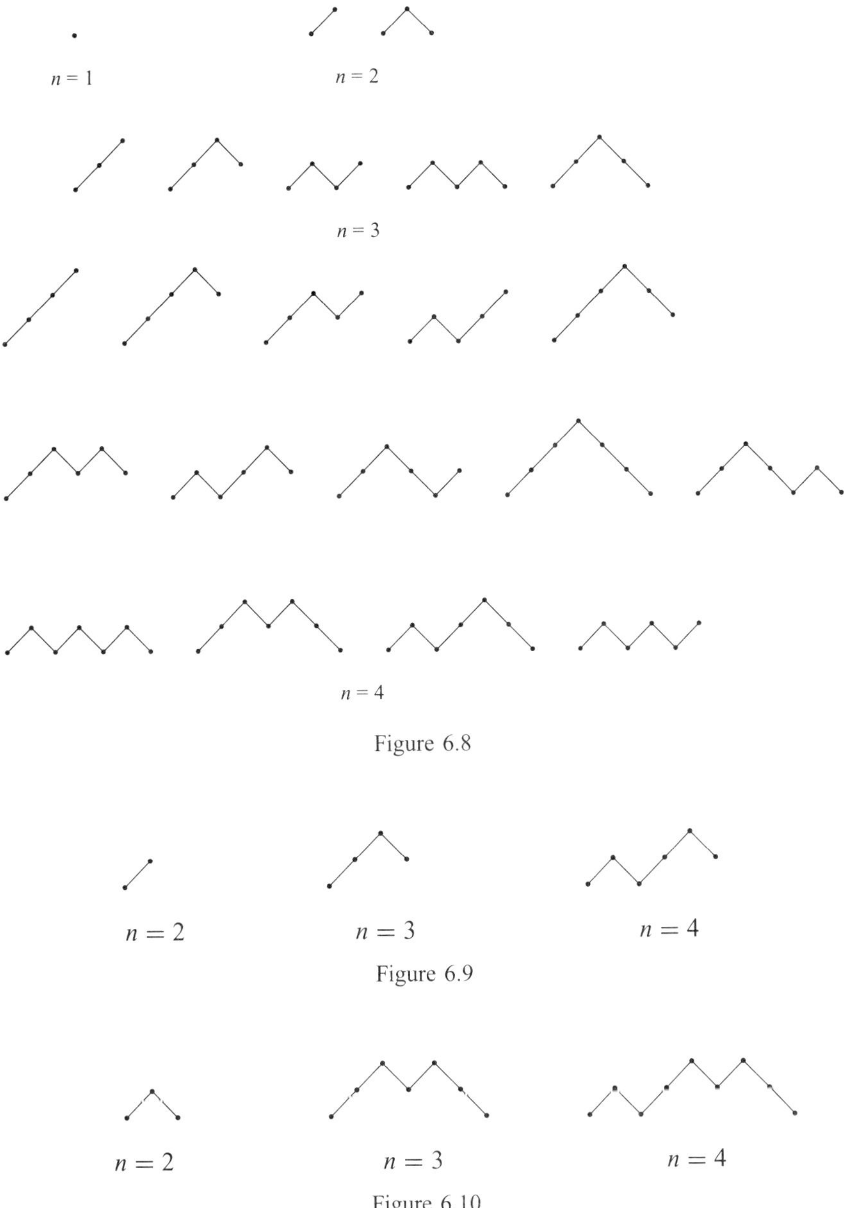

Figure 6.8

Figure 6.9

Figure 6.10

Clearly, this algorithm is reversible: Form the end of a Dyck path (with n upsteps and n downsteps), delete the last sequence of downsteps and the upstep immediately preceding it. This results in an incomplete path with $n - 1$ upsteps.

For example, consider the valid Dyck paths in Figure 6.11. Deleting the portions in bold yields the incomplete paths in Figure 6.12; see Figure 6.13 also.

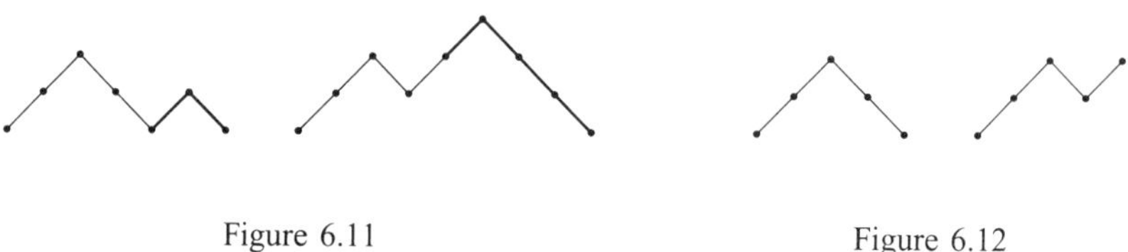

Figure 6.11   Figure 6.12

The next example, developed by Emeric Deutsch, also deals with mountain ranges (or Dyck paths) with a fixed number of peaks. A two-step subpath, where the first upstep is followed by a downstep, is a *peak*. Likewise, a two-step subpath, where the first downstep is followed by an upstep, is a *valley*.

Example 6.7 Find the number of mountain ranges with $n - 1$ peaks such that they do *not* contain three consecutive upsteps or three consecutive downsteps.

Solution   Figure 6.13 shows the various possible mountain ranges, where $1 \leq n \leq 4$. In each case, the answer is a Catalan number.

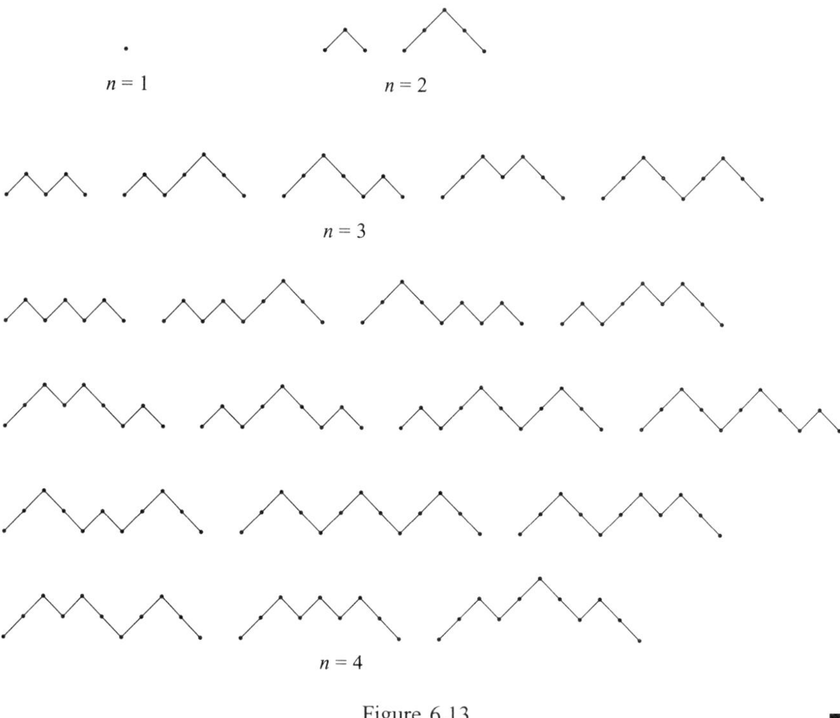

Figure 6.13   ■

The *height* of a peak is the y-coordinate of the right endpoint of its upstep. In 2005, Deutsch proposed a problem asking for the number of mountain ranges with an even number E_n of peaks and an odd number O_n of peaks, both at even height. It can be shown[†] that

$$E_n = \begin{cases} \frac{1}{2}C_n & \text{if } n \text{ is even} \\ \frac{1}{2}\left(C_n + C_{\lfloor n/2 \rfloor}\right) & \text{otherwise} \end{cases}$$

and

$$O_n = \begin{cases} \frac{1}{2}C_n & \text{if } n \text{ is even} \\ \frac{1}{2}\left(C_n - C_{\lfloor n/2 \rfloor}\right) & \text{otherwise} \end{cases}$$

Example 6.8 Suppose we have an infinite supply of pennies and we would like to stack them up in rows. Find the number of different arrangements possible subject to the following conditions:

- The bottom row consists of n pennies, each touching its two neighbors, except the ends, where $n \geq 1$.
- A coin that does not belong to a row sits on the two coins below it.
- Do not distinguish between heads and tails.

Solution   Figure 6.14 shows the possible arrangements, where $1 \leq n \leq 4$. In general, there are C_n such possible arrangements.

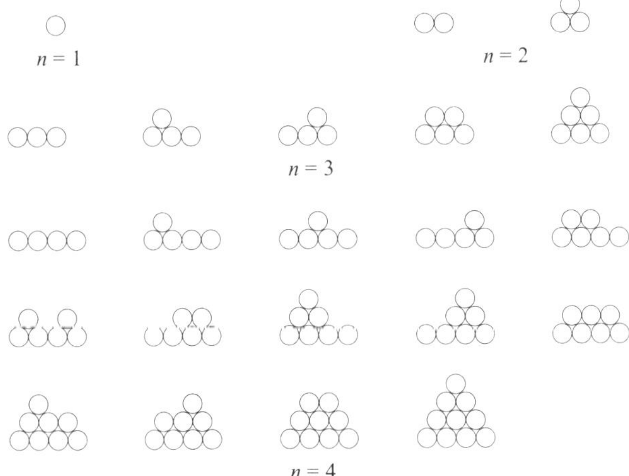

Figure 6.14 Possible Arrangements of Coins for $1 \leq n \leq 4$ ■

[†] See *American Mathematical Monthly* **114** (March 2007) for a solution by O. P. Lossers of Eindhoven University of Technology, The Netherlands.

Example 6.9 Find the number of ways $2n$ people, seated around a round table, can shake hands without their hands crossing, where $n \geq 0$.

Solution   Figure 6.15 shows the various possible handshakes for $0 \leq n \leq 4$. The answer in each case is a Catalan number.

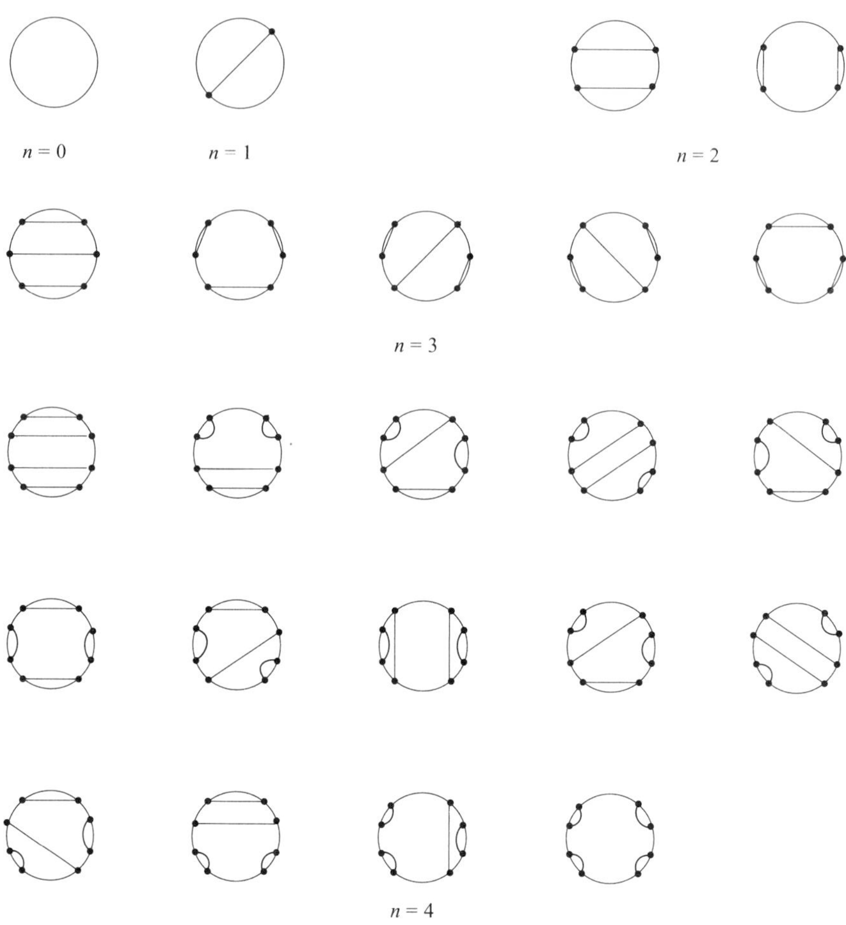

Figure 6.15 Possible Handshakes for $0 \leq n \leq 4$

This problem can be restated as follows: Find the number of ways in which $2n$ points on a circle can be joined by n chords in such a way that no two chords intersect.

Example 6.10 Find the number of sequences of n 1s and n -1s such that every partial sum (from left to right) is nonnegative; that is, the number of $2n$-tuples $a_1 a_2 \cdots a_{2n}$ of 1s and -1s such that $a_1 + a_2 + \cdots + a_k \geq 0$ and $a_1 + a_2 + \cdots + a_{2n} = 0$, where $1 \leq k < 2n$ and $n \geq 0$.

Solution Table 6.2 shows the various possible sequences for $0 \leq n \leq 4$.

Table 6.2

n	Possible Sequences					Count
0	λ					1
1	1 -1					1
2	1-11-1	11-1-1				2
3	1-11-11-1	11-11-1-1	11-1-11-1	1-111-1-1	111-1-1-1	5
4	1-11-11-11-1 111-1-1-11-1 11-1-111-1-1 1-1111-1-1-1 11-1-11-11-1 1-111-1-11-1 1-11-111-1-1 11-11-1-11-1 1-111-11-1-1 1-1111-1-1-1 111-1-1-11-1 11-111-1-1-1 111-1-11-1-1 1111-1-1-1-1					14

Examples 5.6 and 6.2 Revisited

Examples 5.6 and 6.2 Revisited

Notice that this example is essentially the same as Catalan's parenthesization problem in Example 5.6. A 1 corresponds to a left parenthesis and a -1 to a right parenthesis.

It is also the same as Example 6.2. This follows by switching an upstroke with a 1 and a downstroke with -1.

For instance, consider the five mountain ranges corresponding to $n = 3$ in Example 6.2. This transformation translates them into the 6-tuples 1-11-11-1, 1-111-1-1, 11-1-11-1, 11-11-1-1, and 111-1-1-1, respectively; see Table 6.2.

On the other hand, every such n tuple yields a valid mountian range. For example, the 8-tuple 11-11-1-11-1 (see Table 6.2) can be used to construct the mountain range in Figure 6.16.

Figure 6.16

Thus, there is a one-to-one correspondence between the set of mountain ranges in Example 6.2 and the set of $2n$-tuples in Example 6.10.

Example 6.10 leads us naturally to the following example.

Example 6.11 Find the number of sequences $a_0 a_1 \ldots a_{2n}$ of 1s and -1s such that every partial sum $a_0 + a_1 + \cdots + a_k$ is positive and $a_0 + a_1 + \cdots + a_{2n} = 1$.

Solution Every partial sum is positive, the first partial sum a_0 also must be positive. This implies that $a_0 = 1$. Consequently, this problem is the same as Example 6.10 with sequences $1 a_1 a_2 \cdots a_{2n}$ of n 1s and n -1s such that every parital sum $a_0 + a_1 + \cdots + a_k \geq 0$ and $1 \leq k \leq 2n$.

Table 6.3 shows the possible sequences for $0 \leq n \leq 4$. They are the same as the sequences in Table 6.2 with 1 added as a prefix.

Table 6.3

n	Possible Sequences	Count
0	1	1
1	11 -1	1
2	11-11-1 111-1-1	2
3	11-11-11-1 111-11-11 111-1-11-1 11-111-1-1 1111-1-1-1	5
4	11-11-11-11-1 1111-1-1-11-1 111-1-111-1-1 11-1111-1-1-1 111-1-11-11-1 11-111-1-11-1 11-11-111-1-1 111-11-1-11-1 11-111-11-1-1 11-1111-1-1-1 1111-1-1-11-1 111-111-1-1-1 1111-1-11-1-1 11111-1-1-1-1	14

We will now use *Raney's lemma* and a simple combinatorial argument to confirm that there are C_n such sequences. The lemma was discovered in 1959 by George N. Raney (1922–) of Pennsylvania State University. We omit its proof for brevity.[†]

Lemma 6.1 (*Raney's lemma*) Let $a_1 a_2 \ldots a_m$ be a sequence of integers whose sum is 1. Exactly one of its cyclic shifts $a_1 a_2 \ldots a_m, a_2 \ldots a_m a_1, a_3 \ldots a_1 a_2, \ldots, a_m a_1 \ldots a_{m-1}$ has the distinct property that every one of its partial sums is positive.

For example, consider the sequence $2\ 0 - 3\ 3 - 1$. Its cyclic shifts are:

$$
\begin{array}{rrrrr}
2 & 0 & -3 & 3 & -1 \\
-3 & 3 & -1 & 2 & 0 \\
-1 & 2 & 0 & -3 & 3
\end{array}
\qquad
\begin{array}{rrrrr}
0 & -3 & 3 & -1 & 2 \\
3 & -1 & 2 & 0 & -3
\end{array}
$$

[†] See R. L. Graham et al., *Concrete Mathematics*, Addison-Wesley, Massachusetts, 1989.

Exactly one of them has the property that every partial sum is positive, namely, $3 - 1\ 2\ 0 - 3$.

We are now ready to show that the number of $(2n+1)$-tuples in Example 6.11 is C_n. Each $(2n+1)$-tuple consists of $n+1$ 1s and $n-1$s. There are $\frac{(2n+1)!}{(n+1)!n!} = \binom{2n+1}{n}$ such sequences. Every cyclic shift of a $(2n+1)$-tuple also belongs to the same family of tuples. By Raney's lemma, exactly one of the cyclic shifts has the property that every partial sum is positive.

Let N denote the number of $(2n+1)$-tuples such that every partial sum is positive. Since every $(2n+1)$-tuple contributes $2n+1$ cyclic shifts, the total number of $(2n+1)$-tuples in the family is $(2n+1)N$. Thus:

$$(2n+1)N = \binom{2n+1}{n}$$

$$N = \frac{1}{2n+1}\binom{2n+1}{n}$$

$$= \frac{1}{n+1}\binom{2n}{n}$$

$$= C_n$$

as desired. ∎

We will investigate a generalization of Example 6.11 in Chapter 12 and revisit Raney's lemma in the process.

The next example is also essentially the same as Example 6.10.

Example 6.12 Find the number of binary words consisting of n 1s and n 0s such that the number of 1s in each substring, when read from the far left of the substring to the right, is greater than or equal to the number of 0s in the substring. (For example, the binary word 1010 has four such substrings: 1, 10, 101, and 1010.)

Solution Table 6.4 shows the various possible binary words for $0 \le n \le 4$, where λ denotes the null word. ∎

See Table 6.5 also, which shows the possible ways of associating a noncommutative product of $n+1$ operands with n pairs of parentheses.

Table 6.4

n	Possible Binary Words					Count
0	λ					1
1	10					1
2	1100	1010				2
3	111000	110100	110010	101100	101010	5
4	11110000	11101000	11100100	11100010	11011000	14
	11010100	11010010	11001100	11001010	10111000	
	10110100	10110010	10101100	10101010		

Table 6.5

n	Correctly Parenthesized Expressions			Count
0	a			1
1	(ab)			1
2	$((ab)c)$	$(a(bc))$		2
3	$(((ab)c)d)$	$(a(b(cd)))$	$((a(bc))d)$	5
	$((ab)(cd))$	$(a((bc))d)$		

Examples 6.9 and 6.10 Revisited

There is a delightful way we can display a one-to-one correspondence between the set of handshakes in Example 6.9 and the set of binary words in Example 6.12, and hence the set of sequences in Example 6.10.

To see this, suppose there is a (specially trained) bug inside a circle. It travels from the very bottom in the clockwise (or counterclockwise) direction. It calls out a 1 every time it crosses an edge[†] for the first time; otherwise, it calls out a 0. The bug continues like this through every region until it returns to the starting point. (It can cross an edge exactly twice.)

Figure 6.17 shows the various possible handshakes and the corresponding binary numbers, where $0 \le n \le 4$.

Clearly, this algorithm works in the opposite direction as well. For instance, consider the binary word 1110011000. Because it contains five 1s and five 0s, mark ten points on a circle; label them 1 through 10 for convenience, and place a bug inside the circle at the bottom; see Figure 6.18. Since the word begins with 111, the bug must encounter three new edges, after which it can cross only two of

[†] An *edge* is an arc or a chord representing a handshake.

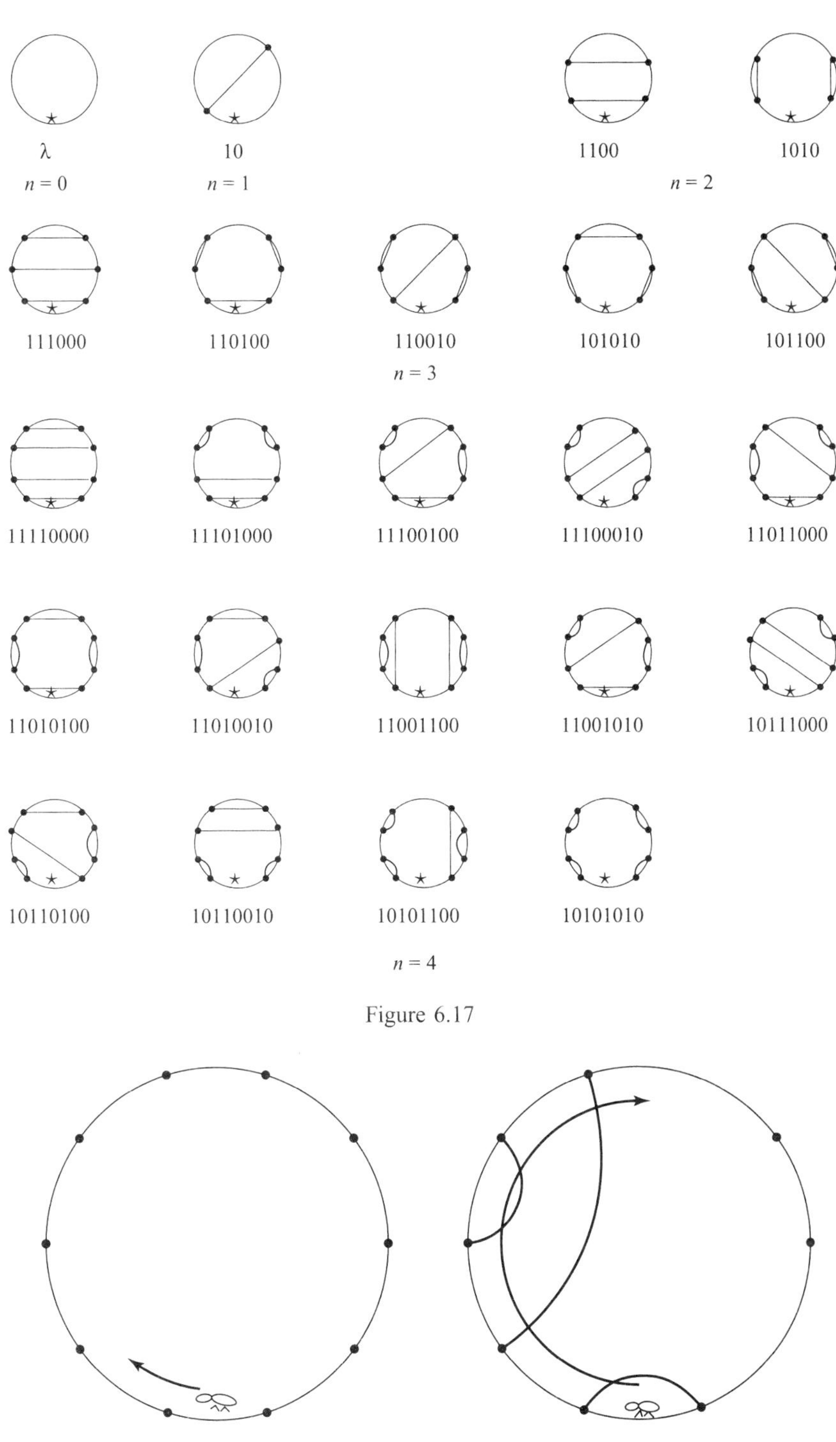

Figure 6.17

Figure 6.18

Figure 6.19

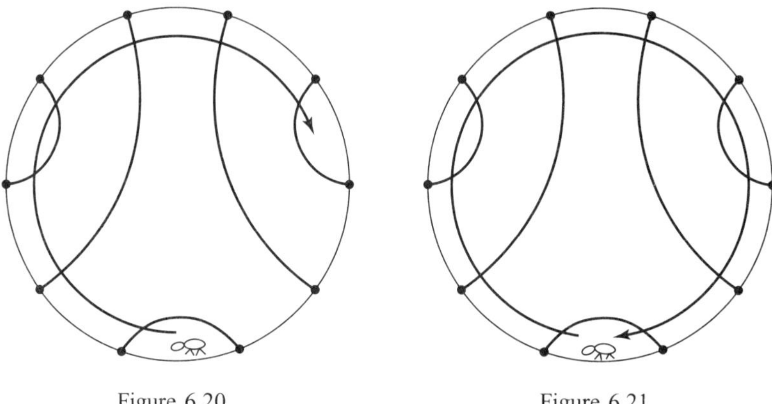

Figure 6.20   Figure 6.21

them; see Figure 6.19. The next two 1s indicate that the bug must cross two new edges. The next three 0s require that the bug now return home, crossing the three edges that have been encountered exactly once; see Figure 6.20. Figure 6.21 shows the resulting handshaking configuration.

The Mountain Ranges Revisited

Changing each 1 into an upstroke ($/$) and each 0 into a downstroke ($\backslash$), the above algorithm also exhibits a bijection between the set of mountain ranges and the set of handshakes.

If we use a left parenthesis and a right parenthesis in lieu of an upstroke and a downstroke, respectively, the algorithm provides a one-to-one correspondence between the set of well-formed sequences of parentheses and the set of handshakes.

The following example, proposed in 1918 by P. Franklin, is directly related to this example.

Example 6.13 Find the number of binary words consisting of $(n + 1)$ 1s and n 0s such that the number of 1s in each substring, when scanned from left to right, is greater than the number of 0s in it, where $n \geq 0$.

Solution   Suppose a word w begins with the bit[†] b. Since b itself is a substring of w, it must satisfy the condition that the number of 1s in b is greater than the number of 0s in it. Consequently, $b = 1$. So every binary word must begin with the bit 1, and is obtained by adding a 1 as a prefix to each binary word in Example 6.12, where $w\lambda = w = \lambda w$.

Table 6.6 shows the resulting binary words for $0 \leq n \leq 4$.

[†]   A *bit* is a contraction of the term *bi*nary dig*it*; it is 0 or 1.

Table 6.6

n	Possible Binary Words					Count
0	1					1
1	110					1
2	11100	11010				2
3	1111000	1110100	1110010	1101100	1101010	5
4	111110000 111101000 111100100 111100010 111011000 111010100 111010010 111001100 111001010 110111000 110110100 110110010 110101100 110101010					14

■

Necklaces

Next we investigate noncyclic binary words consisting of $n + 1$ 1s and n 0s.

Example 6.14 Find the number of binary words consisting of $n + 1$ 1s and n 0s such that *no* binary word is a cyclic shift of another binary word.

Solution   Table 6.7 shows such binary words, where $0 \leq n \leq 4$.   ■

Table 6.7

n	Binary Words					Count
0	1					1
1	110					1
2	11100	10110				2
3	1111000	1110100	1110010	1101100	1101010	5
4	111110000 111101000 111100100 111100010 111011000 111010100 111010010 111001100 111001010 111000110 110110100 110110010 110101100 110101010					14

■

More generally, every such binary word consists of $2n + 1$ bits, of which $n + 1$ are 1s and n are 0s. There are $\frac{(2n+1)!}{(n+1)!n!} = \binom{2n+1}{n}$ such binary words. Each such word produces $2n + 1$ cyclic shifts, so the number of binary words such that no binary word is a cyclic shift of another, is given by

$$\frac{1}{2n + 1} \binom{2n + 1}{n} = C_n$$

The next example deals with permutations and is related to Example 6.6.

Example 6.15 Find the number of sequences of $n - 1$ 1s and any number of -1s such that every parital sum is nonnegative.

Solution   Clearly, the number of -1s in any such sequence is at most $n - 1$, where $n \geq 1$. Table 6.8 shows such sequences for $1 \leq n \leq 4$.   ■

Table 6.8

n	Possible Sequences					Count
1	λ					1
2	1	$1-1$				2
3	11	$11-1$	$1-11$	$1-11-1$	$11-1-1$	5
4	111 $11-11-1$ $1-11-11-1$	$111-1$ $1-111-1$ $11-11-1-1$	$11-11$ $11-1-11$ $1-111-1-1$	$1-111$ $111-1-1-1$ $1-11-11$	$111-1-1$ $11-1-11-1$	14

Example 6.6 Revisited

As can be expected, there is a bijection between the set of sequences in Example 6.15 and the set of paths in Example 6.6. This is obtained by replacing a 1 by an upstep and a -1 by a downstep. This procedure guarantees that the desired conditions in both examples are satisfied.

For example, the sequence $1 - 111 - 1 - 1$ yields the path

and the path

yields the sequence $1 - 111 - 1$.

Next we study $2n$-tuples investigated in 1955 by Eugene P. Wigner[†] of Princeton University.

Example 6.16 Find the number of $2n$-tuples $a_1 a_2 \ldots a_{2n}$ of nonnegative integers a_i such that $a_1 = 1, a_{2n} = 0$, and $|a_i - a_{i-1}| = 1$, where $2 \leq i \leq 2n$.

[†]   Wigner was awarded the Nobel Prize in physics in 1963.

Solution   Clearly, each $2n$-tuple must begin with a 1 and end in a 0. Since $|a_{2n} - a_{2n-1}| = |0 - a_{2n-1}| = 1$, it follows that $a_{2n-1} = 1$. Thus, every $2n$-tuple must terminate in 10. Table 6.9 shows the possible $2n$-tuples, where $0 \leq n \leq 4$.

Table 6.9

n	Possible $2n$-tuples					Count
0	γ					1
1	10					1
2	1010	1210				2
3	1010101	101210	121010	121210	123210	5
4	10101010 10101210 10121010 10121210 10123210 12101010 12101210 12121010 12121210 12321010 12321210 12323210 12123210 12343210					14

Example 6.10 Revisited

Interestingly, there is a bijection between the set of $2n$-tuples in Example 6.16 and those in Example 6.10. To see this, consider the 8-tuple $111 - 11 - 1 - 1 - 1$ in Example 6.10. The sequence of partial sums yields the 8-tuple 12323210, which is a valid 8-tuple in Example 6.16; see Figure 6.22.

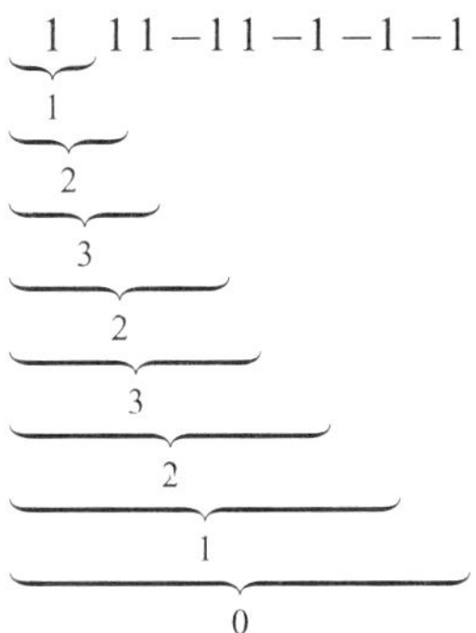

Figure 6.22

Clearly, this procedure is reversible. For instance, consider the 8-tuple $a_1 a_2 \ldots a_8 = 12323210$. Let $b_1 = a_1 = 1$ and $b_i = a_i - a_{i-1}$, where $2 \leq i \leq 2n$. The resulting 8-tuple $b_1 b_2 \ldots b_8 = 111 - 11 - 1 - 1 - 1$ is a valid 8-tuple in Example 6.10.

More generally, let $a_1 a_2 \ldots a_{2n}$ be a valid $2n$-tuple in Example 6.16. Let $b_i = a_i - a_{i-1}$, where $a_0 = 0$ and $1 \leq i \leq 2n$. Since $|a_i - a_{i-1}| = 1$, $b_i = \pm 1$.

Then:

$$\sum_{i=1}^{j} b_i = \sum_{i=1}^{j} (a_i - a_{i-1})$$

$$= a_j - a_0$$

$$= a_j$$

$$\geq 0$$

So every partial sum of the sequence $b_1 b_2 \ldots b_{2n}$ is a $2n$-tuple of 1s and -1s such that every partial sum is nonnegative. Thus, $b_1 b_2 \ldots b_{2n}$ is a valid sequence in Example 6.10.

Conversely, let $b_1 b_2 \ldots b_{2n}$ be a valid $2n$-tuple in Example 6.10. Then $b_1 = 1$ and $b_i = \pm 1$ for every $i \geq 1$. Let $a_j = \sum_{i=1}^{j} b_i$, where $1 \leq j \leq 2n$. Since $\sum_{i=1}^{j} b_i \geq 0$, $a_j \geq 0$. Furthermore, $|a_j - a_{j-1}| = |b_j| = 1$. Thus, the $2n$-tuple $a_1 a_2 \ldots a_{2n}$ is a valid sequence in Example 6.16.

The following example may appear to be different. However, there is no essential difference between it and Example 6.12; it is old wine in a new bottle.

Example 6.17 Find the number of ways the integers 1 through $2n$ can be arranged in two rows of n each such that:

- The numbers in each row increase from left to right; and
- The numbers in each column increase from the top row to the second row.

In other words, find the number of $2 \times n$ matrices

$$\begin{bmatrix} a_{11} & a_{12} & \cdots & a_{1j} & \cdots & a_{1n} \\ a_{21} & a_{22} & \cdots & a_{2j} & \cdots & a_{2n} \end{bmatrix}$$

that can be formed, where $1 \leq a_{ij} \leq 2n, a_{ij} < a_{i,j+1}, a_{1j} < a_{2j}, 1 \leq i \leq 2$, and $1 \leq j \leq n$.

Solution Table 6.10 shows the various possibilities for $0 \leq n \leq 4$. ∎

To see that there is a bijection between the set of binary words of length $2n$ in Example 6.12 and the set of $2 \times n$ arrays in Example 6.17, we shall establish a procedure to translate such an array into a binary word and vice versa. To this end, consider, for example, the array

$$1347$$
$$2568$$

Place a 1 in positions 1, 3, 4, and 7, and a 0 in positions 2, 5, 6, and 8. The resulting binary word is 10110010; it has the desired property that the number of

Table 6.10

n	Possible Arrangements					Count
0	λ					1
1	1 2					1
2	12 34	13 24				2
3	123 456	124 356	125 346	134 256	135 246	5
4	1234 5678	1235 4678	1236 4578	1237 4568	1245 3678	14
	1246 3578	1247 3568	1256 3478	1257 3468	1345 2678	
	1346 2578	1347 2568	1356 2478	1357 2468		

1s in each substring from left to right is greater than or equal to the number of 0s in it.

This process is certainly reversible. For example, consider the binary word 110010 of length 6 from Example 6.12. It contains a 1 in positions 1, 2, and 5; they form the top row. Since the binary word contains a 0 in positions 3, 4, and 6, they form the second row. Thus the desired 2×3 array is

$$125$$
$$346$$

See Table 6.10.

A Delightful Application

This example has a fascinating application. Suppose there are $2n$ soldiers of varying heights. They need to be arranged in two rows of n soldiers each in such a way that the soldiers in each row are standing in ascending order of height from left to right and those in each column are also standing in ascending order of height from the top row to the second row. It follows from Example 6.17 that they can be arranged in C_n different ways.

The next two examples deal with triangular arrays.

Example 6.18 Let $A = (a_{ij})$ be a left-justified triangular array consisting of bits such that $a_{ij} \geq a_{i,j+1}$ and $a_{ij} \geq a_{i+1,j}$, where $1 \leq i \leq n - 1, 1 \leq j \leq n - i$, and $n \geq 1$. Find the number of such triangular arrays.

Solution   The conditions $a_{ij} \geq a_{i,j+1}$ and $a_{ij} \geq a_{i+1,j}$ imply that the elements in each row and in each column form nonincreasing sequences.

For instance, when $n = 3$, i can be 1, 2, or 3. When $i = 1, j = 1$ or 2; when $i = 2, j = 1$; and when $i = 3, j = 0$, which is impossible. Correspondingly, there are five triangular arrays with the given properties:

$$
\begin{array}{ccccc}
0 \;\; 0 & \quad 1 \;\; 0 & \quad 1 \;\; 0 & \quad 1 \;\; 1 & \quad 1 \;\; 1 \\
0 & \quad 0 & \quad 1 & \quad 0 & \quad 1
\end{array}
$$

When $n = 0$, the array is null; and there is exactly one such array. Table 6.11 shows the various possible ways for $1 \leq n \leq 4$. With a little patience, they can be enumerated in a systematic way.

Table 6.11

n	Possible Triangular Arrays					Count
1	null					1
2	0	1				2
3	0 0 0	1 0 0	1 0 1	1 1 0	1 1 1	5
4	0 0 0 0 0 0	1 0 0 0 0 0	1 1 0 0 0 0	1 1 0 1 0 0	1 1 0 1 0 1	14
	1 1 0 1 1 0	1 1 0 1 1 1	1 1 1 0 0 0	1 1 1 1 0 0	1 1 1 1 0 1	
	1 1 1 1 1 0	1 1 1 1 1 1	1 0 0 1 0 0	1 0 0 1 0 1		

Such triangular arrays were studied in 1972 by L. Carlitz of Duke University, Durham, North Carolina, and D. P. Roselle of Louisiana University, Baton Rouge. Carlitz and Roselle showed that there are exactly C_n such triangular arrays, where $n \geq 1$.

The next example also deals with left-justified triangular arrays with similar conditions. They were studied by Carlitz and Richard A. Scoville of Duke University. The number of such arrays is also C_n.

Example 6.19 $A = (a_{ij})$ be a left-justified triangular array consisting of bits such that $a_{ij} \geq a_{i+1,j-1}$ and $a_{ij} \geq a_{i+1,j}$, where $1 \leq i \leq n-1, 1 \leq j \leq n-i$, and $n \geq 1$. Find the number of such triangular arrays.

Solution   The conditions $a_{ij} \geq a_{i+1,j-1}$ and $a_{ij} \geq a_{i+1,j}$ imply that the elements on each southwest diagonal and in each column form nonincreasing sequences.

Table 6.12

n	Possible Triangular Arrays					Count
1	null					1
2	0	1				2
3	0 0 0	0 1 0	1 0 0	1 1 0	1 1 1	5
4	0 0 0 0 0 0	0 0 1 0 0 0	0 1 0 0 0 0	0 1 1 0 0 0	0 1 1 0 1 0	14
	1 0 0 0 0 0	1 0 1 0 0 0	1 1 0 0 0 0	1 1 0 1 0 0	1 1 1 0 0 0	
	1 1 1 0 1 0	1 1 1 1 0 0	1 1 1 1 1 0	1 1 1 1 1 1		

When $n = 1$, there is exactly one such array, namely, the null array. Table 6.12 shows the various possible arrays for $1 \leq n \leq 4$; they also can be listed in a systematic way. Although the array $0\,1\,0$ does not meet the requirements in Example 6.18, notice that it satisfies the conditions in this example.   ∎

The following example was also studied by Carlitz in 1972.

Example 6.20 Find the number of n-tuples $a_1 a_2 \ldots a_n$ of positive integers a_i such that $a_{i+1} \leq a_i \leq n - i + 1$.

Solution   The condition $a_{i+1} \leq a_i$ requires that the sequence $a_1 a_2 \ldots a_n$ be nonincreasing. We could let $n = 0$ by defining the 0-tuple to be the null word λ. Table 6.13 shows the various possible n-tuples for $0 \leq n \leq 4$.   ∎

In this example, suppose we relax the condition that each a_i be positive. Such n-tuples also provide interesting dividends, as the following example shows. It was also studied by Carlitz.

Table 6.13

n	Possible n-tuples					Count
0	λ					1
1	1					1
2	21	11				2
3	321	311	221	211	111	5
4	4321	4311	4221	4211	4111	14
	3321	3311	3221	3211	3111	
	2221	2211	2111	1111		

Example 6.21 Find the number of $(n-1)$-tuples $a_1 a_2 \ldots a_{n-1}$ of nonnegative integers a_i such that $a_{i+1} \leq a_i \leq n - i$.

Solution Table 6.14 shows the various possible such sequences for $0 \leq n \leq 4$.

Table 6.14

n	Possible n-tuples					Count
1	λ					1
2	1	0				2
3	21	20	11	10	00	5
4	321	320	311	310	300	14
	221	220	211	210	200	
	111	110	100	000		

Example 6.10 Revisited

Interestingly, there is a close relationship between the set of sequences $a_1 a_2 \ldots a_{2n}$ of n 1s and n −1s in Example 6.10 and the set of $(n-1)$-tuples $b_1 b_2 \ldots b_{n-1}$ in this example, namely, a one-to-one correspondence.

To see this, consider a sequence $a_1 a_2 \ldots a_{2n}$ of n 1s and n −1s, where every partial sum is nonnegative. Ignore $a_1 = 1$. This yields $a_2 \ldots a_{2n}$, which consists of $n - 1$ 1s and n −1s. Let c_i denote the number of −1s to the left of the ith 1, where $1 \leq i \leq n - 1$. Then the $(n-1)$-tuple $b_1 b_2 \ldots b_{n-1} = c_{n-1} \ldots c_2 c_1$ has the desired properties, where $b_i = c_{n-i}$.

We will now confirm these properties:

- Clearly, every $b_i \geq 0$.
- Reversing the counts of the -1s guarantees that $b_i \geq b_{i+1}$ for every i.
- It remains to show that $b_i \leq n - i$ for every i. Since $a_{i+1} = 1$ has no more than $i - 1$s to its left, it follows that $c_i \leq i$:

$$
\cancel{a_1} \overbrace{a_2 \ldots a_i}^{i-1 \text{ elements}} \; a_{i+1} \ldots a_{2n}
$$

$$
c_1 \ldots c_{i-1} c_i \ldots
$$

So $c_{n-i} \leq n - i$; that is, $b_i \leq n - i$, as desired.

For example, consider the following sequences, where $n = 3$:

	1-11-11-1	11-11-1-1	11-1-11-1	1-111-1-1	111-1-1-1
Ignore leading 1s:	-11-11-1	1-11-1-1	1-1-11-1	-111-1-1	11-1-1-1
Count:	1 2	0 1	0 2	11	00
Reverse:	2 1	1 0	2 0	11	00

This yields the ordered pairs 21, 10, 20, 11, and 00, the same as in Example 6.21; see Table 6.14.

Clearly, the algorithm is reversible. For example, consider the 3-tuple 310. Reversing the steps, we have:

$$
\begin{array}{ll}
\text{3-tuple:} & 310 \\
\text{Reverse:} & 013 \\
\text{Mark three 1s:} & 111
\end{array}
$$

The first 1 must have zero -1s to its left; the second 1 must have one -1 to its left; and the last 1 must have three -1s to its left. This yields the sequence:

$$
1\text{-}11\text{-}1\text{-}11
$$
$$
\text{Insert leading 1:} \quad 11\text{-}11\text{-}1\text{-}11
$$

Now, append a -1 to keep the same number of 1s and -1s. The resulting 8-tuple 11-11-1-11-1 has the desired properties. You may confirm them.

The next example was also studied by Carlitz in 1972.

Example 6.22 Find the number of n-tuples $a_1 a_2 \ldots a_n$ of nondecreasing positive integers a_i such that $a_i \leq i$.

Solution Notice that the definition does not prevent some or all of the elements in an n-tuple from being equal. In other words, repetitions are perfectly acceptable. So the sequences $a_1 a_2 \ldots a_n$ are nondecreasing.

We could let $n = 0$; the corresponding n-tuple is the null word λ. Table 6.15 shows the various possible n-tuples for $0 \leq n \leq 4$. Notice that they are the same as the n-tuples in Table 6.13 in reverse order.

Table 6.15

n	Possible Arrangements					Count
0	λ					1
1	1					1
2	11	12				2
3	111	112	113	122	123	5
4	1111	1112	1113	1114	1122	14
	1123	1124	1133	1134	1222	
	1223	1224	1233	1234		

Example 6.10 Revisited

Although it is not obvious, it is fairly easy to show a bijection between the set of sequences of n 1s and n -1s with nonnegative partial sums in Example 6.10 and the set of n-tuples $a_1 a_2 \ldots a_n$ in Example 6.22, where $a_i \le i$. To this end, write below each 1 in a sequence the number of -1's to its left. Then add a 1 to each count. The resulting n-tuple has the desired properties.

For example, let us apply these steps to the sequence 1-111-11-1-1:

sequence:	1-111-11-1-1
Number of -1s to the left of each 1:	0 11 2
Add 1 to each count:	1 22 3
Resulting 4-tuple:	1223

See Table 6.16, where $n = 3$.

Table 6.16

1-11-11-1	11-11-1-1	11-1-11-1	1-111-1-1	111-1-1-1
0 1 2	00 1	00 2	0 11	000
1 2 3	11 2	11 3	1 22	111

This algorithm is clearly reversible. For example, consider the quadruplet 1223:

4-tuple:	1 2 2 3
Subtract 1 from each:	0 1 1 2
Place a 1 below each:	1 1 1 1
Label them:	$s_1 s_2\ s_3\ s_4$

The first 0 indicates that the sequence always begins with a 1, as we already know. Now we need to locate the -1s. Since there is -1 to the left of s_2 and a -1

to the left of s_3 and two -1s to the left of s_4, insert -1s at the appropriate places:

$$\begin{array}{ll} \text{Labels:} & s_1 \; s_2 s_3 \; s_4 \\ \text{Insert } -1\text{s:} & 1\text{-}11\,1\text{-}11 \end{array}$$

The two remaining -1s did not play any role in the above counts, so they must go at the end, yielding the sequence 1-111-11-1-1.

Example 6.12 Revisited

Using a suitable transformation, we can exhibit a bijection between the set of n-tuples in this example and the set of binary words in Example 6.12. To this end, consider the binary word 111000, where the number of 1s is $\geq$ the number of 0s at each step:

$$\begin{array}{ll} \text{Binary word:} & 111000 \\ \text{Below each 1, enter the number of 0s to its left:} & 000 \\ \text{Add 1 to each bit in the resulting binary word:} & 111 \end{array}$$

Notice that this is one of the 3-tuples in Table 6.15.

Similarly, the binary words 110100, 110010, and 101010 in Table 6.4 yields the 3-tuples 112, 113, 122, and 123, respectively:

$$\begin{array}{cccc} 110100 & 110010 & 101100 & 101010 \\ 00\ \ 1 & 00\ \ 2 & 0\ 11 & 0\ 1\ 2 \\ \hline 11\ \ 2 & 11\ \ 3 & 1\ 22 & 1\ 2\ 3 \end{array}$$

This procedure is also reversible, although a bit tricky. For example, consider the 3-tuple 123:

$$\begin{array}{ll} \text{Given 3-tuple:} & 123 \\ \text{Subtract 1 from each element:} & 012 \end{array}$$

The zero in 012 indicates that the binary word w must begin with a 1; the one 1 indicates that the next bit in w must be a 0; so the next bit must be a 1. Thus far, w looks like this: $w = 101 - --$. Now w must contain two more 0s. But the two 0s cannot go together. So $w = 101010$, as desired.

Example 6.21 Revisited

We now exhibit a bijection between the set of $(n-1)$-tuples $a_1 a_2 \ldots a_{n-1}$ of nonnegative integers in Example 6.21 and the set of n-tuples $b_1 b_2 \ldots b_n$ of positive integers in this example, where $b_i \leq i$.

Append $a_n = 0$ to the $(n-1)$-tuple $a_1 a_2 \ldots a_{n-1}$, where $a_{i+1} \le a_i \le n - i$. Increase each component of the resulting n-tuple $a_1 a_2 \ldots a_{n-1} 0$ by 1. This yields the n-tuple

$$a_1 + 1 \quad a_2 + 1 \quad \ldots \quad a_{n-1} + 1 \quad 1$$

Reverse its order. This yields the n-tuple $b_1 b_2 \ldots b_n$, where $b_i = a_{n-i+1} + 1$.

The n-tuple $b_1 b_2 \ldots b_n$ has the property that $b_i \le i$ for every i. To see this, since $a_i \le n - i$, $a_{n-i+1} \le n - (n - i + 1)$; that is, $b_i - 1 \le i - 1$. So $b_i \le i$, as desired.

Thus every $(n-1)$-tuple in Example 6.21 can be transformed into a a unique n-tuple in Example 6.22.

For example, consider the following pairs:

	21	20	11	10	00
Append 0:	210	200	110	100	000
Add 1 to each component:	321	311	221	211	111
Reverse the order:	123	113	122	112	111

These are the 3-tuples we obtained in Example 6.22.

Clearly, this process is reversible, establishing the desired one-to-one correspondence.

In 2004, E. Deustch slightly modified Carlitz's problem in Example 6.22 as follows: "Find the number of n-tuples $a_1 a_2 \ldots a_n$ of positive integers a_i such that $a_i \le 2i$ and $a_i \le a_{i+1}$." Since $a_1 \le 2, a_1$ can be only 1 or 2. There are 30 triplets satisfying the given conditions:

$$\begin{array}{cccccccccc}
111 & 112 & 113 & 114 & 115 & 116 & 122 & 123 & 124 & 125 \\
126 & 133 & 134 & 135 & 136 & 144 & 145 & 146 & 222 & 223 \\
224 & 225 & 226 & 233 & 234 & 235 & 236 & 244 & 245 & 246
\end{array}$$

In general, there are $\frac{2}{3n+2}\binom{3n+2}{n}$ such n-triplets, as shown in 2006 by Li Zhou of Polk Community College, Winter Haven, Florida.

In the same year, Marc Renault of Shippensburg University, Pennsylvania, generalized the problem even further. He showed that there are exactly

$$C_{n,t} = \frac{1}{n+1}\binom{(t+1)(n+1) - 2}{n}$$

such n-tuples, where $a_i \le ti$ and $a_i \le a_{i+1}$. Notice that $C_{n,1} = C_n$, the nth Catalan number. Interestingly, $C_{n,t}$ counts the number of lattice paths from the origin to (n, tn) that do *not* rise above the line $y = tx$, as noted by the problem editors.

In the following example, we extend the definition to $(n + 1)$-tuples with an extra condition for a_{n+1}. This was also investigated by Carlitz in 1972.

Example 6.23 Find the number of $(n + 1)$-tuples $a_1 a_2 \ldots a_{n+1}$ of nondecreasing positive integers such that $a_i \leq i$ and $a_{n+1} = n + 1$.

Solution   Since $a_{n+1} = n + 1$, every $(n+1)$-tuple ends in $n + 1$. But the elements of the n-tuple $a_1 a_2 \ldots a_n$ must still satisfy the condition that $a_i \leq i$ for every i. Consequently, the various $(n + 1)$-tuples can be obtained by appending $n + 1$ to each n-tuple in Example 6.22. Notice that $w\lambda = w = \lambda w$ for every word w.

Table 6.17 shows the various $(n + 1)$-tuples for $0 \leq n \leq 4$.

Table 6.17

n	Possible Arrangements					Count
0	1					1
1	12					1
2	113	123				2
3	1114	1124	1134	1224	1234	5
4	11115	11125	11135	11145	11225	
	11235	11245	11335	11345	12225	14
	12235	12245	12335	12345		

Example 6.10 Revisited

We now display a bijection between the set of sequences $a_1 a_2 \ldots a_{2n}$ of n 1s and n -1s in Example 6.10 and the set of $(n + 1)$-tuples $b_1 b_2 \ldots b_{n+1}$ in Example 6.23, where $b_i \leq i$ and $b_{n+1} = n + 1$.

In each $2n$-tuple $a_1 a_2 \ldots a_{2n}$, first record the numbers of -1s to the left of each 1. Then add a 1 to each count; this yields the n-tuple $b_1 b_2 \ldots b_n$. Finally, append $n + 1$ to this n-tuple to yield $b_1 b_2 \ldots b_{n+1}$. This $(n + 1)$-tuple has the desired properties:

- Every $b_i \geq 0$
- $b_i \leq i$
- $b_{n+1} = n + 1$

Table 6.18 shows the $2n$-tuples, the counts, the resulting n-tuples, and the final $(n + 1)$-tuples, where $1 \leq n \leq 3$. See Tables 6.18 and 6.17.

Clearly, this algorithm works in the reverse direction as well.

Table 6.18

	$n=1$		$n=2$	
$2n$-tuple :	1-1		1-11-1	11-1-1
Count :	0		0 1	00
Add 1 :	1		1 2	11
Append $n+1$:	12		1 23	113

			$n=3$		
$2n$-tuple :	1-11-11-1	11-11-1-1	11-1-11-1	1-111-1-1	111-1-1-1
Count :	0 1 2	00 1	00 2	0 11	000
Add 1 :	1 2 3	11 2	11 3	1 22	111
Append $n+1$:	1 2 34	11 24	11 34	1 224	1114

Example 6.21 Revisited

We now show that there is a bijection between the set of $(n-1)$-tuples $a_1a_2 \ldots a_{n-1}$ in Example 6.21, where $a_{i+1} \leq a_i \leq n-i$, and the set of $(n+1)$-tuples $b_1b_2 \ldots b_{n+1}$ in Example 6.23, where $b_i \leq i$ and $b_{n+1} = n+1$.

Consider the $(n-1)$-tuple $a_1a_2 \ldots a_{n-1}$. Reverse its order; this yields $a_{n-1} \ldots a_2a_1$. Add 1 to each component. Prefix it with 1 and suffix it with $n+1$. The resulting $(n+1)$-tuple $b_1b_2 \ldots b_{n+1}$ has the properties that $b_i \leq i$ and $b_{n+1} = n+1$.

To see this, notice that $b_i = 1 + a_{n-i+1}$. Because $a_i \leq n-i, a_{n-i+1} \leq n - (n - i + 1)$; that is, $b_i \leq i$. Furthermore, $b_{n+1} = a_{n+1-(n+1)} = a_0 = n + 1$. Thus, the $(n + 1)$-tuple $b_1b_2 \ldots b_{n+1}$ satisfies both properties.

For example, consider the ordered pairs:

	21	20	11	10	00
Reverse the order:	12	02	11	01	00
Add 1 to each component:	23	13	22	12	11
Prefix with 1 and suffix with 4:	1234	1134	1224	1124	1114

See Table 6.17.

This algorithm is certainly reversible, thus establishing the desired bijection.

The next example, although interesting in its own right, is closely related to the foregoing two examples, as we shall see shortly. It was proposed as a problem in 1970 by L. A. Gehami of Glastonbury, Connecticut.

Example 6.24 Find the number of n-tuples $a_1a_2 \ldots a_n$ of nonnegative integers a_i such that every partial sum (from left to right) with k summands is $\geq k$ and their sum equals n, that is, $\sum_{i=1}^{k} a_i \geq k$ and $\sum_{i=1}^{n} a_i = n$.

Solution   First, notice that $0 \le a_i \le n$. Also, the definition does not prevent the components of the n-tuple from being equal.

Table 6.19 shows the various possible n-tuples for $0 \le n \le 4$.

Table 6.19

n	Possible n-tuples					Count
0	λ					1
1	1					1
2	11	20				2
3	111	120	201	210	300	5
4	1111	1120	1201	1210	1300	14
	2011	2101	2110	2020	2200	
	3001	3010	3100	4000		

In 1971, Robert Fray of Florida State University established a one-to-one correspondence between the set of such n-tuples $a_1 a_2 \cdots a_n$ and the set of lattice paths from $(0,0)$ to (n, n) that never fall below the line $y = x$.[†] (See Chapter 9 for a similar set of lattice paths.) He accomplished this by associating the n-tuple to the lattice path connecting the lattice points $(0, 0), (0, s_1), (1, s_1), (1, s_2), (2, s_2), \ldots,$ $(n - 1, s_n), (n, s_n)$, where s_k denotes the partial sum $s_k = a_1 + a_2 + \cdots + a_k$.

Example 6.10 Revisited

There is a systematic way to construct the n-tuples $b_1 b_2 \ldots b_n$ in Example 6.24 from the $2n$-tuples $a_1 a_2 \ldots a_{2n}$ of n 1s and n -1s in Example 6.10.

Let b_1 denote the number of 1s in $a_1 a_2 \ldots a_{2n}$ that precede the first -1. Each remaining b_i denotes the number of 1s between every two consecutive -1's. Then the n-tuple $b_1 b_2 \ldots b_n$ has the desired properties:

- Each $b_i \ge 0$

- $\displaystyle\sum_{i=1}^{k} b_i \ge k$

- $\displaystyle\sum_{i=1}^{n} b_i \ge n$

[†]   For a discussion of such lattice paths, see W. Feller, *An Introduction to Probability Theory and Its Applications*, 2nd ed., Wiley, New York, 1957.

For example, Table 6.20 shows the triplets $b_1b_2b_3$ corresponding to the various 6-tuplets $a_1a_2\ldots a_6$.

Table 6.20

1-11-11-1	11-11-1-1	11-1-11-1	1-111-1-1	111-1-1-1
1 1 1	2 1 0	2 0 1	1 2 0	3 0 0

This algorithm works in the reverse direction also. For example, consider the 4-tuple 2020 in Table 6.19. Since it begins with a 2, the 8-tuple must begin with 11 and the third component must be -1; see Table 6.21. The next element 0 indicates that there are no 1s in between the last -1 and the next -1; see Table 6.22. The next 2 tells us that there are two 1s between the last -1 and the next -1; see Table 6.23.

Table 6.21	Table 6.22	Table 6.23
11-1------	11-1-1----	11-1-111-1 -

The last 0 indicates that there are no 1s between the last -1 and the next -1; so the last entry must be -1, which was obvious:

$$11\text{-}1\text{-}111\text{-}1\text{-}1$$
$$2\quad 0\quad 2\quad 0$$

Example 6.22 Revisited

There is a bijection between the set of n-tuples in this example and that in Example 6.22. To see this, consider the 3-tuple 111 in Table 6.15; it contains three 1s, zero 2s, and zero 3s; correspondingly, we have the 3-tuple 300 in Table 6.19. This procedure is clearly reversible. Thus we have the correspondence

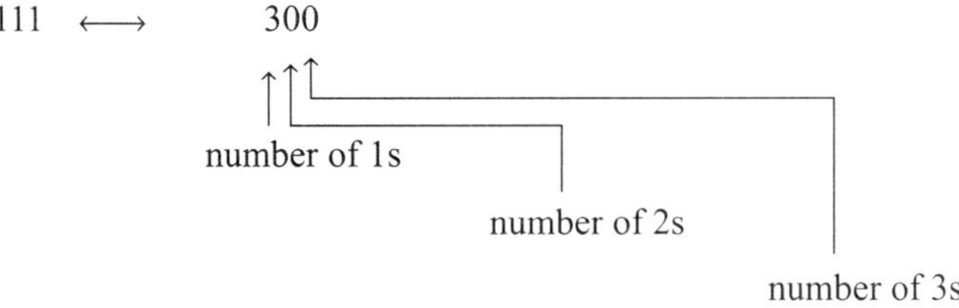

Similarly, we have the following correspondences:

$$112 \longleftrightarrow 210$$
$$113 \longleftrightarrow 201$$
$$122 \longleftrightarrow 120$$
$$123 \longleftrightarrow 111$$

Example 6.21 Revisited

We now claim that there is a one-to-one correspondence between the set of $(n-1)$-tuples $a_1 a_2 \ldots a_{n-1}$ in Example 6.21, where $0 \leq a_{i+1} \leq a_i \leq n - i$, and the set of n-tuples $b_1 b_2 \ldots b_n$ of nonnegative integers b_i in Example 6.24, where $\sum_{i=1}^{k} a_i \geq k$ and $\sum_{i=1}^{n} a_i = n$.

To accomplish this, first append a 0 to each $(n-1)$-tuple $a_1 a_2 \ldots a_{n-1}$. Let b_i denote the number of $(i-1)$'s in the resulting n-tuple $a_1 a_2 \ldots a_{n-1} 0$, where $1 \leq i \leq n$. The n-tuple $b_1 b_2 \ldots b_n$ has the desired properties.

To begin with, clearly, $b_i \geq 0$. Next, suppose that there are exactly j zeros in $a_1 a_2 \ldots a_k$, where $0 \leq j \leq k$. Then $b_1 \geq j$. Since there are $k - j$ nonzero elements in $a_1 a_2 \ldots a_k$, each contributes a 1 to the sum. So

$$\sum_{i=1}^{k} b_i \geq j + (k - j)$$

$$= k$$

Finally, suppose there are exactly m zeros in the n-tuple $a_1 a_2 \ldots a_{n-1} 0$. Then $b_1 = m$. There are $n - m$ nonzeros in the n-tuple, each contributing a 1 to the sum $\sum_{i=1}^{n} b_i$. Thus

$$\sum_{i=1}^{n} b_i = m + (n - m)$$

$$= n$$

as desired.

To illustrate this algorithm, consider the following ordered pairs:

	21	20	11	10	00
Append 0:	210	200	110	100	000
Count the number of 0s, 1s, and 2s:	111	201	120	210	300

These are the same triplets we obtained in Example 6.24; see Table 6.19.

The answers in the next two examples are also Catalan numbers, as we would expect. Catalan enthusiasts can fill in the details.

Example 6.25 Find the number of $(n-1)$-tuples $a_1 a_2 \cdots a_{n-1}$ of positive integers a_i such that $1 \leq a_i \leq 2i$ and $a_i < a_{i+1}$, where $1 \leq i \leq n - 2$.

Solution   Table 6.24 shows the various possible $(n-1)$-tuples, where $1 \leq n \leq 4$.

Table 6.24

n	Possible $(n-1)$-tuples					Count
1	λ					1
2	1	2				2
3	12	13	14	23	24	5
4	123 124 125 126 134 135 136 145 146 234 235 236 245 246					14

Example 6.22 Revisited

We now establish a bijection between the set of n-tuples $a_1 a_2 \ldots a_n$ in Example 6.22, where $1 \leq a_i \leq i$, and the set of $(n-1)$-tuples in this example, where $n \geq 1$. To this end, consider the $(n-1)$-tuple $a_1 a_2 \ldots a_{n-1}$, where $a_1 < a_2 < \cdots < a_{n-1}$ and $a_i \leq 2i$. Let $b_i = a_i - (i-1)$ for each i. Then the n-tuple $c_1 c_2 \ldots c_n = 1 b_1 b_2 \ldots b_{n-1}$ has the property that $1 \leq c_i \leq i$. This correspondence establishes the desired bijection.

For example, consider the 3-tuple 246 in Example 6.25 (see Table 6.24):

$$a_1 a_2 a_3$$

Given 3-tuple:	2 4 6
Subtract $i-1$ from each a_i:	2 3 4
Insert a 1 at the beginning:	1 2 3 4

Notice that 1234 is a valid 4-tuple in Example 6.22.

This correspondence works in the reverse direction also. For example, consider the 4-tuple 1134 in Table 6.15:

$$a_1 a_2 a_3 a_4$$

Given 4-tuple:	1 1 3 4
Delete the leading 1:	1 1 3 4
Add $i-1$ from each a_i:	2 4 5

Notice that 245 is a valid 3-tuple in Example 6.25.

Example 6.26 Find the number of n-tuples $a_1 a_2 \ldots a_n$ of nonnegative integers a_i such that $a_1 = 0$ and $a_{i+1} \leq a_i + 1$, where $1 \leq i \leq n$.

Solution   Table 6.25 shows the various possible n-tuples, where $1 \leq n \leq 4$.

Table 6.25

n	Possible n-tuples					Count
1	0					1
2	00	01				2
3	000	001	010	011	012	5
4	0000	0001	0010	0011	0012	14
	0100	0101	0110	0111	0112	
	0120	0121	0122	0123		

Example 6.10 Revisited

We claim that there is a bijection between the set of sequences of n 1s and n -1s in Example 6.10 and the set of n-tuples in Example 6.26. To see this, we let $b_i = a_i - a_{i+1} + 1$, where $a_{n+1} = 0$. Now, replace each a_i with a 1 followed by b_i -1s. The resulting sequence has the desired property.

For example, consider the 4-tuple $a_1 a_2 a_3 a_4 = 0112$. Then $b_1 = a_1 - a_2 + 1 = 0 - 1 + 1 = 0, b_2 = a_2 - a_3 + 1 = 1 - 1 + 1 = 1, b_3 = 1 - 2 + 1 = 0$ and $b_4 = 2 - 0 + 1 = 3$. Thus, we have:

$$\begin{array}{lcc} & a_1 a_2 & a_3 a_4 \\ \text{Given 4-tuple:} & 0\ 1 & 1\ 2 \\ \text{Convert each } a_i \text{ to a 1:} & 1\ 1 & 1\ 1 \\ \text{Add } b_i -1\text{s immediately after each } b_i: & 1\ 1\ -11 & 1-1-1-1 \end{array}$$

Notice that $11 - 111 - 1 - 1 - 1$ is a valid sequence of four 1s and four -1s in Example 6.10.

The algorithm works in the reverse direction as well. For example, consider the sequence $1 - 111 - 11 - 1 - 1$:

$$\begin{array}{ccc} a_1 & a_2 a_3 & a_4 \\ 1 - 1 & 1\ 1 - 1 & 1 - 1 - 1 \end{array}$$

Since a_1 is followed by one -1, $b_1 = 1$. Thus, $a_1 - a_2 + 1 = 1$. So, $a_1 = a_2$. Because $b_2 = a_2 - a_3 + 1 = 0, a_3 = a_2 + 1$. Similarly, $a_3 = a_4$ and $a_4 = 1$. Thus, $a_1 = a_2, a_2 + 1 = a_3 = a_4 = 1$; so $a_1 = a_2 = 0$ and $a_3 = a_4 = 1$. Consequently, the corresponding 4-tuple is 0011; see Table 6.23.

Example 6.27 Find the number of $(n-1)$-tuples $a_1 a_2 \ldots a_{n-1}$ of integers $a_i \leq 1$ such that every partial sum is nonnegative, where $n \geq 2$.

Solution Table 6.26 shows the various possible $(n-1)$-tuples with the desired property, where $2 \leq n \leq 4$.

Table 6.26

n	Possible $(n-1)$-tuples					Count
2	0	1				2
3	00	01	$1-1$	10	11	5
4	000	001	$01-1$	010	011	14
	$1-10$	$1-11$	$10-1$	100	101	
	$11-1$	110	111	$11-2$		

Example 6.28 Find the number of n-tuples $a_1a_2\ldots a_n$ of integers $a_i \geq -1$ such that every partial sum is nonnegative and $a_1 + a_2 + \cdots + a_n = 0$.

Solution Table 6.27 shows the various possible n-tuples, where $1 \leq n \leq 4$.

Table 6.27

n	Possible n-tuples					Count
1	0					1
2	00	$1-1$				2
3	000	$01-1$	$10-1$	$1-10$	$2-1-1$	5
4	0000	$001-1$	$01-10$	$010-1$	$1-100$	14
	$100-1$	$10-10$	$1-11-1$	$11-1-1$	$02-1-1$	
	$20-1-1$	$2-10-1$	$2-1-10$	$3-1-1-1$		

We now turn to an example about flipping coins, which is very closely related to Examples 6.2, 6.10, and 6.12.

Example 6.29 Suppose that when a coin is tossed is $2n$ times it falls heads n times and tails n times. Find the number of outcomes so that after each toss the number of heads that have occurred is greater than or equal to the number of tails that have occurred.

Solution Table 6.28 shows the various possible outcomes, where $0 \leq n \leq 4$.

This example is basically the same as Example 6.12 (or 6.10). The correspondence H$\leftrightarrow$1 and T$\leftrightarrow$0 provides a clear bijection between the two sets.

The following example deals with finite sequences of positive integers.

Table 6.28

n	Possible Outcomes					Count
0	λ					1
1	HT					1
2	HHTT	HTHT				2
3	HHHTTT	HHTHTT	HHTTHT	HTHHTT	HTHTHT	5
4	HHHHTTTT HHHTHTTT HHHTTHTT HHHTTTHT HHTHHTTT HHTHTHTT HHTHTTHT HHTTHHTT HHTTHTHT HTHHHTTT HTHHTHTT HTHHTTHT HTHTHHTT HTHTHTHT					14

Example 6.30 Find the number of n-tuples $a_1a_2 \ldots a_n$ of positive integers a_i such that:

- $i \leq a_i \leq n$; and
- If $i \leq j \leq a_i$, then $a_j \leq a_i$.

Solution   Since $n \leq a_n \leq n$, it follows that every n-tuple ends in n. Table 6.29 shows the various possible n-tuples, where $0 \leq n \leq 4$. Notice, for example, that 233 is not such a triplet. This is so, since when $i = 1$, we have $1 \leq j \leq 2$; when $j = 2, a_j = 3$, and $a_j \not\leq a_i$. Likewise, 2334 is not a valid 4-tuple.

Table 6.29

n	Possible n-tuples					Count
0	λ					1
1	1					1
2	12	22				2
3	123	133	223	323	333	5
4	1234 1244 1334 1344 1444 2234 2244 3234 3334 3344 4234 4244 4334 4444					14

Example 6.31 Find the number of n-tuples $a_1a_2 \ldots a_n$ of nonnegative integers $a_i \leq n - i$ such that if $i < j, a_i, a_j > 0$, and $a_{i+1} = a_{i+2} = \cdots = a_{j-1} = 0$, then $a_i + i < a_j + j$; see Figure 6.23, where $a_i, a_j > 0$.

Solution   Since $0 \leq a_n \leq n - n$, it follows that every such n-tuple must end in 0. Table 6.30 lists the various n-tuples with the desired properties, where $0 \leq n \leq 4$.

$$\text{subscript:} \qquad i \qquad j$$
$$\dots a_i \underbrace{0 \dots 0}_{\text{zeros}} a_j \dots$$

Figure 6.23

When $n = 3$, for example, $0 \le a_i \le 3 - i$. So $0 \le a_1 \le 2, 0 \le a_2 \le 1$, and $a_3 = 0$. These conditions yield five triplets:

$$000 \quad 010 \quad 100 \quad 110 \quad 200$$

Table 6.30

n	Possible n-tuples					Count
0	λ					1
1	0					1
2	00	10				2
3	000	010	100	110	200	5
4	0000	0010	0100	0110	0210	14
	1000	1010	2000	2010	2100	
	2110	3000	3100	3210		

Example 6.32 Find the number of $(n + 2)$-tuples $a_0 a_1 a_2 \dots a_n a_{n+1}$ of integers $a_i \ge 2$ such that $a_0 = 1 = a_{n+1}$ and $a_i \mid (a_{i-1} + a_{i+1})$, where $1 \le i \le n$ and $x \mid y$ means x is a factor of y.

Solution   Because $a_i \mid (a_{i-1} + a_{i+1})$, each element a_i divides the sum of its adjacent neighbors. Notice that if the n-tuple $a_1 a_2 \dots a_n$ has the given properties, so does its reverse $a_n a_{n-1} \dots a_1$. This cuts down the task of finding such n-tuples by 50 percent. Table 6.31 lists such sequences for $0 \le n \le 4$.

Table 6.31

n	Possible n-tuples					Count
0	λ					1
1	121					1
2	1231	1321				2
3	12341	12531	13231	13521	14321	5
4	123451	123741	125341	125831	127531	14
	132341	132531	135231	135721	138521	
	143231	143521	147321	154321		

The next example counts the number of subsets of the set $W = \{0, 1, 2, \ldots\}$ of whole numbers with certain properties.

Example 6.33 Find the number of subsets S of W such that:

- $0 \in S$; and
- If $k \in S$, then $k + n, k + n + 1 \in S$, where $n \geq 0$.

Solution   Table 6.32 lists the possible subsets for $0 \leq n \leq 4$.

Table 6.32

n	Possible Subsets of S				Count
0	W				1
1	W				1
2	W	$W - \{1\}$			2
3	W	$W - \{1\}$	$W - \{2\}$	$W - \{1,2\}$	$W - \{1,2,5\}$	5
4	W $W - \{1,3\}$ $W - \{1,2,3,6\}$	$W - \{1\}$ $W - \{2,3\}$ $W - \{1,2,3,7\}$	$W - \{2\}$ $W - \{1,2,3\}$ $W - \{1,2,3,11\}$	$W - \{3\}$ $W - \{1,2,6\}$ $W - \{1,2,3,6,7\}$	$W - \{1,2\}$ $W - \{2,3,7\}$	14

We will explore additional occurrences of Catalan numbers in the following chapters.

7

The Ubiquity of Catalan Numbers II

In the previous chapter, we encountered Catalan numbers in many quite unexpected places and studied their close relationships. This chapter continues the investigation of their ubiquitous occurrence.

The following example deals with random walks of a particle on the positive x-axis.

Example 7.1 A particle at the origin moves n units to the right and n units to the left along the x-axis, one unit at a time. After moving $2n$ units, the particle returns to the origin. Find the number of such moves so that the number of moves to the left does not exceed the number of moves to the right.

Solution Figure 7.1 shows the various possible moves for $0 \leq n \leq 4$, where we have distorted the figures to avoid overlapping line segments.

These figures clearly resemble the mountain ranges in Figure 6.3, especially if the mountains are rotated through 90 degrees in the counterclockwise direction. Consequently, switching an upstroke with a unit line segment in the postive direction and a downstroke with a unit line segment in the negative direction establishes a one-to-one correspondence between the set of mountain ranges (or Dyck paths) in Example 6.2.

A bijection can also be seen between the set of Dyck paths in Example 6.2 and the set of sequences of n 1s and n -1s in Example 6.10 (or the binary words in

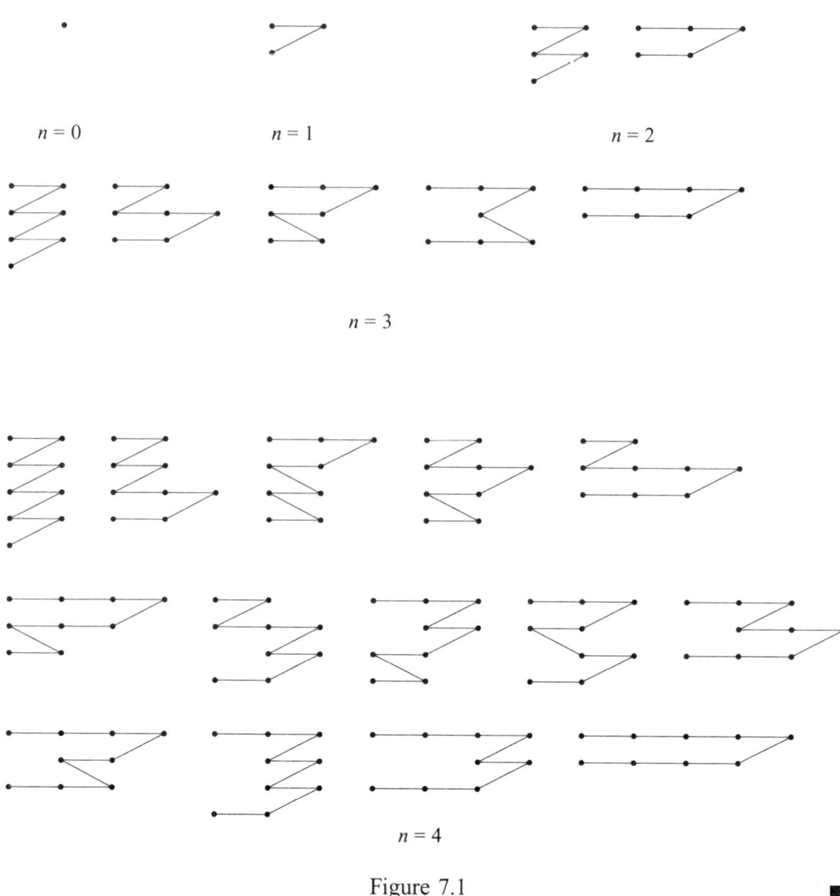

Figure 7.1

Example 6.12). This follows by the following correspondence:

$$1 \quad \leftrightarrow \quad \text{unit distance to the right}$$
$$-1 \quad \leftrightarrow \quad \text{unit distance to the left}$$

Although these two bijections are obvious, I exhibit a less trivial bijection with the set of $(n-1)$-tuples $a_1 a_2 \ldots a_{n-1}$ of nonnegative integers a_i in Example 6.21.

Example 6.21 Revisited

To realize the bijection, consider any path of the particle; see Figure 7.2, where $n = 5$. Ignore the first step to the right. Below each positive step, record the number of negative steps to its left; see Figure 7.3. Read the $n - 1$ counts in the reverse order to get the desired $(n-1)$-tuple; see Figure 7.4.

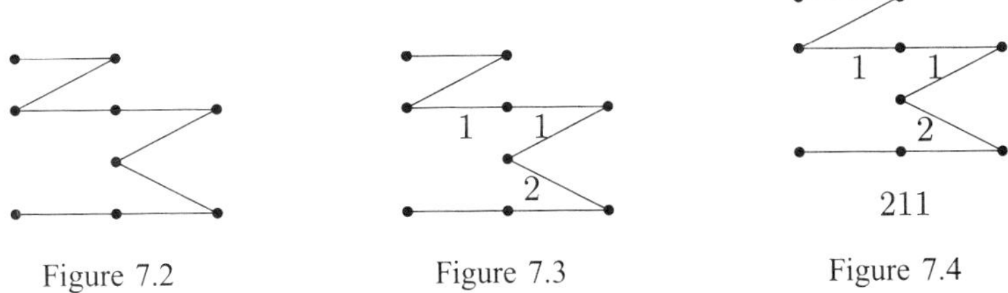

Figure 7.2   Figure 7.3   Figure 7.4

Figure 7.5 shows the various random walks of the particle and the corresponding $(n-1)$-tuples; where $1 \leq n \leq 3$.

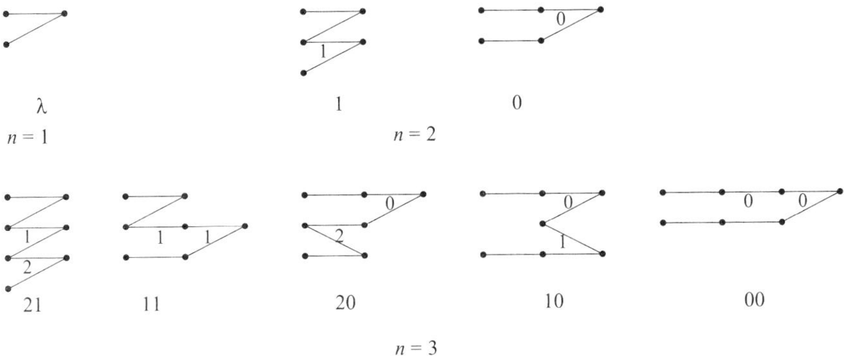

Figure 7.5

This algorithm is clearly reversible. For instance, consider the triplet 310. Reversing it, we get 013. Each element is a count of the number of negative steps to the left of a positive move. Since 013 is a 3-tuple, the path must consist of four positive steps, including the initial positive step.

We start with the initial step; see Figure 7.6. The first count 0 indicates that there are zero negative moves to the left of the second positive move; see Figure 7.7.

The next count 1 shows that the particle has made a negative move before the next positive move. So add a negative step and then a positive one; see Figure 7.8.

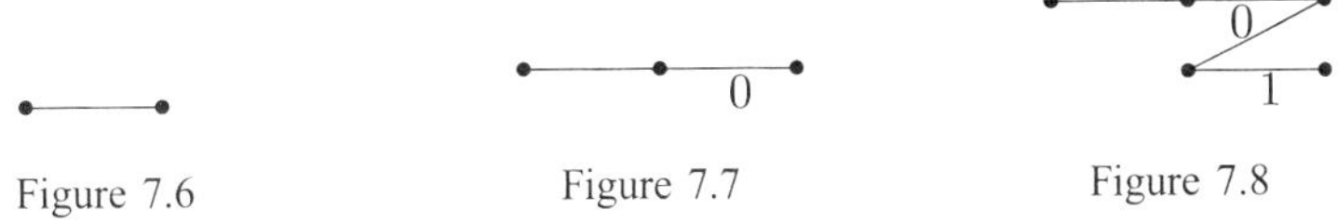

Figure 7.6   Figure 7.7   Figure 7.8

The final count 3 indicates a total of three negative steps before the next positive step. So add two negative steps, followed by a positive step; see Figure 7.9.

By now, the path contains four positive moves, but only three negative moves. Now, add a negative move, so the particle will return home; see Figure 7.10.

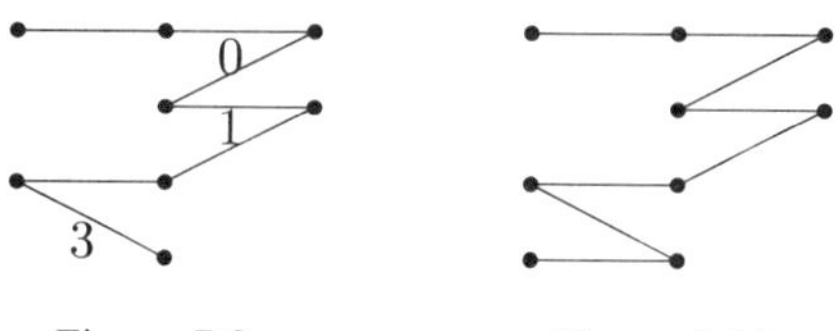

Figure 7.9   Figure 7.10

Clearly, the same algorithm can be adapted in a straightforward way to show a bijection between the set of mountain ranges in Example 6.2 and the set of $(n-1)$-tuples in Example 6.21.

The next example was proposed as a problem in 1985 by N. K. Krier of Colorado State University and F. R. Bernhart of Rochester Institute of Technology. The solution is a composite solution given by J. Biasotti, P. W. Lindstrom, R. Parris, and W. P. Wardlaw in 1986.

Example 7.2 Find the number N_n of n-tuples $a_1 a_2 \ldots a_n$ of positive integers a_i such that $a_1 = 1$ and $a_{i+1} \leq a_i + 1$, where $1 \leq i \leq n$. Find also the sum S_n of all such n-tuples.

Solution   Let $A_n = \{(a_1 a_2 \ldots a_n) | a_1 = 1 \text{ and } a_{i+1} \leq a_i + 1\}$. Let $A_{n,k}$ be the subset of A_n with $a_n = k$. Let $N_{n,k}$ be the number of elements in $A_{n,k}$ and $S_{n,k}$ the sum of the elements in $A_{n,k}$.
Then:

$$N_n = \sum_{k=1}^{n} N_{n,k} = N_{n+1,1}$$

$$N_{n,k} = 0 \quad \text{if } k < 1 \quad \text{or} \quad k > n$$

$$S_n = \sum_{k=1}^{n} S_{n,k} = S_{n+1,1}$$

$$S_{n,k} = 0 \quad \text{if } k < 1 \quad \text{or} \quad k > n$$

Furthermore, both $N_{n,k}$ and $S_{n,k}$ can be defined recursively:

$$N_{n,n} = 1$$

$$N_{n+1,k} = N_{n+1,k+1} + N_{n,k-1}, \quad 1 \leq k \leq n \tag{7.1}$$

$$S_{n,n} = n!$$

$$S_{n+1,k} = \frac{k}{k+1} S_{n+1,k+1} + k S_{n,k-1}, \quad 1 \leq k \leq n \tag{7.2}$$

These follow because the correspondence between $(a_1 a_2 \ldots a_n k)$ in $A_{n+1,k}$ and $(a_1 a_2 \ldots a_n k + 1)$ in $A_{n+1,k+1}$ is one-to-one, except when $a_n = k - 1$.

It is easy to verify that these recursive definitions are satisfied by

$$N_{n,k} = \frac{k}{2n-k}\binom{2n-k}{n} \quad \text{and} \quad S_{n,k} = \frac{n!}{2^{n-k}N_{n,k}}$$

where $0 \le k \le n$. Consequently,

$$N_n = N_{n+1,1} = \frac{1}{2n+1}\binom{2n+1}{n+1} = \frac{1}{n+1}\binom{2n}{n}$$

$$= C_n$$

and

$$S_n = S_{n+1,1} = \frac{n!}{2^n}\binom{2n}{n}$$

$$= 1 \cdot 3 \cdot 5 \cdots (2n-1)$$

as desired. $\blacksquare$

K. L. Bernstein of MITRE Corporation, Bedford, Massachusetts, provided a delightful alternate solution, using two well-known combinatorial identities:

$$b\binom{a+b}{a} = (a+1)\binom{a+b}{a+1}$$

$$\sum_{j=1}^{d}\binom{c-1+j}{c} = \binom{c+d}{c+1}$$

Notice that the required sum S_n can be written as

$$S_n = \sum_{a_1=1}^{a_0+1} a_1 \sum_{a_2=1}^{a_1+1} a_2 \cdots \sum_{a_n=1}^{a_{n-1}+1} u_n$$

where $a_0 = 0$. The rightmost sum can be written as

$$\sum_{a_n=1}^{a_{n-1}+1} a_n = \sum_{a_n=1}^{a_{n-1}+1} a_n\binom{a_n}{0} = 1\sum_{a_n=1}^{a_{n-1}+1}\binom{1-1+a_n}{0}$$

$$= 1\sum_{a_n=1}^{a_{n-1}+1}\binom{0+a_n}{0} = 1\sum_{a_n=1}^{a_{n-1}+1}\binom{0+a_n}{0+1}$$

$$= 1\binom{2+a_{n-1}}{2}$$

Continuing like this successively with the remaining rightmost summations, we get:

$$S_n = 1 \cdot 3 \cdot 5 \cdots (2m-1) \sum_{a_1=1}^{a_0+1} a_1 \sum_{a_2=1}^{a_1+1} a_2 \cdots \sum_{a_{n-m}=1}^{a_{n-m-1}+1} a_{n-m} \binom{2m + a_{n-m}}{2m}$$

where $0 \leq m \leq n$. When $m = n$, this yields:

$$S_n = 1 \cdot 3 \cdot 5 \cdots (2n-1) \binom{2n + a_0}{2n}$$

$$= 1 \cdot 3 \cdot 5 \cdots (2n-1) \binom{2n + 0}{2n}$$

$$= 1 \cdot 3 \cdot 5 \cdots (2n-1)$$

as obtained earlier.

To evaluate N_n, we use the identity

$$\sum_{j=1}^{a+1} \frac{j+1}{m} \binom{2m+j}{m-1} = \frac{a+1}{m+1} \binom{2m+2+a}{m} \tag{7.3}$$

Beginning with the innermost sum, a repeated application of formula (7.3) yields

$$N_n = \sum_{a_1=1}^{a_0+1} \sum_{a_2=1}^{a_1+1} \cdots \sum_{a_{n-k}=1}^{a_{n-k-1}+1} \frac{a_{n-k}+1}{k} \binom{2k + a_{n-k}}{k-1}$$

where $1 \leq k \leq n$. When $k = n$, this yields

$$N_n = \frac{a_0+1}{n} \binom{2n + a_0}{n-1} = \frac{1}{n}\binom{2n}{n-1}$$

$$= \frac{1}{n+1}\binom{2n}{n}$$

$$= C_n$$

as expected.   ∎

Permutations and Catalan Numbers

The next two examples deal with permutations and cycles, but first a few definitions for clarity.[†]

A *permutation f* of the elements of the set $S = \{a_1, a_2, \cdots, a_n\}$ is a bijection on S. Let $f(a_i) = b_i$ for each i. Then f is written as

$$f = \begin{pmatrix} a_1 & a_2 & \cdots & a_n \\ b_1 & b_2 & \cdots & b_n \end{pmatrix}$$

using the bilevel notation. In particular, suppose $f(a_i) = a_{i+1}$ for $1 \le i \le n-1$ and $f(a_n) = a_1$. Then, we use a compact *cyclic* notation to identify f:

$$f = (a_1 \, a_2 \, \cdots \, a_n)$$

This permutation f cyclically shifts every element a_i by one position to the right. It is a *cycle* of length n.

For example, let $S = \{1, 2, 3, 4, 5\}$. Then

$$(1\,3\,4\,2\,5) = \begin{pmatrix} 1 & 2 & 3 & 4 & 5 \\ 3 & 5 & 4 & 2 & 1 \end{pmatrix}$$

Notice that

$$(1\,3\,4\,2\,5) = (3\,4\,2\,5\,1) = (4\,2\,5\,1\,3) = (2\,5\,1\,3\,4) = (5\,1\,3\,4\,2)$$

An element a_i is left *fixed* by a permutation f if $f(a_i) = a_i$. For example, consider the permutation g on the set S:

$$g = (1\,4\,3\,5) = \begin{pmatrix} 1 & 2 & 3 & 4 & 5 \\ 4 & 2 & 5 & 3 & 1 \end{pmatrix}$$

It leaves the element 2 fixed: $g(2) = 2$. The permutation g is a cycle of length 4.

A cycle of length 2 is a *transposition*. For example, $(3\,5)$ is a transposition:

$$(3\,5) = \begin{pmatrix} 1 & 2 & 3 & 4 & 5 \\ 1 & 2 & 5 & 4 & 3 \end{pmatrix}$$

Because cycles are permutations, they can be multiplied, where the generic term *multiplication* means composition. Since the composition of two bijections

---

[†] For a detailed discussion, see J. B. Fraleigh, *A First Course in Abstract Algebra*, 7th ed., Addison-Wesley, Reading, Massachusetts, 2003.

is a bijection, it follows that the product of two permutations (and hence two cycles) is also a permutation.

For example, consider the cycles $(1\,4\,3\,5)$ and $(3\,2\,5)$ of the set S. Then:

$$(1\,4\,3\,5)(3\,2\,5) = \begin{pmatrix} 1 & 2 & 3 & 4 & 5 \\ 4 & 2 & 5 & 3 & 1 \end{pmatrix} \begin{pmatrix} 1 & 2 & 3 & 4 & 5 \\ 1 & 5 & 2 & 4 & 3 \end{pmatrix}$$

$$= \begin{pmatrix} 1 & 2 & 3 & 4 & 5 \\ 4 & 5 & 3 & 2 & 1 \end{pmatrix}$$

$$= (1\,4\,2\,5)$$

and

$$(3\,2\,5)(1\,4\,3\,5) = \begin{pmatrix} 1 & 2 & 3 & 4 & 5 \\ 1 & 5 & 2 & 4 & 3 \end{pmatrix} \begin{pmatrix} 1 & 2 & 3 & 4 & 5 \\ 4 & 2 & 5 & 3 & 1 \end{pmatrix}$$

$$= \begin{pmatrix} 1 & 2 & 3 & 4 & 5 \\ 4 & 1 & 2 & 3 & 5 \end{pmatrix}$$

$$= (1\,4\,3\,2)$$

In general, the product of two cycles need not be a cycle.

Two cycles are *disjoint* if they have no common elements. For example, the cycles $(1\,3\,5)$ and $(2\,4)$ are disjoint.

Notice that the cycle $(1\,4\,3\,5)$ can be considered the product of two cycles: $(1\,4\,3\,5) = (1\,4\,3\,5)(2)$. Although it is customary to drop the cycle (2), we will need it in our discussions.

Since the multiplication of disjoint cycles is commutative, the order of the factors in such a product does not affect the outcome. For example, $(1\,3\,5)(2\,4) = (2\,4)(1\,3\,5)$.

We now ready to present the two examples on permutations and Catalan numbers.

Example 7.3 Find the number of permutations α of the set $\{1, 2, \ldots, 2n\}$ with n transpositions such that the permutation $\omega\alpha$ is the product of $n + 1$ cycles, where $\omega = (1\,2\,\ldots\,2n)$ and $n \geq 1$.

Solution Table 7.1 enumerates the various possible such permutations α, where $1 \leq n \leq 4$. Notice, for instance, that $(1\,2\,3\,4)\cdot(1\,3)(2\,4) = (1\,4\,3\,2)$, so $(1\,3)(2\,4)$ is not such a permutation. Likewise, $(1\,2)(3\,5)(4\,6)(7\,8)$ is also not such a permutation.

Table 7.1

n	Permutation α	$\omega\alpha$	Count
1	$(1\,2)$	$(1\,2) \cdot (1\,2) = (1)(2)$	1
2	$(1\,2)(3\,4)$	$(1\,2\,3\,4) \cdot (1\,2)(3\,4) = (1)(2\,4)(3)$	
	$(1\,4)(2\,3)$	$(1\,2\,3\,4) \cdot (1\,4)(2\,3) = (1\,3)(2)(4)$	2
3	$(1\,2)(3\,4)(5\,6)$	$(1\,2\,3\,4\,5\,6) \cdot (1\,2)(3\,4)(5\,6) = (1)(2\,4\,6)(3)(5)$	
	$(1\,2)(3\,6)(4\,5)$	$(1\,2\,3\,4\,5\,6) \cdot (1\,2)(3\,5)(4\,6) = (1)(2\,6)(3\,5)(4)$	
	$(1\,4)(2\,3)(5\,6)$	$(1\,2\,3\,4\,5\,6) \cdot (1\,4)(2\,3)(5\,6) = (1\,3)(2)(4\,6)(5)$	5
	$(1\,6)(2\,3)(4\,5)$	$(1\,2\,3\,4\,5\,6) \cdot (1\,6)(2\,3)(4\,5) = (1\,3\,5)(2)(4)(6)$	
	$(1\,6)(2\,5)(3\,6)$	$(1\,2\,3\,4\,5\,6) \cdot (1\,6)(2\,5)(3\,4) = (1\,5)(2\,4)(3)(6)$	
4	$(1\,2)(3\,4)(5\,6)(7\,8)$	$(1\,2\,3\,4\,5\,6\,7\,8) \cdot (1\,2)(3\,4)(5\,6)(7\,8) = (1)(2\,4\,6\,8)(3)(5)(7)$	
	$(1\,2)(3\,4)(5\,8)(6\,7)$	$(1\,2\,3\,4\,5\,6\,7\,8) \cdot (1\,2)(3\,4)(5\,8)(6\,7) = (1)(2\,4)(3)(5\,7)(6)$	
	$(1\,2)(3\,6)(4\,5)(7\,8)$	$(1\,2\,3\,4\,5\,6\,7\,8) \cdot (1\,2)(3\,6)(4\,5)(7\,8) = (1)(2\,6\,8)(3\,5)(4)(7)$	
	$(1\,2)(3\,8)(4\,5)(6\,7)$	$(1\,2\,3\,4\,5\,6\,7\,8) \cdot (1\,2)(3\,8)(4\,5)(6\,7) = (1)(2\,8)(3\,5\,7)(4)(6)$	
	$(1\,2)(3\,8)(4\,7)(5\,6)$	$(1\,2\,3\,4\,5\,6\,7\,8) \cdot (1\,2)(3\,8)(4\,7)(5\,6) = (1)(2\,8)(3\,7)(4\,6)(5)$	
	$(1\,2)(3\,8)(4\,7)(5\,6)$	$(1\,2\,3\,4\,5\,6\,7\,8) \cdot (1\,2)(3\,8)(4\,7)(5\,6) = (1)(2\,8)(3\,7)(4\,6)(5)$	
	$(1\,4)(2\,3)(5\,6)(7\,8)$	$(1\,2\,3\,4\,5\,6\,7\,8) \cdot (1\,4)(2\,3)(5\,6)(7\,8) = (1\,3)(2)(4\,6\,8)(5)(7)$	14
	$(1\,6)(2\,3)(4\,5)(7\,8)$	$(1\,2\,3\,4\,5\,6\,7\,8) \cdot (1\,6)(2\,3)(4\,5)(7\,8) = (1\,3\,5)(2)(4)(6\,8)(7)$	
	$(1\,6)(2\,5)(3\,4)(7\,8)$	$(1\,2\,3\,4\,5\,6\,7\,8) \cdot (1\,6)(2\,5)(3\,4)(7\,8) = (1\,5)(2\,4)(3)(6\,8)(7)$	
	$(1\,8)(2\,3)(4\,7)(5\,6)$	$(1\,2\,3\,4\,5\,6\,7\,8) \cdot (1\,8)(2\,3)(4\,7)(5\,6) = (1\,3\,7)(2)(4\,6)(5)(8)$	
	$(1\,8)(2\,5)(3\,4)(6\,7)$	$(1\,2\,3\,4\,5\,6\,7\,8) \cdot (1\,8)(2\,5)(3\,4)(6\,7) = (1\,5\,7)(2\,4)(3)(6)(8)$	
	$(1\,8)(2\,7)(3\,4)(5\,6)$	$(1\,2\,3\,4\,5\,6\,7\,8) \cdot (1\,8)(2\,7)(3\,4)(5\,6) = (1\,7)(2\,4\,6)(3)(5)(8)$	
	$(1\,8)(2\,7)(3\,6)(4\,5)$	$(1\,2\,3\,4\,5\,6\,7\,8) \cdot (1\,8)(2\,7)(3\,6)(4\,5) = (1\,7)(2\,6)(3\,5)(4)(8)$	
	$(1\,8)(2\,3)(4\,5)(6\,7)$	$(1\,2\,3\,4\,5\,6\,7\,8) \cdot (1\,8)(2\,3)(4\,5)(6\,7) = (1\,3\,5\,7)(2)(4)(6)(8)$	

We now turn to the next example on permutations and Catalan numbers. They were studied in 1985 by D. M. Jackson of the University of Waterloo, Ontario, Canada, while he was a visiting scholar at Massachusetts Institute of Technology. According to Richard P. Stanley of Massachusetts Institute of Technology, they could have been investigated much earlier.

Example 7.4 Find the number of ordered pairs of permutations α and β of the set $\{1, 2, 3, \ldots, n\}$ with a total of $n + 1$ cycles such that $\alpha\beta = (1\,2\,3\,\ldots\,n)$.

Solution Table 7.2 shows the various possible such ordered pairs of permutations, where $1 \leq n \leq 4$. ∎

Next we pursue the well-known *ballot problem*, studied by French mathematician Joseph Louis François Bertrand. Again, it may appear to be different; but it is exactly the same as Examples 6.10 and 6.12, except for the wording.

Example 7.5 (*The Ballot Problem*) Two candidates, A and B, are running for an office. Each person gets n votes and the votes are counted one at a time. Find the

Table 7.2

n	α	β	$\alpha\beta$	Count
1	(1)	(1)	(1)	1
2	(1)(2)	(1 2)	(1 2)	2
	(1 2)	(1)(2)	(1 2)	
3	(1)(2)(3)	(1 2 3)	(1 2 3)	5
	(1 2 3)	(1)(2)(3)	(1 2 3)	
	(1 2)(3)	(1 3)(2)	(1 2 3)	
	(1 3)(2)	(1)(2 3)	(1 2 3)	
	(1)(2 3)	(1 2)(3)	(1 2 3)	
4	(1)(2)(3)(4)	(1 2 3 4)	(1 2 3 4)	14
	(1 2 3 4)	(1)(2)(3)(4)	(1 2 3 4)	
	(1 2)(3)(4)	(1 3 4)(2)	(1 2 3 4)	
	(1 3)(2)(4)	(1 4)(2 3)	(1 2 3 4)	
	(1 4)(2)(3)	(1)(2 3 4)	(1 2 3 4)	
	(1)(2 3)(4)	(1 2 4)(3)	(1 2 3 4)	
	(1 2 4)(3)	(1 2)(3 4)	(1 2 3 4)	
	(1)(2 4)(3)	(1 2)(3 4)	(1 2 3 4)	
	(1 2)(3 4)	(1 3)(2)(4)	(1 2 3 4)	
	(1 4)(2 3)	(1)(2 4)(3)	(1 2 3 4)	
	(1 3 4)(2)	(1)(2 3)(4)	(1 2 3 4)	
	(1 2 3)(4)	(1 4)(2)(3)	(1 2 3 4)	
	(1)(2)(3 4)	(1 2 3)(4)	(1 2 3 4)	
	(1)(2 3 4)	(1 2)(3)(4)	(1 2 3 4)	

number of ways the $2n$ votes can be counted in such a way that at each count the number of votes received by A is greater than or equal to the number of votes received by B. ∎

We puruse the ballot problem in a slightly different setting. Suppose we would like to arrange the six integers a_1, a_2, a_3, b_1, b_2, and b_3 in a row such that $a_i \prec b_i$, $a_i \prec a_{i+1}$, and $b_i \prec b_{i+1}$ for each i, where $x \prec y$ means x *precedes* y. There are $5 = C_3$ such possible arrangements:

$$
\begin{array}{cccccc}
a_1 & a_2 & a_3 & b_1 & b_2 & b_3 \\
a_1 & a_2 & b_1 & a_3 & b_2 & b_3 \\
a_1 & a_2 & b_1 & b_2 & a_3 & b_3 \\
a_1 & b_1 & a_2 & a_3 & b_2 & b_3 \\
a_1 & b_1 & a_2 & b_2 & a_3 & b_3
\end{array}
$$

Joseph Louis François Bertrand (1822–1900), the son of a writer of popular scientific articles and books, was born in Paris. At the age of eleven, he unofficially began attending classes at the École Polytechnique. In 1838, at the age of sixteen, he earned two degrees, one in the arts and the other in science. A year later, he received his doctorate for his work in thermomechanics and published his first paper. In 1841, he became professor at the Collège Saint-Louis. Subsequently, he taught at the Lycée Henry IV, the École Normale Supérieure, the École Polytechnique, and finally at the Collège de France until his death.

Bertrand wrote many popular textbooks and made important contributions to applied mathematics, analysis, differential geometry, probability, and theoretical physics.

This amusing problem was generalized in 1968 by E. Just of Bronx Community College, New York. The solution, by E. Andersen of the University of Oslo, Norway, uses induction.

Example 7.6 Find the number of ways the $2n$ items $a_1, a_2, \ldots, a_n, b_1, b_2, \ldots, b_n$ can be arranged in a row such that $a_i \prec b_i, a_i \prec a_{i+1}$ and $b_i \prec b_{i+1}$ for each i, where $1 \leq i \leq n - 1$.

Solution   Table 7.3 shows the possible lists for $0 \leq n \leq 4$.

Table 7.3

n	Possible Ordered Lists			Count
0	λ			1
1	$a_1 b_1$			1
2	$a_1 a_2 b_1 b_2$	$a_1 b_1 a_2 b_2$		2
3	$a_1 a_2 a_3 b_1 b_2 b_3$	$a_1 a_2 b_1 a_3 b_2 b_3$	$a_1 a_2 b_1 b_2 a_3 b_3$	5
	$a_1 b_1 a_2 a_3 b_2 b_3$	$a_1 b_1 a_2 b_2 a_3 b_3$		
4	$a_1 a_2 a_3 a_4 b_1 b_2 b_3 b_4$	$a_1 a_2 a_3 b_1 a_4 b_2 b_3 b_4$	$a_1 a_2 a_3 b_1 b_2 a_4 b_3 b_4$	14
	$a_1 a_2 a_3 b_1 b_2 b_3 a_4 b_4$	$a_1 a_2 b_1 a_3 a_4 b_2 b_3 b_4$	$a_1 a_2 b_1 a_3 b_2 a_4 b_3 b_4$	
	$a_1 a_2 b_1 a_3 b_2 b_3 a_4 b_4$	$a_1 a_2 b_1 b_2 a_3 a_4 b_3 b_4$	$a_1 a_2 b_1 b_2 a_3 b_3 a_4 b_4$	
	$a_1 b_1 a_2 a_3 a_4 b_2 b_3 b_4$	$a_1 b_1 a_2 a_3 b_2 a_4 b_3 b_4$	$a_1 b_1 a_2 a_3 b_2 b_3 a_4 b_4$	
	$a_1 b_1 a_2 b_2 a_3 a_4 b_3 b_4$	$a_1 b_1 a_2 b_2 a_3 b_3 a_4 b_4$		

To find the number of such ordered lists, notice that there are $\binom{2n}{n}$ arrangements with $a_i \prec a_{i+1}$, and $b_i \prec b_{i+1}$; they are the different ways of choosing n items from $2n$ distinct items. From these, we extract all *permissible* arrangements, where $a_i \prec b_i$. Let $P(n)$ denote the number of permissible arrangements.

Let $q(n,j)$ denote the number of nonpermissible arrangements with b_1 in the jth position. Clearly, $q(n, 1) = \binom{2n-1}{n-1}$. It follows by induction that $q(n,j) = \binom{2n-j}{n-j}$. So $q(n) = q(n+1, 2) = \binom{2n}{n-1}$. Thus

$$p(n) = \binom{2n}{n} - q(n)$$

$$= \binom{2n}{n} - \binom{2n}{n-1}$$

$$= C_n \qquad \blacksquare$$

A Special Case

As a special case, let $a_i = 2i - 1$ and $b_i = 2i$, where $1 \le i \le n$. Then there are C_n different permuations of the set $\{1, 2, \ldots, 2n\}$ such that:

- The odd integers occur in ascending order.
- The even integers occur in ascending order.
- The integer $2i - 1$ occurs before $2i$.

Table 7.4 shows such permutations for $0 \le n \le 4$.

Table 7.4

n	Permutations					Count
0	λ					1
1	12					1
2	1324	1234				2
3	135246	132546	132456	123546	123456	5
4	13572468	13527468	13524768	13524678	13257468	14
	13254768	13254678	13245768	13245678	12357468	
	12354768	12354678	12345768	12345678		

Examples 6.10 and 6.22 Revisited

In this example, suppose we change each odd number in a permutation into 1 and each even number into -1. For example, this transformation converts the permutations 1324 and 1234 into $11 - 1 - 1$ and $1 - 11 - 1$, respectively; they are valid permutations in Example 6.10.

Clearly, this algorithm works in the reverse direction also. For example, changing 1s into successive odd numbers and -1s into successive even numbers, the sequence $1 - 11 - 11 - 11 - 1$ becomes 12345678, a valid 8-tuple in Example 6.22.

Thus, there is a bijection between the set of sequences in Example 6.10 (and hence Example 6.12) and the set of permutations in Table 7.4.

Example 7.6 Revisited

By slightly modifying this algorithm, we can establish a bijection between the set of $2n$-tuples in Example 7.6 and the set of permutations in Table 7.4. By changing each a_i to $2i-1$ and each b_i to $2i$, we can recover all permutations from Example 7.6.

For example, we have:

$$\begin{aligned}
a_1b_1 &\leftrightarrow 12 \\
a_1a_2b_1b_2 &\leftrightarrow 1324 \\
a_1b_1a_2b_2 &\leftrightarrow 1234
\end{aligned}$$

To see the correspondence in the opposite direction, it suffices to change $2i - 1$ to a_i and $2i$ to b_i:

$$\begin{aligned}
13572468 &\leftrightarrow a_1a_2a_3a_4b_1b_2b_3b_4 \\
13527468 &\leftrightarrow a_1a_2a_3b_1a_4b_2b_3b_4
\end{aligned}$$

Example 6.22 Revisited

We show that there is a one-to-one correspondence between the set of these ordered lists and the set of n-tuples in Example 6.22. To realize this, keep a count of the number of b_js to the left of each a_i. Add one to each count. The resulting n-tuple $c_1c_2\ldots c_n$ has the property that $c_k \leq k$ for every k.

For example, consider the list $a_1b_1a_2a_3b_2a_4b_3b_4$:

$$a_1b_1a_2a_3b_2a_4b_3b_4$$

Count the b_js to the left of each a_i:	0	1 1	2
Add 1 to each count:	1	2 2	3

Notice that the 4-tuple $c_1c_2c_3c_4 = 1223$ has the desired property; see Table 6.15.

Clearly, this process is reversible. For example, consider the 4-tuple $c_1c_2c_3c_4 = 1124$:

$$
\begin{array}{lcccc}
c_1c_2c_3c_4: & 1\ 1 & 2 & 4 \\
\text{Subtract 1 from each:} & 0\ 0 & 1 & 3 \\
\text{Place an } a_i \text{ below each:} & a_1a_2 & a_3 & a_4 \\
\text{Insert } b_i\text{s:} & a_1a_2b_1a_3b_2b_3a_4b_4
\end{array}
$$

So the corresponding ordered list is $a_1a_2b_1a_3b_2b_3a_4b_4$; see Table 7.3.

An Interesting Application

As an interesting application, suppose that an undergraduate student wants to take n mathematics courses $m_1, m_2, \ldots, m_n$ and n computer science courses $c_1, c_2, \ldots, c_n$, where m_i is a prerequisite of m_{i+1}, c_i is a prerequisite of c_{i+1}, and m_i is a prerequisite of c_i. Then there are C_n ordered ways the student can take these $2n$ courses.

Example 6.20 Revisited

Next we show that there is a bijection between the set of ordered lists in Example 7.6 and the set of n-tuples in Example 6.20. To this end, it suffices to read the above n-tuples $c_1c_2 \ldots c_n$ to yield $c_nc_{n-1} \ldots c_1$, as illustrated here:

$$
\begin{array}{lccccc}
\text{Ordered list:} & a_1b_1b_1a_2a_3b_2a_4b_4 \\
\text{Count the } b_j\text{s to the left of each } a_i: & 1 & 2\ 2 & 3 \\
\text{Reverse the order:} & 3 & 2\ 2 & 1
\end{array}
$$

Notice that the n-tuple 3221 satisfies the properties in Example 6.20; see Table 6.13. Clearly, this algorithm is reversible.

Example 6.23 Revisited

Interestingly, there is a bijection between the set of ordered lists in Example 7.6 and the set of $(n + 1)$-tuples $x_1x_2 \ldots x_{n+1}$ in Example 6.23, where $x_i \leq i$ and $x_{n+1} = n + 1$. To establish this correspondence, first we append a_{n+1} at the end of each list. Then, as before, we keep a count of the number of b_js to the left of each a_i. Add 1 to each count. The resulting $(n + 1)$-tuple $c_1c_2 \ldots c_{n+1}$ has the desired properties.

For example, consider the word $a_1b_1a_2a_3b_2a_4b_3b_4$:

$$
\begin{array}{lcccccc}
\text{Given word:} & a_1b_1a_2a_3b_2a_4b_3b_4 \\
\text{Append } a_5: & a_1b_1a_2a_3b_2a_4b_3b_4a_5 \\
\text{Take a count:} & 0 & 1\ 1 & 2 & 4 \\
\text{Add 1 to each count:} & 1 & 2\ 2 & 3 & 5
\end{array}
$$

Clearly, the algorithm is reversible.

The following revised version of the ballot problem was proposed in 1969 by K. Steffen of Johannes Gutenberg University, Mainz, Germany. A neat solution was provided by H. Lass of California Institute of Technology, but we omit it for convenience:

> Consider $2n$ items $a_1, a_2, \ldots, a_n, b_1, b_2, \ldots, b_n$ arranged in a row such that $a_i \prec b_i$ and $b_i \prec b_{i+1}$, where $1 \le i \le n-1$. An *inversion* is a pair (a_i, b_i) such that $a_i \not\prec b_i$. Let v_k denote the number of arrangements with k inversions, where $0 \le k \le n$. Then $v_0 = v_1 = \cdots = v_n = C_n$.

For example, let $n = 3$. Table 7.5 shows the various arrangements with k inversions, where $0 \le k \le 3$. Notice that $v_0 = v_1 = v_2 = v_3 = 5 = C_3$.

Table 7.5

a_1	a_2	a_3	b_1	b_2	b_3
a_1	a_2	b_1	a_3	b_2	b_3
a_1	a_2	b_1	b_2	a_3	b_3
a_1	b_1	a_2	a_3	b_2	b_3
a_1	b_1	a_2	b_2	a_3	b_3

With Zero Inversions

b_1	a_1	a_2	a_3	b_2	b_3
b_1	a_1	a_2	b_2	a_3	b_3
b_1	a_1	a_2	b_2	a_3	b_3
a_1	b_1	b_2	a_2	a_3	b_3
a_1	b_1	a_2	b_2	b_3	a_3

With One Inversion

b_1	a_1	b_2	a_2	a_3	b_3
b_1	a_1	a_2	b_2	b_3	a_3
b_1	a_1	b_2	a_2	a_3	b_3
a_1	b_1	b_2	a_2	b_3	a_3
a_1	b_1	b_2	b_3	a_2	a_3

With Two Inversions

b_1	a_1	b_2	a_2	b_3	a_3
b_1	a_1	b_2	b_3	a_2	a_3
b_1	b_2	a_1	a_2	b_3	a_3
b_1	b_2	a_1	b_3	a_2	a_3
b_1	b_2	b_3	a_1	a_2	a_3

With Three Inversions

The following example is directly related to Example 7.6.

Example 7.7 Find the number of binary words consisting of $(n + 1)$ *a*s and n *b*s that can be formed such that the number of *a*s in every subword, when scanned from the far left to right, is greater than the number of *b*s in it.

Solution   Table 7.6 shows the various possible binary words, where $0 \le n \le 4$.   ∎

Notice that this is basically the same as Example 6.13.

Example 6.22 Revisited

There is a bijection between the set of binary words in Example 7.7 and the set of n-tuples $a_1 a_2 \ldots a_n$ in Example 6.22, where $1 \le a_i \le i$. To see this, first ignore the

Table 7.6

n	Possible Binary Words					Count
0	$a\lambda = a$					1
1	aab					1
2	$aaabb$	$aabab$				2
3	$aaaabbb$	$aaababb$	$aaabbab$	$aabaabb$	$aababab$	5
4	$aaaaabbbb$ $aaaababbb$ $aaaabbabb$ $aaaabbbab$ $aaabaabbb$ $aaababab b$ $aaababbab$ $aaabbaabb$ $aaabbabab$ $aabaaabbb$ $aabaababb$ $aabaabbab$ $aababaabb$ $aabababab$					14

leading symbol a. Then count the number of bs to the left of each a and increase each count by 1. The resulting n-tuple $c_1 c_2 \ldots c_n$ has the property that $1 \le c_i \le i$.

For example, consider the word $aaabbab$:

$$
\begin{array}{ll}
\text{Given word:} & aaabbab \\
\text{Delete the leftmost } a: & \cancel{a}aabbab \\
\text{Take a count:} & 00\quad 2 \\
\text{Add 1 to each count:} & 11\quad 3
\end{array}
$$

The triplet 113 has the desired property; see Table 6.15.

Clearly, this process is reversible.

Example 6.21 Revisited

We now find a one-to-one correspondence between the set of binary words in Example 7.7 and the set of $(n-1)$-tuples $a_1 a_2 \ldots a_{n-1}$ of nonnegative integers a_i such that $a_{i+1} \le a_i \le n - i$. First, ignore the two leading as; this leaves $n - 1$ as in the word. Keep a count of the number of bs to the left of each a. Reverse their order. The resulting $(n-1)$-tuple $c_1 c_2 \ldots c_{n-1}$ has the property that $c_{i+1} \le c_i \le n - i$.

For example, consider the binary word $aabaabbab$ in Example 7.7:

$$
\begin{array}{ll}
\text{Binary word:} & aa\,baabbab \\
\text{Delete the two leading } a\text{s:} & \cancel{a}\cancel{a}baabbab \\
\text{Count the } b\text{s to the left of each } a: & 11\quad 3 \\
\text{Reverse the order:} & 31\quad 1
\end{array}
$$

The triplet $c_1 c_2 c_3 = 311$ has the property that $c_{i+1} \le c_i \le n - i$.

To reverse the process, consider the triplet 221:

Triplet:	2 21
Reverse the order:	1 22
Place an *a* below each:	*a aa*
Using the counts, place *b*s to the left of each *a*:	*babaa*
Insert the two leading *a*s:	*aababaa*
Append two *b*s (to get four *b*s):	*aababaabb*

Thus the binary word corresponding to 221 is *aababaabb*.

Mountain Ranges Revisited

There is an obvious bijection between the set of binary words in Example 7.7 and the set of mountain ranges in Example 6.2. To see this, first drop the leading *a*. Then replace each *a* with an upstroke and each *b* with a downstroke. Clearly, this process is reversible.

For example, Figure 7.11 shows the binary words and the corresponding mountain ranges for $0 \leq n \leq 3$.

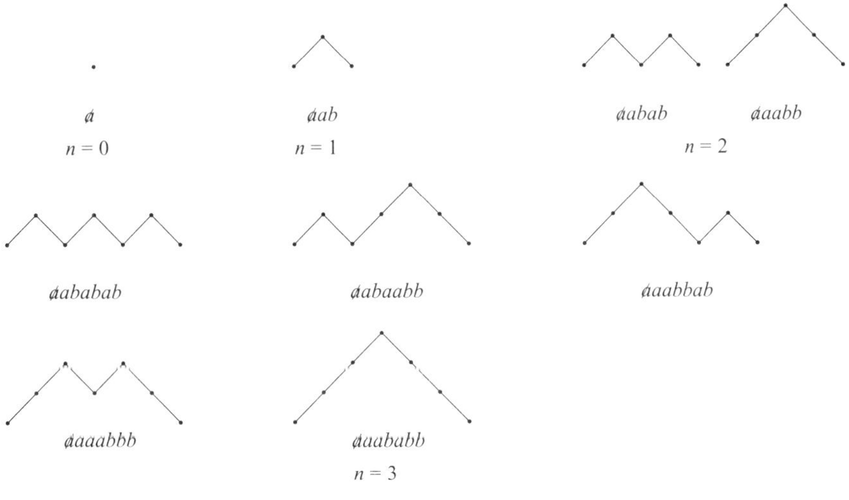

Figure 7.11

This algorithm also implies an obvious correspondence between the set of binary words and the set of linear paths of the particle in Example 7.1. See, for example, Figure 7.12.

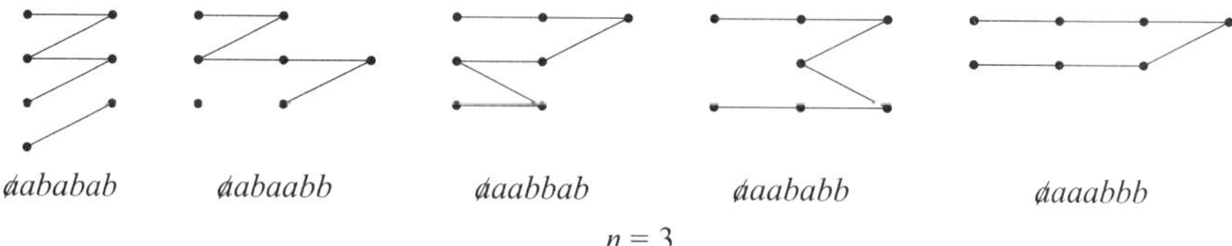

Figure 7.12

Multisets

Before I present the next application of Catalan numbers to permutations, I introduce the concept of a multiset. Unlike an ordinary set, a *multiset* is a set with repeated elements. For example, $\{a, a, b, b, b\}$ and $\{a, a, a, b, b, b, c, c, c, c\}$ are multisets. Using the exponential notation, these sets are often written as $\{a^2, b^3\}$ and $\{a^3, b^3, c^4\}$, respectively. It is interesting to note that combinations with repetitions are multisets.

We are now ready to present the example of multisets.

Example 7.8 Find the number of permutations $a_1 a_2 \ldots a_{2n}$ of the multiset $\{1^2, 2^2, \ldots, n^2\}$ such that:

- The first occurrences of the integers $1, 2, \ldots, n$ are in ascending order.
- There is no subsequence of the form *abab*, where $a \neq b$.

Solution   Table 7.7 shows the possible permutations of the multiset for $0 \leq n \leq 4$.

Table 7.7

n	Permutations					Count
0	λ					1
1	11					1
2	1221	1122				2
3	123321	122331	122133	112332	112233	5
4	12344321	12334421	12332441	12332144	12234431	14
	12233441	12233144	12213443	12213344	11234432	
	11233442	11233244	11223443	11223344		

■

Example 5.6 Revisited

Next we exhibit a bijection between the set of permutations in Example 7.8 and the set of well-formed sequences in Example 5.6.

To this end, consider, for example, the well-formed sequence (()())() with four pairs of parentheses. Count the number of left parentheses, from left to right, and assign the count as the label for each left parenthesis and the corresponding right parenthesis:

$$(\; (\;) \; (\;) \;) \; (\;)$$
$$1 \; 2 \; 2 \; 3 \; 3 \; 1 \; 4 \; 4$$

The resulting sequence 12233144 is a valid permutaion of the multiset $\{1^2, 2^2, 3^3, 4^2\}$; see Table 7.7. Since the sequence of parentheses is well formed, this procedure guarantees that the resulting sequence of labels will not contain a subsequence of the form *abab*.

This process is clearly reversible. For example, consider the permutation 1221 3344. Replace the first occurrence of each integer with a left parenthesis and the second occurrence with a right parenthesis:

$$1 \; 2 \; 2 \; 1 \; 3 \; 3 \; 4 \; 4$$
$$(\; (\;) \;) \; (\;) \; (\;)$$

The resulting sequence (())()() is a well-formed one.

Example 6.12 Revisited

Although it follows from the preceding discussion that there is a bijection between the set of permutations in Example 7.8 and the set of binary words in Example 6.12, we take a direct approach to exhibit the same relationship.

Consider, for example, the binary word 10110100. Count the numbers of 1s from left to right. Save each count as the label for the corresponding 1:

$$\begin{array}{ll} \text{Binary word:} & 1 \; 0 \; 1 \; 1 \; 0 \; 1 \; 0 \; 0 \\ \text{Count:} & 1 \quad\;\; 2 \; 3 \quad\; 4 \end{array}$$

If a 0 follows a 1, then keep the same label for 0 as the one for 1:

$$\begin{array}{ll} \text{Binary word:} & 1 \; 0 \; 1 \; 1 \; 0 \; 1 \; 0 \; 0 \\ \text{Count:} & 1 \; 1 \; 2 \; 3 \; 3 \; 4 \; 4 \end{array}$$

If there is another 0 following a 1, then subtract a 1 from its count to find the label of the new 0; if this label is already used twice, then keep subtracting a 1 until we get a label not used more than once:

$$\begin{array}{ll} \text{Binary word:} & 1 \; 0 \; 1 \; 1 \; 0 \; 1 \; 0 \; 0 \\ \text{Count:} & 1 \; 1 \; 2 \; 3 \; 3 \; 4 \; 4 \; \not{3} \\ & 1 \; 1 \; 2 \; 3 \; 3 \; 4 \; 4 \; 2 \end{array}$$

This results in the valid permutation 11233442; see Table 7.7.

Likewise, the binary word 11011000 yields 12234431:

$$\begin{array}{ll}\text{Binary word:} & 1\ 1\ 0\ 1\ 1\ 0\ 0\ 0 \\ \text{Count:} & 1\ 2\ 2\ 3\ 4\ 4\ 3\ \not{2} \\ & 1\ 1\ 2\ 3\ 3\ 4\ 4\ 1\end{array}$$

To see that this algorithm is reversible, consider the permutation 11233442 of the multiset $\{1^2, 2^2, 3^2, 4^2\}$. Beginning with the second element in this sequence, subtract each from the previous element. If a subtraction results in a negative number, then record a 0:

$$\begin{array}{ll}\text{Permutation:} & 1\ 1\ 2\ 3\ 3\ 4\ 4\ 2 \\ \text{Save the leading 1:} & 1 \\ \text{Subtract:} & 1\ 0\ 1\ 1\ 0\ 1\ 0\ 0\end{array}$$

(Here, we have recorded $2-4$ as 0.) This results in the valid binary word 10110100, as expected.

Likewise, the permutation 11234432 yields the binary word 10111000:

$$\begin{array}{ll}\text{Permutation:} & 1\ 1\ 2\ 3\ 4\ 4\ 3\ 2 \\ \text{Save the leading 1:} & 1 \\ \text{Subtract:} & 1\ 0\ 1\ 1\ 1\ 0\ 0\ 0\end{array}$$

(Here, we have recorded $3-4$ and $2-3$ as 0s.)

Noncrossing Matchings

The next example is fascinating in its own right.

Example 7.9 Find the number of ways n semicircles can be arranged on a horizontal line in such a way that no two semicircles intersect.

Solution   Figure 7.13 shows the various possibilities for $0 \leq n \leq 4$.

Interestingly, this example can be translated into Example 6.12 or 6.10 by a simple transformation and vice versa. For example, consider the arrangement and scan it from left to right. Enter a 1 each time the left side of a semicircle is encountered and a 0 when the right side of a semicircle is encountered. Thus, the array yields the binary word 110100. Clearly, this process is reversible. Consequently, there is a bijection between the set of such semicircles and the set of binary words in Example 6.12.

As an added bonus, the semicircle problem can be considerd exactly the same as Catalan's parenthesization problem. To see this, consider once again the array . Cut each semicircle into two halves by a vertical cut. Think of the left half as a left parenthesis and the right half as a right parenthesis. This procedure yields the sequence (( )( )). This procedure is also clearly reversible, establishing a bijection

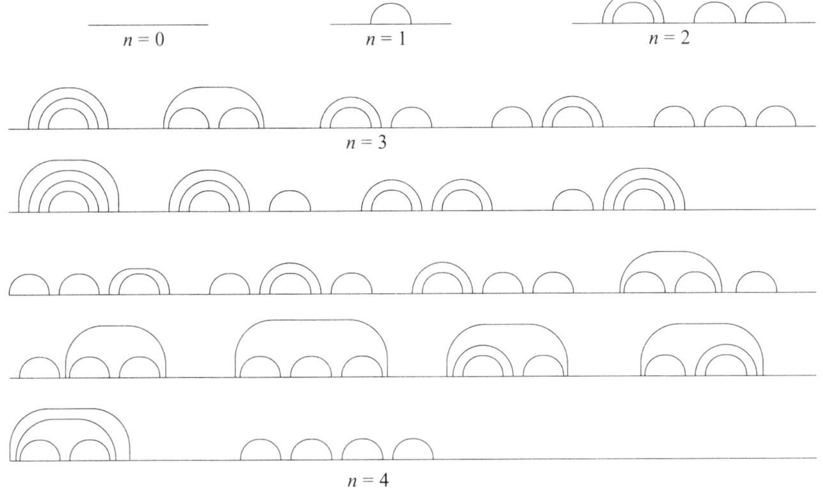

Figure 7.13

between the set of semicircles in Example 7.8 and the set of well-formed sequences of parentheses. Thus, the desired answer is once again the Catalan number C_n.

Notice that the semicircle problem can be restated as follows: Find the number of ways $2n$ points on a horizontal line can be joined by nonintersecting arcs, where each arc connects two points and lies above the line, and $n \geq 0$.

See Figure 7.14, where $n = 3$

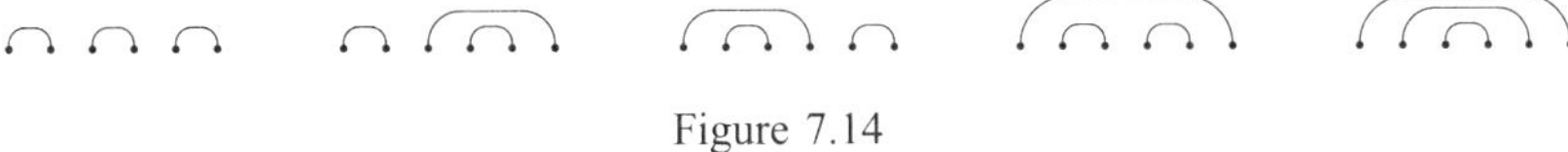

Figure 7.14

Example 7.8 Revisited

Because there is a one-to-one correspondence between the set of semicircles in Example 7.9 and the set of well-formed sequences of parentheses (or the binary words in Example 6.12), it follows that there is a bijection between the set of semicircles and the set of permutations in Example 7.8.

For example, consider the array of four semicircles. Scan it from left to right. Enter a count for each left end of a semicircle and keep the same count for its right end:

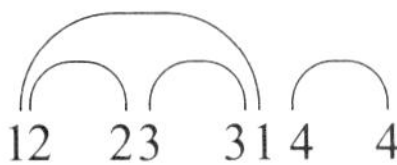

The resulting sequence 12233144 is a permutation of the multiset $\{1^2, 2^2, 3^2, 4^2\}$ with the desired properties.

Likewise, the arrangement

generates the permutation 12332441.

Clearly, the algorithm is reversible.

The following example is directly related to the semicircle problem in Example 7.9.

Example 7.10 Find the number of ways n disjoint circles (or rings), not necessarily of the same size, can be arranged on a line.

Solution   Figure 7.15 shows the various ways n circles can be placed as desired, where $0 \leq n \leq 4$. Notice that the circles can be concentric, and two or more circles can be placed inside a larger circle.

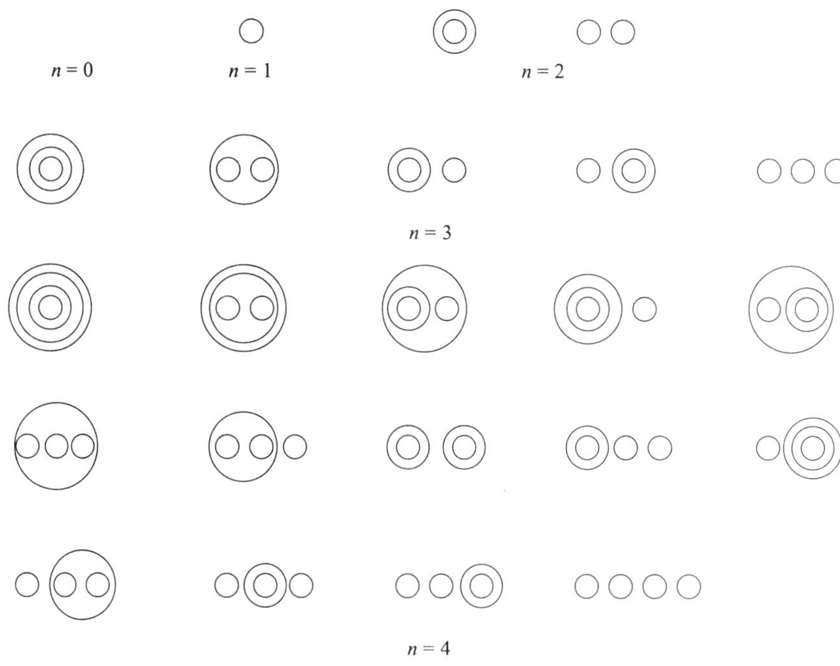

Figure 7.15

Clearly, there is a bijection between the set of semicircles in Example 7.8 and the set of circles in this example. It is achieved by completing the semicircles into circles and by taking the upper halves of the circles in Example 7.9.

An Application to Computer Science

Next we pursue an interesting application of Catalan numbers to computer science, based on an example by S. Even of Israel Institute of Technology. But first we define some basic vocabulary from data structures in computer science.

A *stack*, a very important abstract data structure in computer science, is used for storing data. Just as a stack of trays in a cafeteria or a stack of dinner plates at a buffet table can grow or shrink in size, so does a stack. We can add new trays only to the top of the stack and remove trays from the top. Likewise, a new item can be inserted only at the top of a stack and an item can be deleted only from the top. See Figure 7.16.

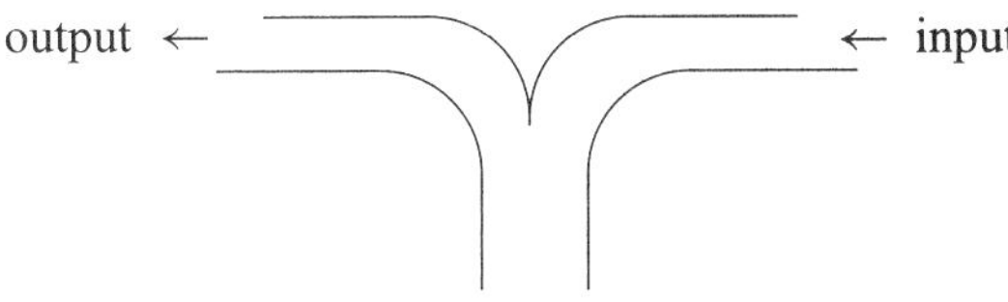

Figure 7.16 A Stack

In computer science, arrays are used to implement the stack structure; see Figure 7.17. The last item inserted is the first item that can be deleted; accordingly, unlike a queue, which is a *first-in-first-out* (FIFO) data structure, a stack is a *last-in-first-out* (LIFO) data structure.

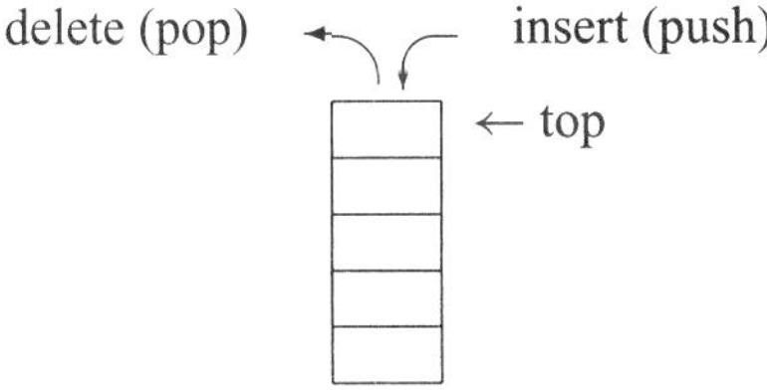

Figure 7.17 A Stack

Insertion and deletion are the two important stack operations; they are known as *pushing* and *popping*, respectively. Both take place at the top of the stack. Notice that we can push an item onto an empty stack but cannot pop from an empty stack. We can always pop an item from a nonempty stack.

We are now ready to present the application.

Example 7.11 Using a stack, determine the number of different permutations of the ordered list of integers $1, 2, 3, \ldots, n$.

Solution   To get a better understanding of the problem, let us consider a few special cases.

Case 1 Let $n = 1$. So the input list consists of one element, namely, $n = 1$.

We can push 1 onto the empty stack (see Figure 7.18). Now, pop it. The stack is now empty, and there are no more input values. So we are done and the output is the permutation 1.

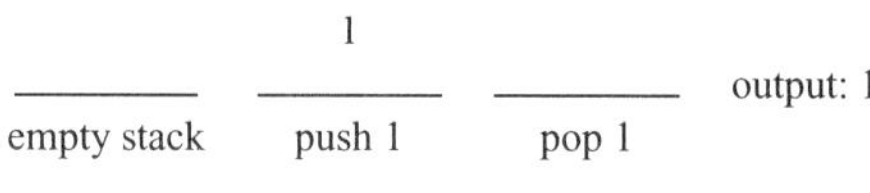

Figure 7.18

Case 2 Let $n = 2$, so the input list is $\{1,2\}$.

There are two pemutations of these two elements. They are 12 and 21, both can be obtained using a stack:

- To obtain the permutation 12, first push 1 onto the stack; then pop it. Now, push 2 onto the stack and then pop it. The resulting output is 12 (see Figure 7.19):

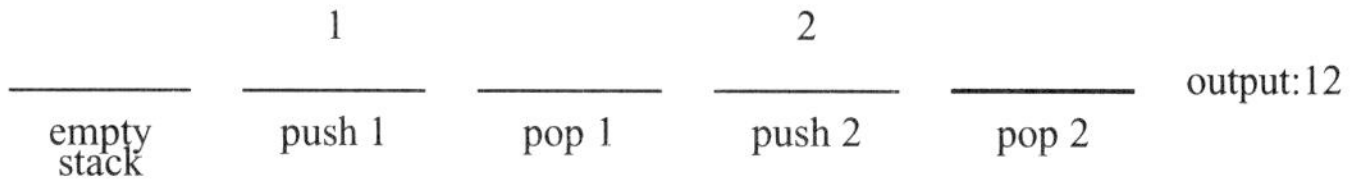

Figure 7.19

- To obtain the permutation 21, first push 1 onto the stack; then push 2; then pop 2; and then pop 1. The resulting output is 21 (see Figure 7.20).

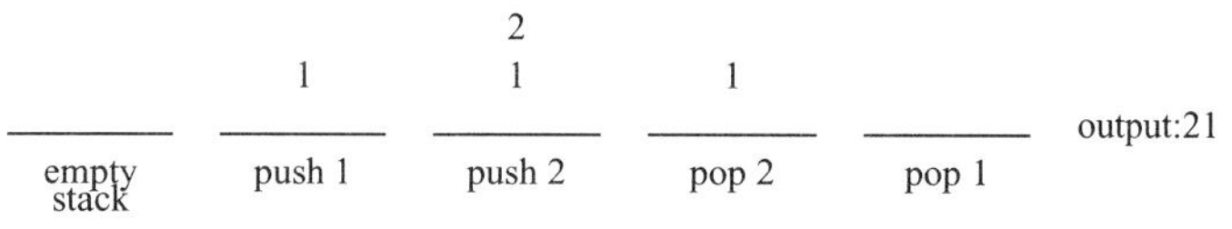

Figure 7.20

Case 3 Let $n = 3$, so the input list is $\{1,2,3\}$.

Although there are $3! = 6$ different permutations of three elements, only five of them can be realized using a stack. Those permutations and the corresponding steps are displayed in Figure 7.21.

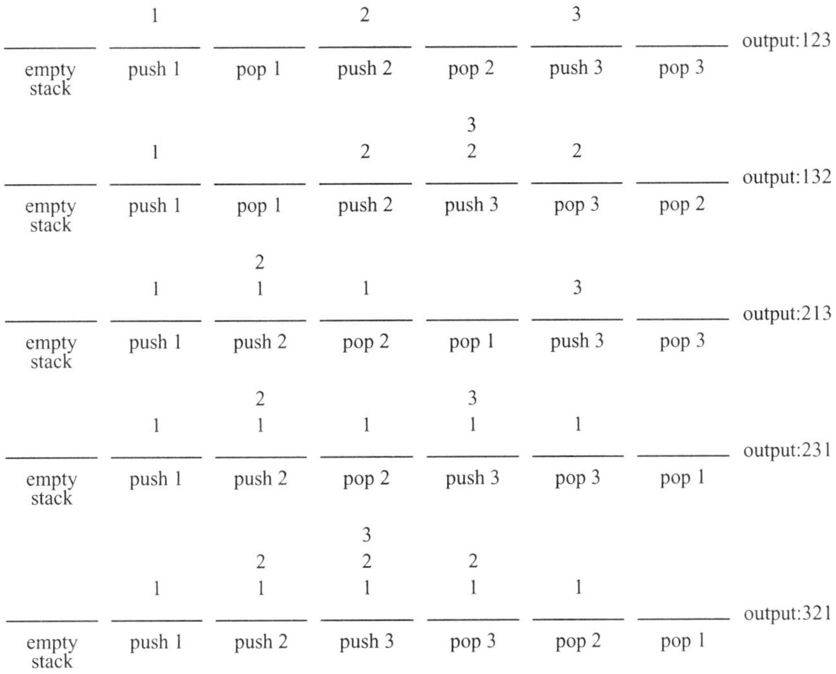

	1		2		3		
empty stack	push 1	pop 1	push 2	pop 2	push 3	pop 3	output:123

	1		2	3 2	2		
empty stack	push 1	pop 1	push 2	push 3	pop 3	pop 2	output:132

	1	2 1	1		3		
empty stack	push 1	push 2	pop 2	pop 1	push 3	pop 3	output:213

	1	2 1	1	3 1	1		
empty stack	push 1	push 2	pop 2	push 3	pop 3	pop 1	output:231

	1	2 1	3 2 1	2 1	1		
empty stack	push 1	push 2	push 3	pop 3	pop 2	pop 1	output:321

Figure 7.21

Notice that the permutation 312 cannot be generated using a stack. To see this, to have 3 as the first element in the output, the first element pushed off the stack must be 3. Before this push, the stack must look as in Figure 7.22.

Now pop 3 (see Figure 7.23). From this stack, there is no way of popping 1 without first popping the 2. Consequently, the permutation 312 cannot be generated using the stack structure.

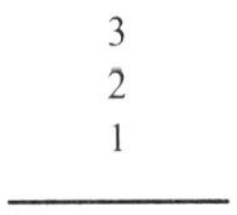

3
2
1   2
1

Figure 7.22 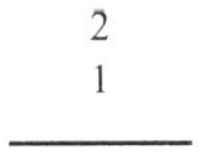   Figure 7.23

Case 4 Let $n = 4$, so the input list is $\{1,2,3,4\}$.

Of the $4! = 24$ possible permutations, only 14 of them can be realized using a stack. They are listed in increasing order of the position of 1 in Table 7.8. (This observation will come in handy later when we develop a formula for the number of permutations that can be generated.)

Table 7.8

1234	2134	2314	2341
1243	2143	3214	2431
1324			3241
1342			3421
1432			4321

Notice, for example, that the permutation 1423 cannot be generated using the stack structure. This is so since to have 1 as the first element in the output, first we must push 1 and then pop it. Then we must push 2, 3, and 4 onto the stack. Now, we pop 4. Thus far we have 14-- in the output. Now, 3 is sitting at the top of the stack; so we cannot pop 2. Consequently, we cannot produce the permutation 1423 using the stack.

Let P_n denote the number of possible permutations of the elements $1, 2, \ldots, n$ using the stack structure. Let i denote the position of 1 in a valid permuation generated by the stack, where $1 \leq i \leq n$. Then P_{i-1} denotes the number of valid permutations that have $i - 1$ elements to the left of i and P_{n-i-1} denotes the number of valid permutations that have $n - i - 1$ elements to the right of i, where we define $P_0 = 1$.

Using the multiplication and addition principles, this yields the recurrence relation

$$P_n = \sum_{i=1}^{n} P_{i-1} P_{n-i-1}$$
$$= P_0 P_{n-1} + P_1 P_{n-2} + \cdots + P_{n-1} P_0$$

where $P_0 = 1$. This is exactly the same as Segner's recurrence relation (5.6) encountered in Chapter 5. Consequently, $P_n = C_n$ for every $n \geq 0$. ∎

It is interesting to note that this permutation problem is practically the same as Catalan's parenthesization problem and the binary word problem in Example 6.12. To see this similarity, consider the permutation 2431 in Table 7.8. To generate it, first we push 1, push 2, pop 2, push 3, push 4, pop 4, pop 3, and then pop 1. So we performed the stack operations push and pop in the following order to generate it:

push, push, pop, push, push, pop, pop, pop

This order of operations outputs a unique permutation. Replace each push with a left parenthesis (and a pop with a right parenthesis). This results in the well-formed sequence (()(())) with four pairs of parentheses.

On the other hand, suppose we replace each push with a 1 and each pop with a 0. This transformation yields the binary word 11011000 with four 1s and four 0s, where the number of 1s in each partial substring from the far left to right is greater than or equal to the number of 0s in it.

Because both procedures are reversible, they establish a bijection between the set of permutations and the set of well-formed sequences, and the set of permutations and the set of binary words with the added restriction.

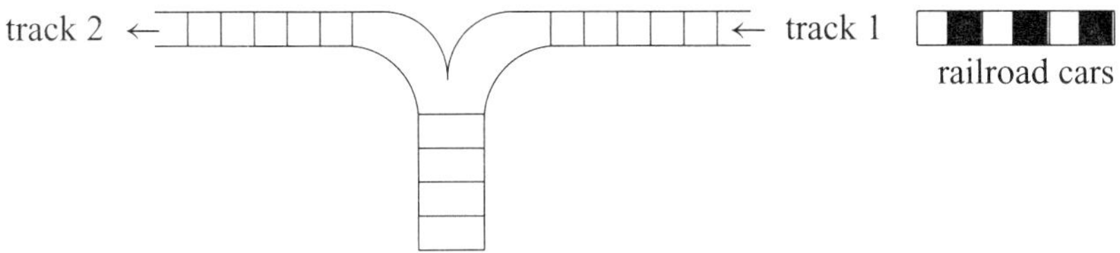

Figure 7.24 Railroad Tracks

Before we leave this permutation problem, we examine a practical application. Suppose there are n railroad cars on track 1 (see Figure 7.24) and they need to be moved to track 2 using a special track for the switching. Each car on track 1 must be moved to the switching track before it can be moved to track 2, and only one car at a time can be moved from the switching track to track 2. To find the number of ways the n cars can be moved from track 1 to track 2 by using the switching track is exactly the same as the above permutation problem.

The next two examples also deal with restricted permutations.

Example 7.12 Find the number of permutations $a_1 a_2 \ldots a_n$ of the set $\{1, 2, \ldots, n\}$ such that $i < j < k$ does *not* imply that $a_i > a_j > a_k$; that is, the conditions $i < j < k$ and $a_i > a_j > a_k$ do *not* hold simultaneously.

Solution   Table 7.9 shows the possible permutations for $0 \le n \le 4$.

Table 7.9

n	Permutations					Count
0	λ					1
1	1					1
2	12	21				2
3	123	132	213	231	312	5
4	1234	1243	1324	1342	1423	14
	2134	2143	2314	2341	2413	
	3124	3142	3412	4123		

Such restricted permutations were originally conceived by J. M. Hammersley of Oxford University. However, the first combinatorial proof that the number of such permuations is C_n was given in 1978 by Douglas G. Rogers of Crozley Green, United Kingdom.

Example 7.13 Find the number of permutations $a_1 a_2 \ldots a_n$ of the set $\{1, 2, \ldots, n\}$ such that $i < j < k$ does *not* imply that $a_j > a_k > a_i$; that is, the conditions $i < j < k$ and $a_j > a_k > a_i$ do *not* hold simultaneously.

Solution   Table 7.10 shows the possible permutations for $0 \le n \le 4$.

Table 7.10

n	Permutations					Count
0	λ					1
1	1					1
2	12	21				2
3	123	132	213	231	321	5
4	1234	1243	1324	1342	1432	14
	2134	2143	2314	2341	2431	
	3214	3241	3421	4321		

Example 7.8 Revisited

Interestingly, Examples 7.8 and 7.13 are closely related. To see this, suppose we delete the first occurrences of the integers 1 through n in the permutation $a_1 a_2 \ldots a_{2n}$ in Example 7.8. The resulting sequence is a permutation of the set $\{1, 2, \cdots, n\}$ satisfying the conditions in Example 7.13.

For example, consider the permutation 12234431; see Table 7.7. Deleting the first occurrences of the integers yields $\cancel{1}2\cancel{2}2\cancel{3}\cancel{4}431 = 2431$, which is a valid permutation in Example 7.13; see Table 7.9. Likewise, the permutation 12213443 yields $\cancel{1}2\cancel{2}1\cancel{3}\cancel{4}43 = 2143$; see Tables 7.7 and 7.10.

Staircase Tessellations

Next we turn to a delightful occurrence of Catalan numbers in tessellations.

Example 7.14 Find the number of tilings of the staircase shape that can be made with n rectangles.

Solution   Figure 7.25 shows the possible tilings for $0 \le n \le 4$.

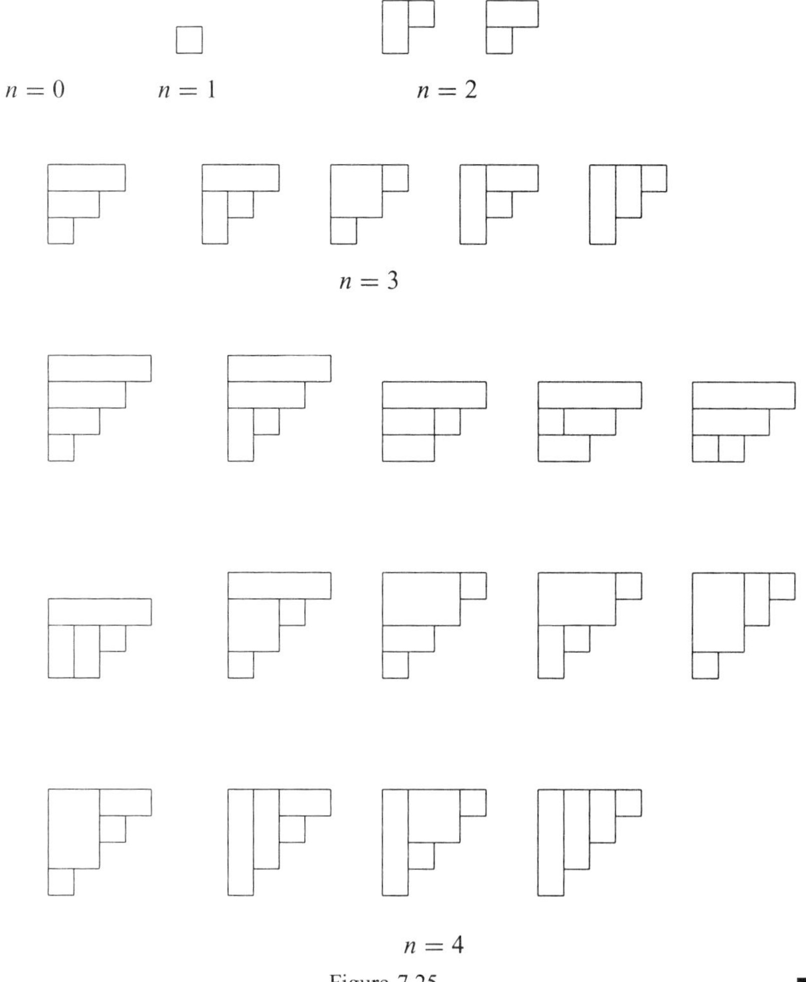

Figure 7.25

The next four examples deal with pairing points on a horizontal line subject to certain restrictions.

Nonnested Matchings

Example 7.15 Find the number of ways $2n$ distinct points on a horizontal line can be connected by n arcs above the line, such that each arc connects exactly two points and no arc lies entirely inside another arc.

Solution   Figure 7.26 shows the various possible pairings, where $0 \le n \le 4$.

Figure 7.26

Example 6.12 Revisited

There is a bijection between the set of arcs in Example 7.15 and the set of binary words in Example 6.12. To see this, consider the arcs in Figure 7.27. Label each left end of an arc with a 1 and its corresponding right end with a 0; see Figure 7.28. The resulting binary word 11010010 has the desired property: The number of 1s in each substring, when scanned from left to right, is greater than or equal to the number of 0s in it.

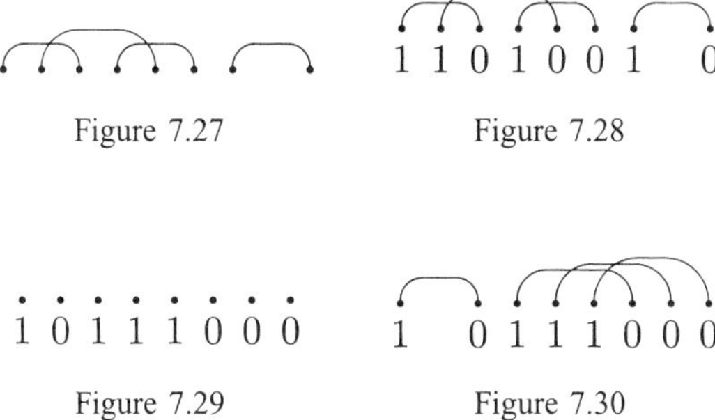

Figure 7.27   Figure 7.28

Figure 7.29   Figure 7.30

Clearly, this algorithm is reversible. To see this, consider the binary word 10111000. First, mark eight points in a row and label them 1, 0, 1, 1, 1, 0, 0, and 0 in that order; see Figure 7.29. Now, draw an arc from each 1 to the next (available) 0 to its right; see Figure 7.30. The resulting arcs have the desired properties.

Example 7.16 Find the number of ways n distinct points on a horizontal line can be connected by noncrossing arcs that lie above the line in such a way that if two arcs share a common endpoint A, then A must be a left endpoint of both arcs.

Solution Figure 7.31 shows the various possible configurations for $0 \leq n \leq 4$.

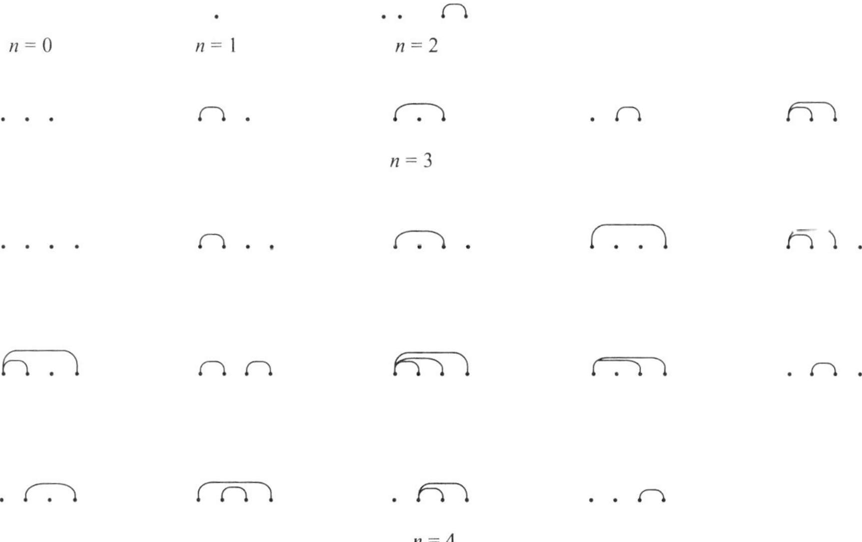

Figure 7.31

Example 7.17 Find the number of ways n distinct points on a horizontal line can be connected by noncrossing arcs above the line such that no two adjacent points are connected and the right endpoints of the arcs are all distinct.

Solution   Figure 7.32 shows the possible ways n points can be connected, as desired, where $1 \le n \le 5$.

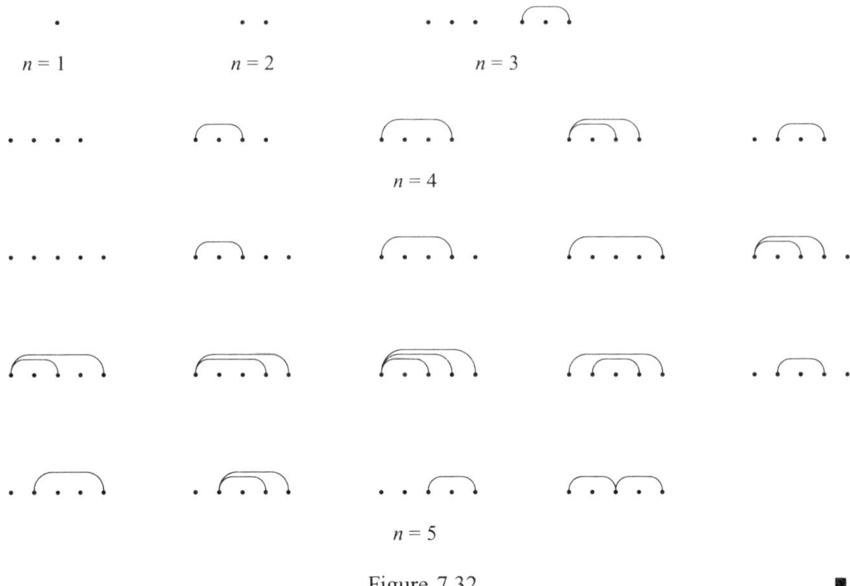

Figure 7.32

The next example is also due to E. Deutsch.

Example 7.18 Find the number of ways any number of distinct points on a horizontal line can be connected by nonintersecting arcs above the line in such a way that:
- The sum of the number of arcs and the number of isolated points is $n - 1$; and
- No isolated point lies below an arc.

Solution   Figure 7.33 shows the various possibilities for $1 \le n \le 4$. ■

The next example shows that Catalan numbers can occur in many other situations also. It was proposed as a problem in 1973 by Carlitz. The solution is based on one given by him in 1974.

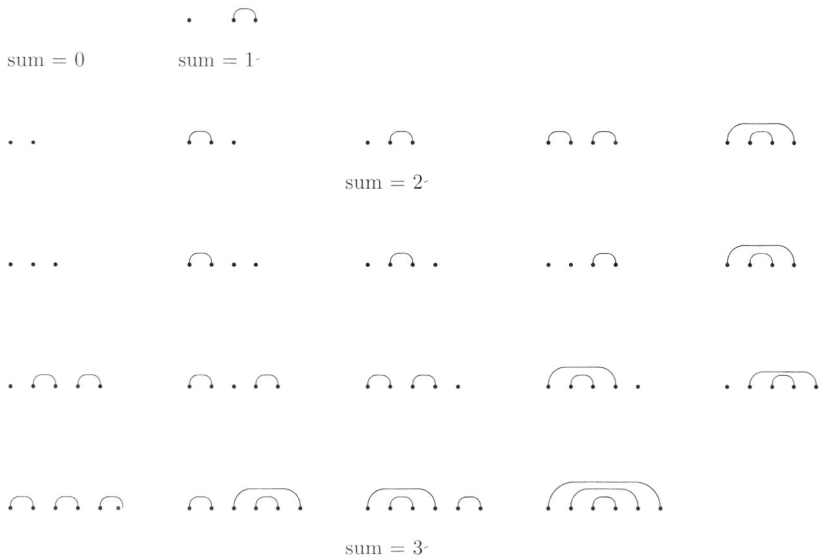

sum = 0
sum = 1
sum = 2
sum = 3

Figure 7.33 ■

Example 7.19 Let

$$u_n = \frac{\alpha^{n+1} - \beta^{n+1}}{\alpha - \beta}$$

where $\alpha + \beta = \alpha\beta = z$. Find the coefficient $C(n, k)$ such that

$$z^n = \sum_{k=1}^{n} C(n, k) u_{n-k+1}$$

where $n \geq k \geq 1$.

Solution Notice that

$$\begin{aligned} z &= \sum_{k=1}^{1} C(1, k) u_{2-k} \\ &= C(1, 1) u_1 \\ &= C(1, 1)(\alpha + \beta) \\ &= C(1, 1) z \end{aligned}$$

So $C(1, 1) = 1$ and hence $z = u_1$.

Similarly, z^2, z^3, z^4, etc. can be expressed as linear combinations of u_1, u_2, u_3, etc.:

$$z = u_1$$

$$z^2 = u_2 + u_1$$

$$z^3 = u_3 + 2u_2 + 2u_1$$

$$z^4 = u_4 + 3u_3 + 5u_2 + 5u_1$$

$$z^5 = u_5 + 4u_4 + 9u_3 + 14u_2 + 14u_1$$

Notice that

$$(\alpha + \beta)u_k = (\alpha + \beta) \cdot \frac{\alpha^{k+1} - \beta^{k+1}}{\alpha - \beta}$$

$$= \frac{\alpha^{k+2} - \beta^{k+2}}{\alpha - \beta} + \alpha\beta \cdot \frac{\alpha^k \beta^k}{\alpha - \beta}$$

$$= u_{k+1} + (\alpha + \beta)u_{k-1}$$

$$= u_{k+1} + u_k + (\alpha + \beta)u_{k-2}$$

$$\vdots$$

Continuing like this, we get:

$$(\alpha + \beta)u_k = u_{k+1} + u_k + u_{k-1} + \cdots + u_2 + (\alpha + \beta)u_0$$

But $u_0 = 1$ and $\alpha + \beta = z = u_1$. So

$$(\alpha + \beta)u_k = u_1 + u_2 + \cdots + u_k + u_{k+1}$$

Thus

$$z^{n+1} = (\alpha + \beta)z^n$$

$$\sum_{j=1}^{n+1} C(n+1, j)u_{n-j+2} = \sum_{k=1}^{n} C(n, k)(\alpha + \beta)u_{n-k+1}$$

$$= \sum_{k=1}^{n} C(n,k) \sum_{j=1}^{n-k+2} u_j$$

$$= \sum_{j=1}^{n+1} u_j \sum_{k=1}^{n-j+2} C(n,k)$$

$$= \sum_{j=1}^{n+1} u_{n-j+2} \sum_{k=1}^{j} C(n,k)$$

Consequently, $C(n,k)$ satifies the recurrence relation

$$C(n+1,j) = \sum_{k=1}^{j} C(n,k) \tag{7.4}$$

In particular, $C(2,1) = C(1,1) = 1$ and

$$C(2,2) = \sum_{k=1}^{2} C(1,k)$$

$$= C(1,1) + C(1,2)$$

$$= 1 + 0$$

$$= 1$$

If we know the values of $C(2,1)$ and $C(2,2)$, we can use the recurrence relation (7.4) to compute $C(n,j)$, where $n \geq 3 : C(n,j)$ is the sum of the first j values of $C(n-1,j)$. The various values of $C(n,j)$ are summarized in Figure 7.34.

```
1
1   1
1   2   2
1   3   5   5
1   4   9   14   14
1   5   14  28   42   42
```

Figure 7.34

The recurrence relation (7.4) can also be written as

$$C(n,j) = \sum_{k=1}^{j-1} C(n-1,k) + C(n-1,j)$$

$$= C(n,j-1) + C(n-1,j)$$

Consequently, each element $C(n,j)$ in the triangular array can also be obtained by adding its northern and western neighbors. For example, $42 = 28 + 14$; see Figure 7.34.

8

Trees and Catalan Numbers

This chapter presents some delightful occurrences of Catalan numbers found in the study of trees in graph theory. In the interest of clarity and precision, we begin the discussion by defining some basic graph-theoretic vocabulary.

Basic Graph Terminology

A *graph G* is a nonempty set of points, called *vertices* or *nodes*, together with line segments or arcs, called *edges*, joining them. Let V denote the set of vertices and E the set of edges. The graph G can be considered the ordered pair $(V, E) : G = (V, E)$. An edge between vertices v and w is denoted by $v - w$ or $\{v, w\}$.

For example, the graph in Figure 8.1 has four vertices—A, B, C, and D— and seven edges. It has edges with the same vertices; such edges are *parallel edges*. The graph in Figure 8.2 has an edge at Q that starts from and ends at the same vertex; such an edge is a *loop*.

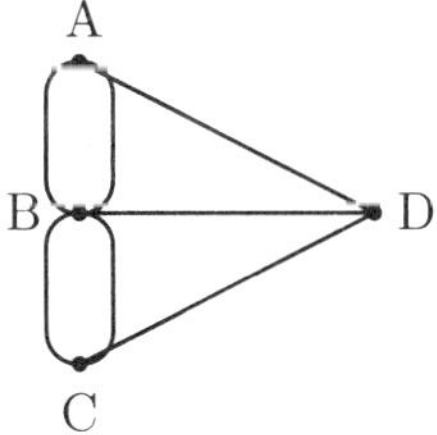

Figure 8.1

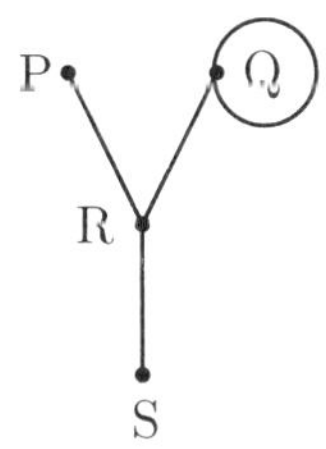

Figure 8.2

A loop-free graph that contains no parallel edges is a *simple graph*. A simple graph with m vertices that contains exactly one edge between every two distinct

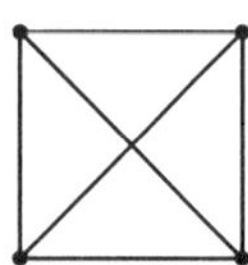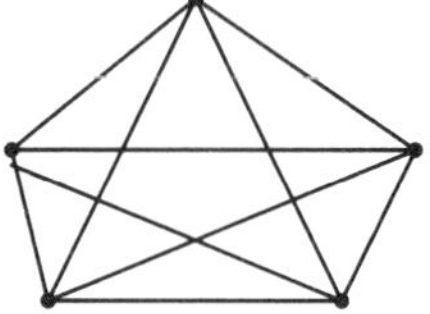

Figure 8.3

vertices is a *complete graph*, denoted by K_m. Figure 8.3 shows the complete graphs K_4 and K_5.

Let $H = (V', E')$ be a graph such that $V' \subseteq V$ and $E' \subseteq E$. Then H is a *subgraph* of G. For example, the graph in Figure 8.4 is a subgraph of the graph in Figure 8.1.

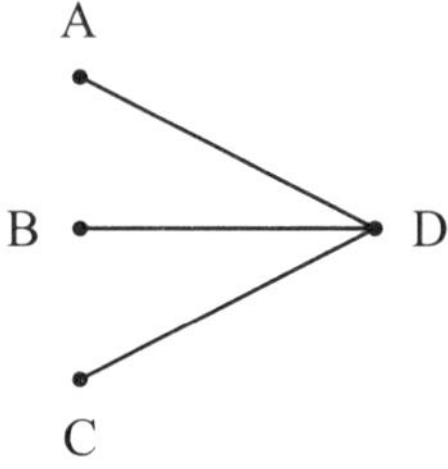

Figure 8.4

A *path* between two vertices v_0 and v_n is a sequence $v_0 - v_1 - v_2 - \cdots - v_n$ of vertices v_i and edges connecting them. The *length* of the path is n, the number of edges in it. A graph is *connected* if there is a path between every two distinct vertices.

For example, the graph in Figure 8.5 is connected. But the graph in Figure 8.6 is not connected; it contains an *isolated vertex*.

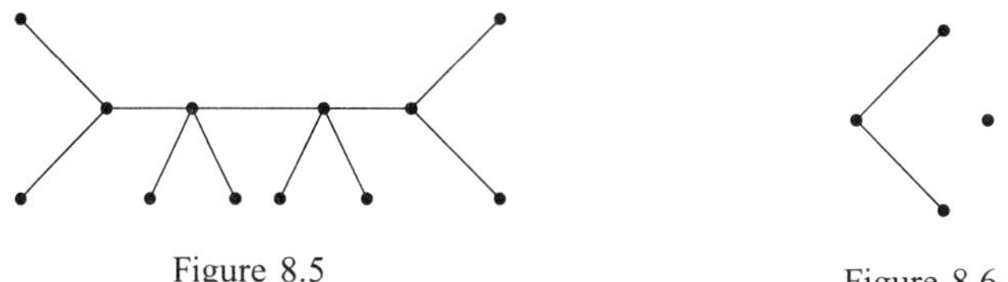

Figure 8.5

Figure 8.6

A *cycle* is a path with the same endpoints; it contains no repeated vertices. A graph is *acyclic* if it contains no cycles. A connected, acyclic graph is a *tree*. A connected graph with n vertices is a tree if and only if it has exactly $n - 1$ edges.[†] For example, the graph in Figure 8.7 is a tree, but the one in Figure 8.8 is not.

[†]   See T. Koshy, *Discrete Mathematics with Applications*, 2004.

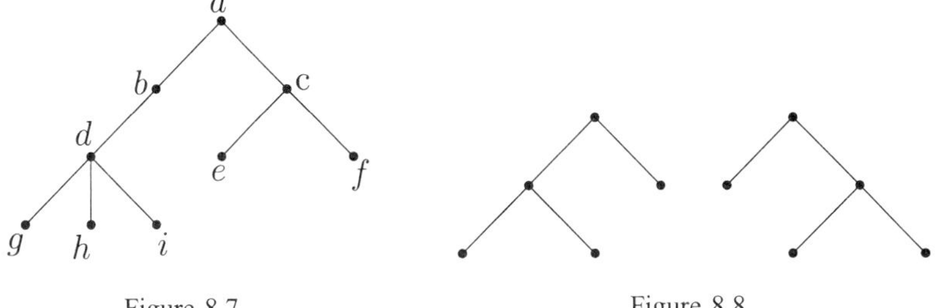

Figure 8.7   Figure 8.8

The tree in Figure 8.7 contains a specially designated vertex, called the *root*. It is a rooted tree, rooted at a. A *forest* is a set of trees; see Figure 8.8.

An Ordered Rooted Tree

An *ordered rooted tree* is a rooted tree in which the vertices at each level are ordered as the first, second, third, and so on. As ordered trees, the two trees in Figure 8.8 are different.

The basic terminology of (rooted) trees reflects that of a family tree. Let T be a tree with root v_0. Let $v_0 - v_1 - v_2 - \ldots - v_n$ be the path from v_0 to v_n. Then:

- v_{i-1} is the parent of v_i, where $i \geq 1$.
- v_i is a child of v_{i-1}, where $i \geq 1$.
- The vertices $v_0, v_1, v_2, \ldots, v_{n-1}$ are *ancestors* of v_n.
- The *descendants* of a vertex v are those vertices for which v is an ancestor.
- A vertex v with no children is a *leaf* or a *terminal vertex*.
- A vertex that is not a leaf is an *internal vertex*.
- The *degree* of a vertex is the number of edges meeting at the vertex.
- The *subtree* rooted at v consists of v, its descendants, and all its edges.

For example, consider the tree T in Figure 8.7. Vertex b is a parent of vertex d. Vertex d has three children: g, h, and i. Vertex h has three ancestors: d, b, and a. Vertex c has two descendants, e and f. Vertices g, h, and i are leaves. Vertices b, d, and c are internal vertices. The degree of vertex b is two and that of vertex d is four. The tree in Figure 8.9 is a subtree of T rooted at b.

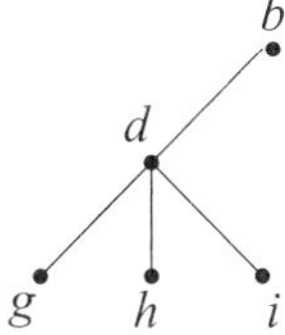

Figure 8.9

Binary Tree

An ordered rooted tree is a *binary tree* if each vertex has at most two children. For example, the tree in Figure 8.10 is a binary tree; its *left subtree* is the binary tree rooted at b (see Figure 8.11) and its *right subtree* is the binary tree rooted at d (see Figure 8.11). The tree in Figure 8.9 is not a binary tree.

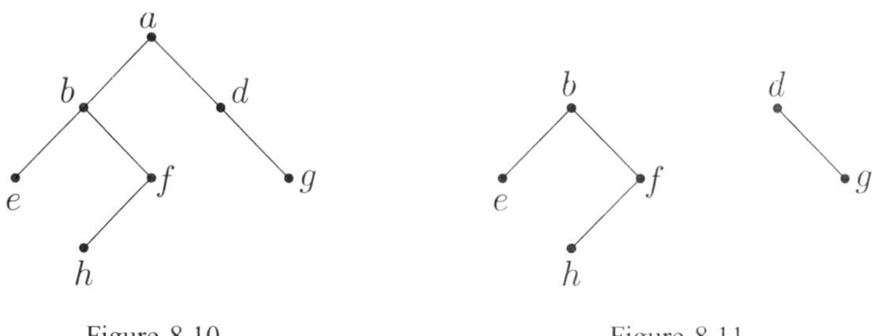

Figure 8.10   Figure 8.11

The next example shows the close relationship between binary trees and Catalan numbers.

Example 8.1   Find the number of binary trees with n vertices.

Solution   Figure 8.12 shows the various possible binary trees for $0 \le n \le 4$.

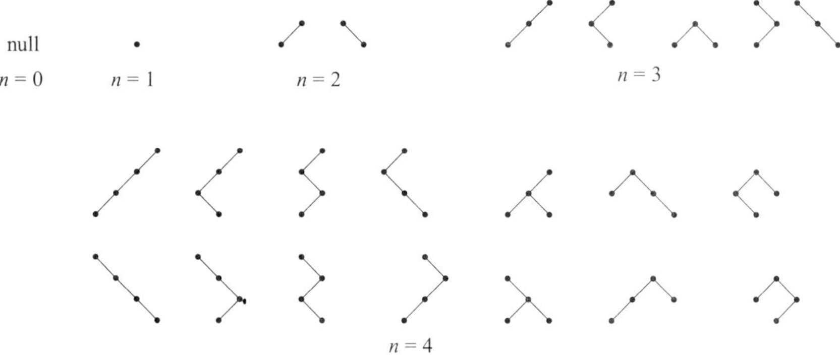

Figure 8.12 Binary Trees with n Vertices, where $0 \le n \le 4$ ■

Using these data, we conjecture that there are C_n binary trees with n vertices. The following theorem confirms our conjecture.

Theorem 8.1   The number of binary trees with n vertices is C_n.

Proof   Let b_n denote the number of binary trees with n vertices, where $n \ge 0$. Since the empty tree is the only binary tree with 0 vertices, $b_0 = 1$.

When $n = 1$, the tree is made up of the root, and there is exactly one such binary tree; so $b_1 = 1$.

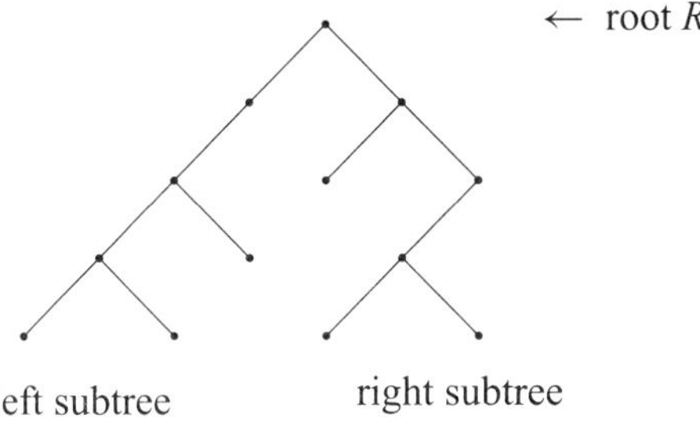

Figure 8.13

We proceed to find a recurrence relation satisfied by b_n. To this end, consider a binary tree T with n vertices; see Figure 8.13. The root of the tree has $n - 1$ descendants. Suppose it has i descendants on the left subtree and hence $n - i - 1$ descendants on the right subtree, where $0 \le i \le n - 1$. By definition, there are b_i binary trees with i vertices and b_{n-i-1} binary trees with $n - i - 1$ vertices. By the multiplication principle, there are $b_i b_{n-i-1}$ binary trees with i descendants on their left subtrees and hence $n - i - 1$ descendants on their right subtrees. Thus, by the addition principle, the total number of binary trees with n vertices is given by

$$b_n = \sum_{i=0}^{n-1} b_i b_{n-i-1}$$

$$= b_0 b_{n-1} + b_1 b_{n-2} + \cdots + b_{n-1} b_0$$

where $b_0 = 1$. This is exactly the same as Segner's recurrence relation with the same initial condition, so $b_n = C_n$.

Thus there are exactly C_n binary trees with n vertices (and hence $n - 1$ edges).

For example, there are exactly

$$C_{10} = \frac{1}{11} \binom{20}{10}$$

$$= 16{,}796$$

binary trees with 10 vertices.

Next, we investigate the number of a special class of binary trees. However, first a definition.

Full Binary Trees

A *full binary tree* is a binary tree in which each internal vertex has exactly two children. For example, the trees in Figure 8.8 are full binary trees, whereas the binary tree in Figure 8.10 is not.

Before we investigate the number of full binary trees with n vertices, let us make an important observation: A full binary tree with l leaves has $n = 2l - 1$ vertices.* Consequently, the number of vertices in a full binary tree is an odd positive integer.

Example 8.2 Conjecture the number of full binary trees with n vertices.

Solution   Figure 8.14 shows the various possible full binary trees with One, Three, Five, and Seven vertices.

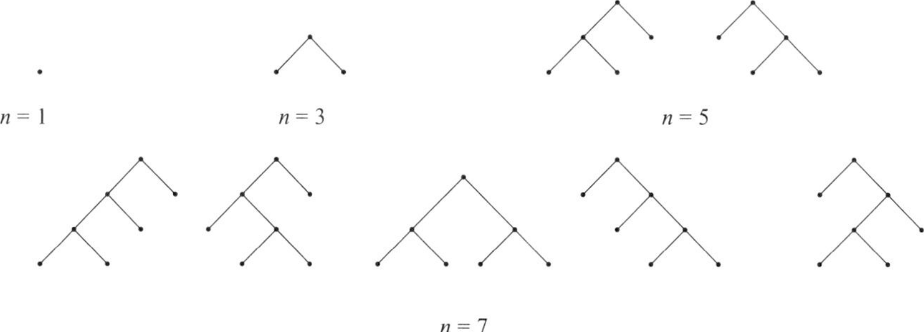

Figure 8.14 Full Binary Trees

Let b_n denote the number of full binary trees with n vertices. Then:

$$b_1 = 1 = C_{\frac{1-1}{2}}$$

$$b_3 = 1 = C_{\frac{3-1}{2}}$$

$$b_5 = 2 = C_{\frac{5-1}{2}}$$

$$b_7 = 5 = C_{\frac{7-1}{2}}$$

So, we conjecture that $b_n = C_{\frac{n-1}{2}}$, where n is odd.   ∎

The next theorem confirms this conjecture by setting up a suitable one-to-one correspondence.

Theorem 8.2 The number of full binary trees with n vertices is $C_{\frac{n-1}{2}}$, where n is odd.

---

* See T. Koshy, *Discrete Mathematics with Applications*, Elsevier, Burlington, Massachusetts, 2004.

Proof   Let F denote the set of full binary trees with n vertices and B the set of binary trees with $\frac{n-1}{2}$ vertices. We establish a bijection between the sets F and B.

Let T be a full binary tree with n vertices; it has $l = \frac{n+1}{2}$ leaves (see Figure 8.15). Delete all its leaves; the resulting tree is a binary tree T' with $n - l = n - \frac{n+1}{2} = \frac{n-1}{2}$ vertices (see Figure 8.16). This procedure clearly produces different binary trees corresponding to different full binary trees.

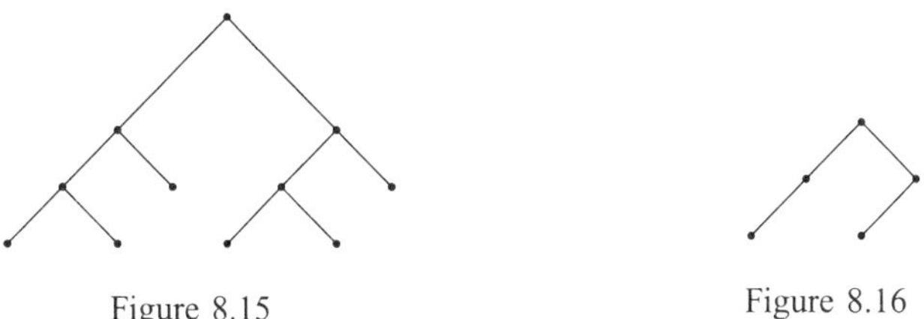

Figure 8.15   Figure 8.16

On the other hand, consider a binary tree T' (see Figure 8.17). Now, attach two leaves to each leaf of T' and one leaf to each internal vertex with exactly one child (see Figure 8.18). The resulting tree T is clearly a full binary tree.

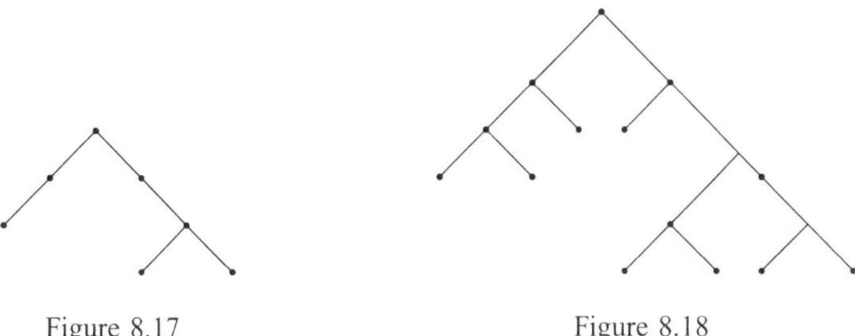

Figure 8.17   Figure 8.18

It remains to show that if T' has v vertices, then T has $2v + 1$ vertices. To this end, we shall use the fact that the number of leaves l in a binary tree equals one more than the number of internal vertices i_2 with exactly two children. (This fact can be established by induction on l.) So $l - i_2 = 1$.

Let l denote the number of leaves in T', i_1 the number of internal vertices with exactly one child, i_2 the number of internal vertices with exactly two children, and v the number of vertices. Let L, I_1, I_2, and V denote the corresponding numbers for the full binary tree T. Then:

$$v = l + i_1 + i_2$$

$$L = i_1 + 2l$$

$$I_1 = 0$$

$$I_2 = v$$

$$\begin{aligned}
V &= I_2 + I_1 + L \\
&= 2l + v + i_1 \\
&= 3l + 2i_1 + i_2 \\
&= 2(l + i_1 + i_2) + (l - i_2) \\
&= 2v + (l - i_2) \\
&= 2v + 1
\end{aligned}$$

Thus, the full binary tree has $2v + 1$ vertices, as desired.

To sum up, the algorithm establishes a bijection between the set of full binary trees with n vertices and the set of binary trees with $\frac{n-1}{2}$ vertices. Thus:

Number of full binary trees with n vertices = number of binary trees

$$\text{with } \frac{n-1}{2} \text{ vertices}$$

$$= C_{\frac{n-1}{2}}$$

$$= \frac{2}{n+1} \binom{n-1}{\frac{n-1}{2}} \qquad \blacksquare$$

For instance, there are

$$C_{11} = \frac{2}{24} \binom{22}{11}$$

$$= 58{,}786$$

full binary trees with 23 vertices.

Example 6.10 Revisited

We exhibit a bijection between the set of full binary trees with $2n + 1$ vertices and the set of sequences of n 1s and $n - 1$s in Example 6.10, using the following recursive *preorder tree traversal algorithm* of a binary tree:[†]

> Visit the root.
> Traverse the left subtree in preorder.
> Traverse the right subtree in preorder.

[†]   See T. Koshy, *Discrete Mathematics with Applications*, 2004.

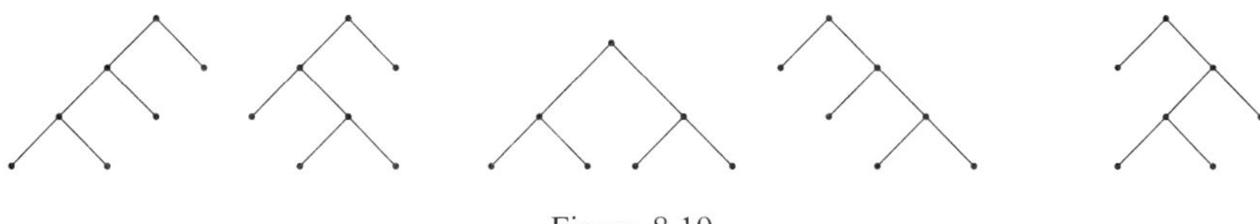

Figure 8.19

Every time a left edge is traversed, we record a 1 and every time a right edge is traversed, we record a -1.

For example, consider the full binary trees in Figure 8.19. Using the preorder traversal, they yield the sequences $111 - 1 - 1 - 1, 11 - 11 - 1 - 1, 11 - 1 - 11 - 1, 1 - 11 - 11 - 1$, and $1 - 111 - 1 - 1$, respectively. These are valid sequences in Example 6.10.

Clearly, this algorithm is reversible. Thus, the desired bijection follows.

Now we introduce another family of binary trees.

Planted Trivalent Binary Trees

A binary tree is *planted trivalent* if the degree of its root is one and that of each internal vertex is three. For example, the trees in Figure 8.20 are planted trivalent binary trees.

Figure 8.20 Planted Trivalent Binary Trees

Planted trivalent binary trees, studied by English mathematician Arthur Cayley, are often drawn upward for convenience. This convention will come in handy shortly.

By deleting the root of a planted trivalent binary tree, we get an ordinary binary tree; by attaching a new root at the existing root of a binary tree, we get a planted trivalent binary tree. Thus there is a bijection between the set of planted trivalent binary trees with n vertices and the set of binary trees with $n - 1$ vertices. Thus:

$$\begin{pmatrix} \text{Number of planted trivalent} \\ \text{binary trees with n vertices} \end{pmatrix} = \text{number of binary trees with } n - 1 \text{ vertices}$$

$$= C_{n-1}$$

$$= \frac{1}{n}\binom{2n - 2}{n - 1}$$

Arthur Cayley (1821–1895) was born in Richmond, England. At fourteen, he entered King's College, London. His teachers, recognizing his superb mathematical talents, encouraged him to be a mathematician.

At seventeen, Cayley entered Trinity College, Cambridge, where he was rated in a class by himself, "above the first."By age twenty-five, he had published twenty-five articles, the first one at age twenty.

In 1846, he left his position at Cambridge to study law and became a successful lawyer. While practicing law, he published more than 200 articles in mathematics. Dissatisfied with law, Cayley rejoined the faculty at Cambridge University in 1863. He pursued his mathematical interests until his death.

Polygonal Triangulations and Planted Trivalent Binary Trees

Earlier we established that there is a bijection between the set of triangulations of an n-gon and the set of well-formed sequences of parentheses with n pairs, each containing C_n elements.

Interestingly, there is also a one-to-one correspondence between polygonal triangulations and planted trivalent binary trees. We demonstrate this delightful relationship using the hexagonal triangulation in Figure 5.7. Place a dot next to each label, below the base, and inside each triangle. (Strategically locate them so as to get an aesthetically pleasing diagram at the end of this procedure; see

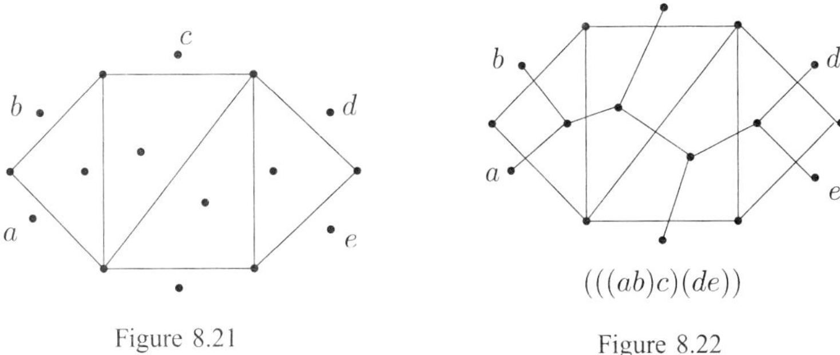

Figure 8.21

Figure 8.22

Figure 8.21). Join every two dots if they are separated by exactly one side of a triangle. Figure 8.22 shows the ensuing diagram. The resulting graph is a planted trivalent binary tree with five leaves; see Figure 8.23. It is fascinating to observe how the development of the planted trivalent binary tree parallels the building up of the infix expression $(((ab)c)(de))$. The planted trivalent binary tree can be redrawn to a normal-looking and aesthetically more interesting one, as in Figure 8.24.

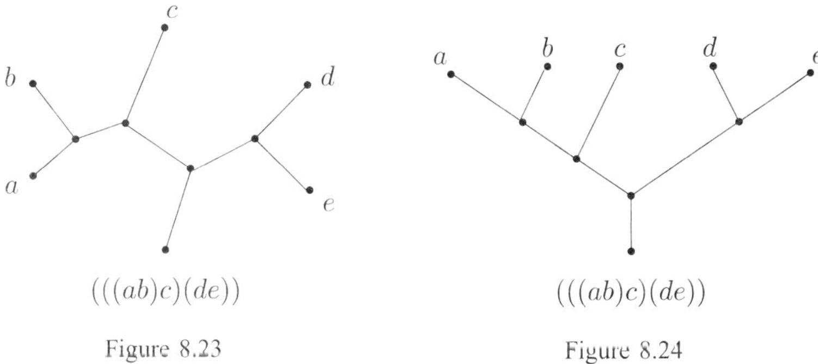

Figure 8.23

Figure 8.24

More generally, this algorithm constructs a planted trivalent binary tree with $n-1$ leaves corresponding to every triangulation of a convex n-gon, where $n \geq 3$.

Can we reverse this process? In other words, does there exist a polygonal dissection into triangles corresponding to each planted trivalent binary tree? Fortunately, the answer is yes, and we illustrate the reverse of the algorithm.

To this end, consider the planted trivalent binary tree in Figure 8.25; it has five leaves, labeled a through e. Place a dot between every two leaves. Join the dots by line segments in such a way that each line segment crosses an edge of the planted trivalent binary tree exactly once. This results in Figure 8.26. Deleting the original planted trivalent binary tree yields a unique triangulation of a hexagon, as Figure 8.27 shows.

Obviously, this technique applied to any planted trivalent binary tree with $n-1$ leaves yields a unique triangulation of a convex n-gon, where $n \geq 3$.

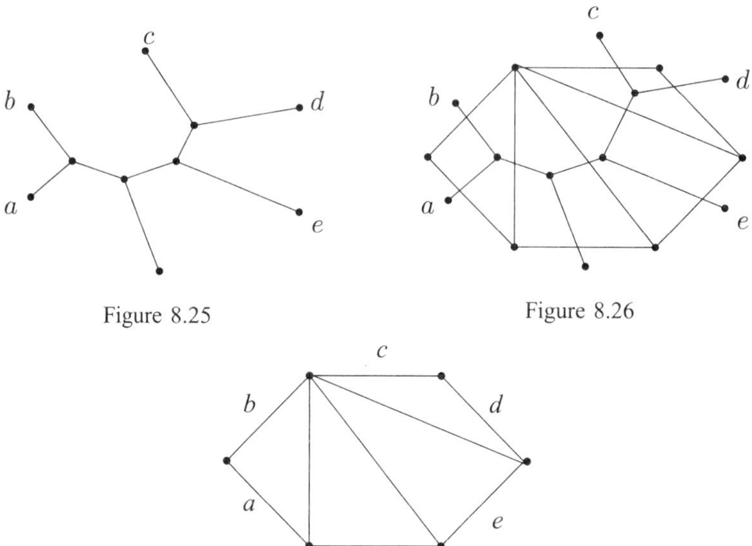

Figure 8.25

Figure 8.26

Figure 8.27 A Hexagonal Triangulation

Thus there is a bijection between the set of triangulations of a convex n-gon and the set of planted trivalent binary trees with $n-1$ leaves. Consequently, we have:

$$\begin{pmatrix} \text{Number of triangulations} \\ \text{of a convex } n\text{-gon} \end{pmatrix} = \begin{pmatrix} \text{number of planted trivalent} \\ \text{binary trees with } n-1 \text{ leaves} \end{pmatrix}$$

$$= C_n$$

where $n \geq 3$.

Thus, it follows from our discussions that:

$$\begin{pmatrix} \text{Number of triangulations} \\ \text{of a convex } n\text{-gon} \end{pmatrix} = \begin{pmatrix} \text{number of well-formed sequences} \\ \text{of parentheses with } n \text{ pairs} \end{pmatrix}$$

$$= \begin{pmatrix} \text{number of trivalent binary trees} \\ \text{with } n-1 \text{ leaves} \end{pmatrix}$$

$$= C_n$$

Figure 8.28 shows the various triangular dissections of a convex n-gon, and the corresponding trivalent binary trees and correctly parenthesized expressions with $n-1$ operands, where $n \geq 3$.

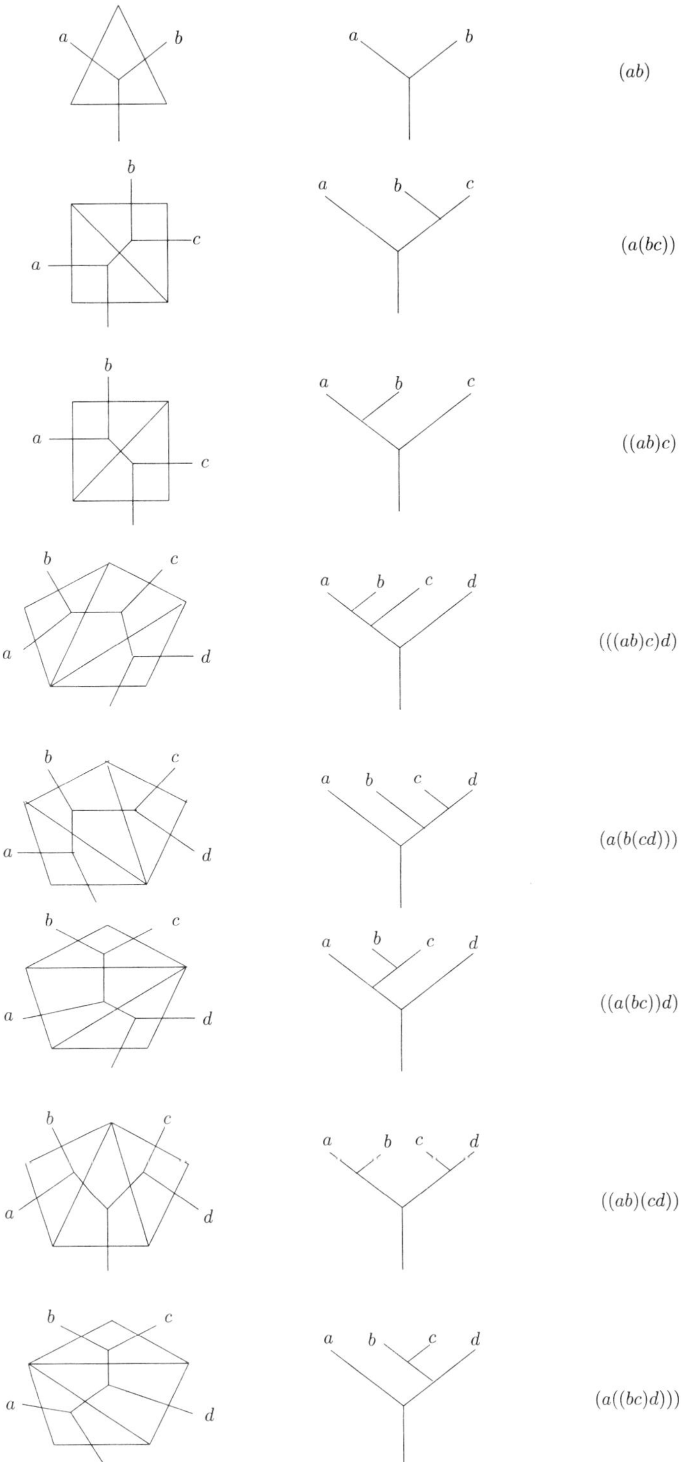

Figure 8.28

Triangulations, Parenthesized Expressions, and Binary Numbers

Consider the polygonal disection and the infix expression $(((ab)c)(de))$ in Figure 8.22; the expression contains exactly four pairs of parentheses.

For convenience, let us call the binary operation multiplication. Suppose we drop all right parentheses. Then we get the expression $(((abc(de$. Although the right parentheses are missing, the expression makes sense. Let's see why. As we scan the expression from left to right, by inserting a right parenthesis every time we encounter two consecutive operands, we can recover the original parenthesized expression.

Suppose we replace each left parenthesis with a 1 and an operand with a 0. Then the left-parenthesized expression $(((abc(de$ yields a unique binary number 111000100.

Consequently, the dissection in Figure 8.22 and the corresponding planted trivalent binary tree in Figure 8.24 can be uniquely represented by the binary number 111000100.

Clearly, this algorithm can be applied to any convex n-gon and to the corresponding planted trivalent binary tree and the left-parenthesized expression.

Triangulations and Prefix Expressions

Interestingly, the left-parenthesized expression $(((abc(de$ should remind us of the prefix expression $* * *abc * de$. Since this technique can be generalized to any convex n-gon, it follows that there is a bijection between the set of triangulations of a convex n-gon and the set of prefix expressions consisting of n binary operators (or $n+1$ binary operands). Consequently, there are exactly C_n valid prefix expressions consisting of n binary operands, where $n \geq 2$.

For instance, there is exactly one prefix expression with one multiplication $*$: $*ab$; there are two prefix operations with two multiplications: $*ab * c$ and $* * abc$.

Binary Number of a Planted Trivalent Binary Tree

Returning to the binary number 111000100 of the left-parenthesized expression $(((abc(de$, it is worth noting that the binary number can be read directly from the trivalent binary tree in Figure 8.24. The algorithm generating it was developed by Polish mathematician Jan Lukasiewicz.

Lukasiewicz's Algorithm

Label each leaf with a 0 and each internal vertex with a 1. Think of a bug crawling up the trunk and around the entire tree, and then returning to the ground. See the

Jan Lukasiewicz (Wu-cash-AY-rich) (1878–1956), a Polish logician and philosopher, was born in Lvov. Son of a captain in the Austrian army, he studied mathematics and philosophy, and in 1902 earned his Ph.D. in philosophy from the University of Lvov. He taught there for the next five years. In 1915, he accepted an invitation to teach at the University of Warsaw. In 1919, he served as minister of Education in independent Poland, and returned to the university as professor from 1920 to 1939, serving twice as its rector. Exiled to Belgium, Lukasiewicz accepted a professorship at the Royal Irish Academy, Dublin, from 1946 until his death.

A "resourceful and imaginative scholar"and "a gifted and inspiring teacher,"he was one of the founding fathers of the Warsaw School of Logic, a member of several scientific societies, and recipient of numerous honors.

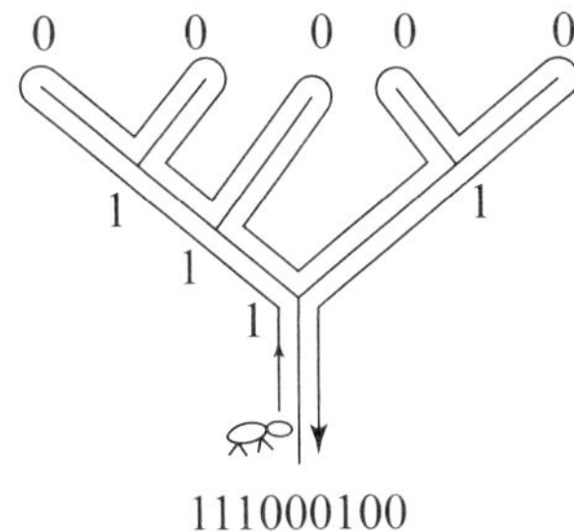

Figure 8.29

dotted path in Figure 8.29. Each time the bug visits a new vertex, it calls out that vertex's label.

This algorithm yields the binary number 111000100, the same number obtained directly from the left-parenthesized expression. This should not come as a surprise, because each internal vertex in the tree represents a left parenthesis, that is, a binary operator.

The planted trivalent binary trees in Figure 8.29 and their binary numbers are given in Figure 8.30.

A Delightful Byproduct

Consider the five possible parenthesized expressions and the corresponding binary numbers:

$$((ab)(cd)) \quad 1100100$$

$$(((ab)c)d) \quad 1110000$$

$$(a(b(cd))) \quad 1010100$$

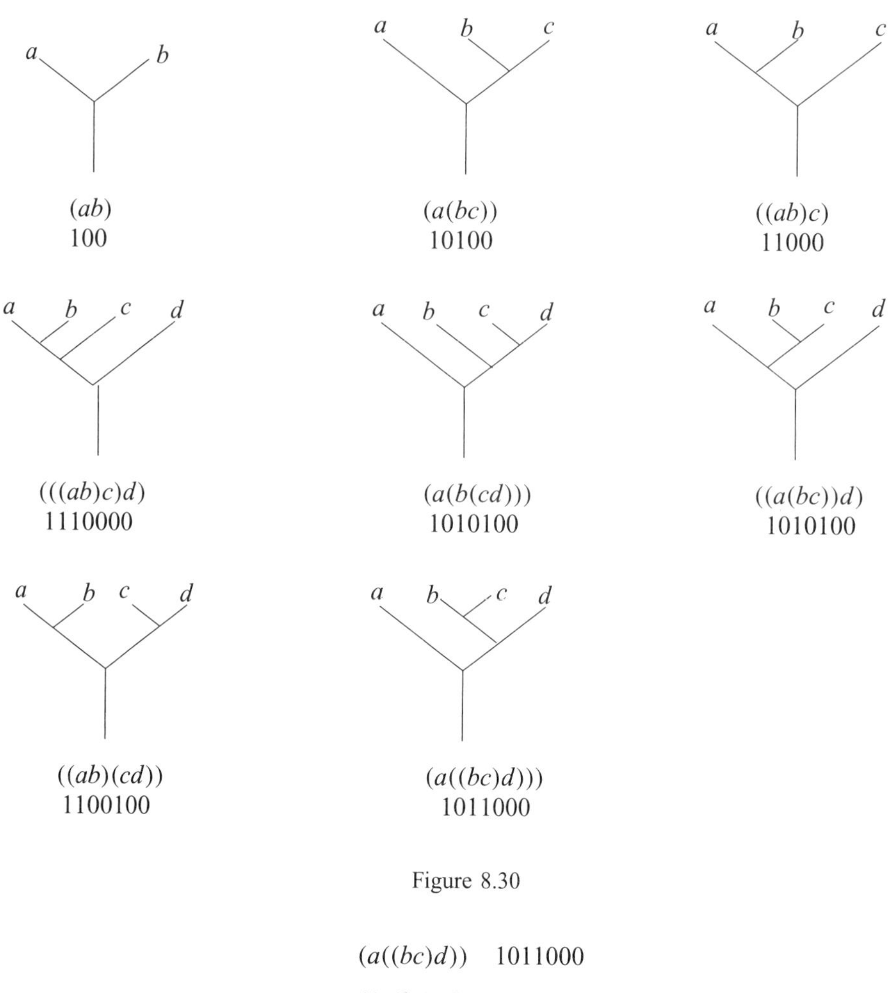

Figure 8.30

$$(a((bc)d)) \quad 1011000$$
$$((a(bc))d) \quad 1101000$$

Suppose we delete the rightmost 0 from each. Then we get five binary numbers:

$$110010$$
$$111000$$
$$101010$$
$$101100$$
$$110100$$

Each consists of three 1s and three 0s such that the number of 1s is $\geq$ the number of 0s in each substring, as we observed in Example 6.12.

This technique is reversible and can be extended to any convex n-gon (or parenthesized expression with n pairs of parentheses). It establishes a bijection between the set of triangulations of a convex n-gon and the set of binary words consisting n 1s and n 0s such that the number of 1s is $\geq$ the number of 0s in each partial sum. Consequently, there are exactly C_n such binary words.

Catalan Numbers and the Chessboard

We now turn to another occurrence of Catalan numbers, studied by Forder in 1961.

Example 8.3 Find the number of paths a rook can take from the upper left-hand corner to the lower right-hand corner on an $n \times n$ chessboard without crossing over the main southeast diagonal.

Solution Figure 8.31 shows the various possible paths for $1 \leq n \leq 5$.

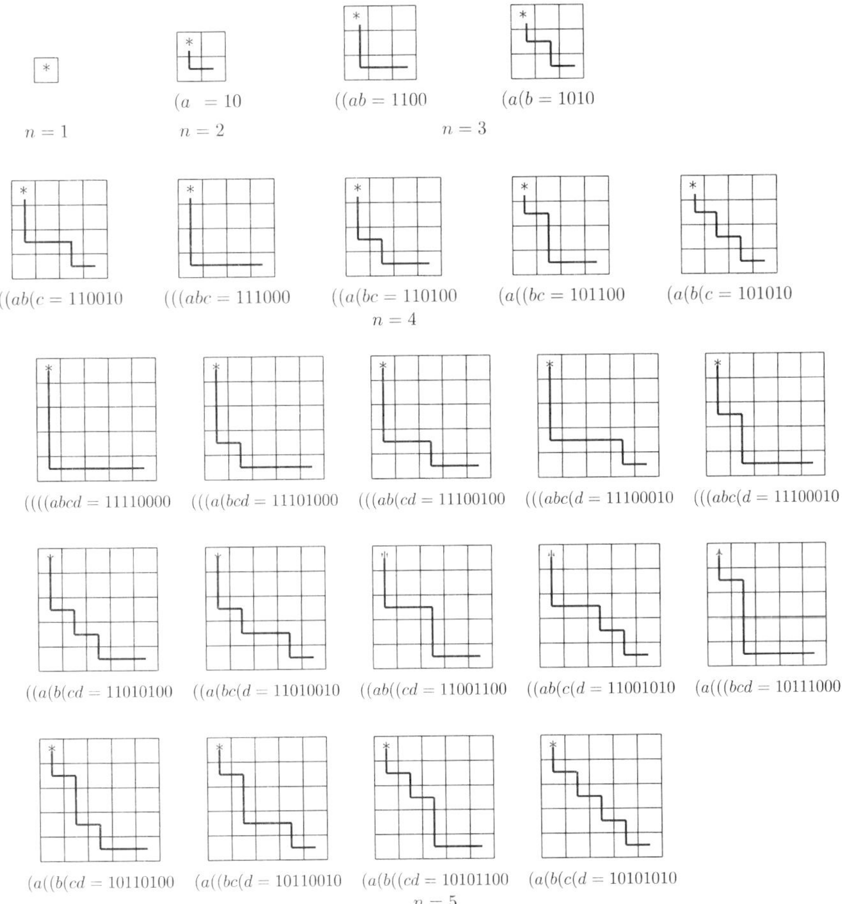

Figure 8.31

Interestingly, the left-parenthesized expressions and hence their corresponding binary numbers can be employed to identify the rook's various paths. For example, consider the five possible paths on a 4×4 chessboard. They can be represented by the expressions $((ab(c = 110010, (((abc = 111000, ((a(bc = 110100, (a((bc = 101100,$ and $(a(b(c = 101010,$ where a left parenthesis (or a 1) indicates moving down to the cell below and a letter (or a 0) indicates moving to the right cell.

More generally, this procedure establishes a bijection between the set of possible paths of the rook from the northwest corner to the southeast corner on an $n \times n$ chessboard and the set of binary words consisiting of n 1s and n 0s such that the number of 1s is $\geq$ the number of 0s in each substring (see Example 6.12). Accordingly, there are exactly C_n possible paths the rook can take.   ∎

An Added Dividend

The numbers of paths from the upper left-hand corner to any square on or below the main diagonal on an $n \times n$ chessboard can be employed to construct a Pascal-like triangular array, *Catalan's triangle*, as Forder did; see Table 8.1.

Table 8.1 Catalan's Triangle

n \ j	0	1	2	3	4	5	6
0	1						
1	1	1					
2	1	2	2				
3	1	3	5	5			
4	1	4	9	14	14		
5	1	5	14	28	42	42	
6	1	6	20	48	90	132	132

Catalan numbers

This triangular array has several interesting properties:

- Each row begins with a 1.
- Each row ends in a Catalan number.
- Each entry $C(n,j)$ in row n and column j can be obtained by adding the element to its left and the element just above it, where $n, j \geq 1$; that is, $C(n,j) = C(n,j-1) + C(n-1,j)$. For example, $14 = 5 + 9$; see the curly arrow in Table 8.1.
- The sum of the first r elements in row $n - 1$ equals the rth element in row n. For example, $1 + 4 + 9 + 14 = 28$.
- The last two elements in each row, except row 0, are equal.

Accordingly, each element $C(n,j)$ in Catalan's triangle can be defined recursively, as follows, where $n \geq 0$:

$$C(n,j) = \begin{cases} 1 & \text{if } j = 0 \\ n & \text{if } j = 1 \\ C(n,j-1) + C(n-1,j) & \text{if } 1 \leq j \leq n \\ 0 & \text{if } j > n \end{cases}$$

The recursive formula $C(n,j) = C(n,j-1) + C(n-1,j)$ makes sense because there are exactly two different ways for the rook to reach the square (n,j), through the square $(n,j-1)$ to its left or through the square $(n-1,j)$ above it. There are $C(n-1,j)$ paths to the square $(n,j-1)$ and $C(n-1,j)$ paths to the square $(n-1,j)$. It follows by the addition principle that $C(n,j) = C(n,j-1) + C(n-1,j)$, where $1 \leq j \leq n$.

In particular,

$$\begin{aligned} C(n,n) &= C(n,n-1) + C(n-1,n) \\ &= C(n,n-1) + 0 \\ &= C(n,n-1) \end{aligned}$$

Thus, the last two entries in each row n are equal, where $n \geq 1$. This is true since there is only one way for the rook to move to square (n,n), namely, through the square $(n,n-1)$.

An Explicit Formula for $C(n, j)$

There is an interesting explicit formula for $C(n,j)$:

$$C(n,j) = \frac{(n+j)!(n-j+1)}{(n+1)!j!} \tag{8.1}$$

$$= \frac{n-j+1}{n+1} \binom{n+j}{j} \tag{8.2}$$

where $n > j$.

For example,

$$\begin{aligned} C(5,3) &= \frac{8!(5-3+1)}{6!3!} \\ &= 28 \end{aligned}$$

as desired.

To confirm formula (8.1), it suffices to show that $\frac{(n+j)!(n-j+1)}{(n+1)!j!}$ satisfies the above initial conditions and the recurrence relation. To this end, notice that when $j = 0$:

$$\frac{(n+j)!(n-j+1)}{(n+1)!j!} = \frac{(n+0)!(n-0+1)}{(n+1)!0!}$$

$$= \frac{(n+1)!}{(n+1)!}$$

$$= 1$$

When $j = 1$:

$$\frac{(n+j)!(n-j+1)}{(n+1)!j!} = \frac{(n+1)!(n-1+1)}{(n+1)!1!}$$

$$= \frac{(n+1)!n}{(n+1)!}$$

$$= n$$

and

$$C(n,j-1) + C(n-1,j) = \frac{(n+j-1)!(n-j+2)}{(n+1)!(j-1)!} + \frac{(n+j-1)!(n-j)}{n!j!}$$

$$= \frac{(n+j-1)![(n-j+2)j + (n+1)(n-j)]}{(n+1)!j!}$$

$$= \frac{(n+j-1)!(n-j+1)(n+j)}{(n+1)!j!}$$

$$= \frac{(n+j)!(n-j+1)}{(n+1)!j!}$$

$$= C(n,j)$$

In particular,

$$C(n,n) = \frac{(n+n)!(n-n+1)}{(n+1)!n!}$$

$$= \frac{(2n)!}{(n+1)!n!}$$

$$= C_n$$

as expected. Thus $C(n,n) = C_n = C(n,n-1)$.

Ordered Rooted Trees and Catalan Numbers

The next example shows another close relationship between ordered rooted trees and Catalan numbers.

Example 8.4 Find the number of ordered rooted trees with n edges.

Solution   Figure 8.32 shows the various possible ordered rooted trees with $0 \leq n \leq 3$ edges.

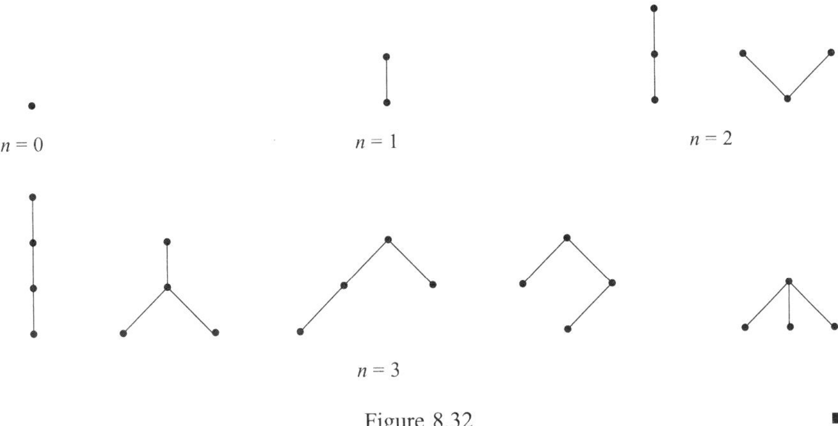

Figure 8.32

Using this example, we conjecture that there are C_n ordered rooted trees with n edges. The next theorem confirms this using a special one-to-one correpondence between the set of ordered rooted trees with n vertices and the set of planted trivalent binary trees with $n - 1$ vertices, developed by F. R. Bernhart of the Rochester Institute of Technology.

Before we illustrate Bernhart's algorithm, we need to demonstrate that every ordered rooted tree can be transformed into one with edges pointing either upward, to the right, or both. To this end, consider the trivalent binary tree with six leaves in Figure 8.33. At each vertex of degree three, the right edge is drawn as a horizontal edge and the left edge is drawn upward. See Figure 8.34.

Bernhart's Algorithm

Now prune the tree in Figure 8.34 using Bernhart's algorithm. To this end, compress each horizontal edge to the vertex on the left. If there is a vertex of degree three to the right of a vertical edge, it is merged with the vertex on the left. Save all vertical edges.

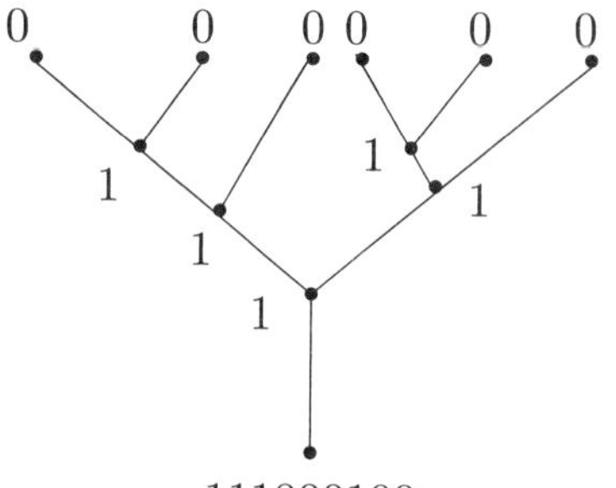

Figure 8.33 A Planted Trivalent
Binary Tree with 6 Leaves

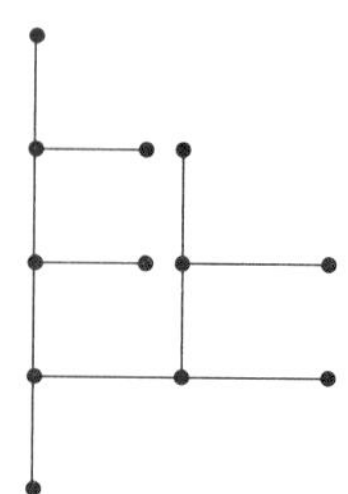

Figure 8.34 A Planted Trivalent
Binary Tree with Horizontal and
Vertical Edges

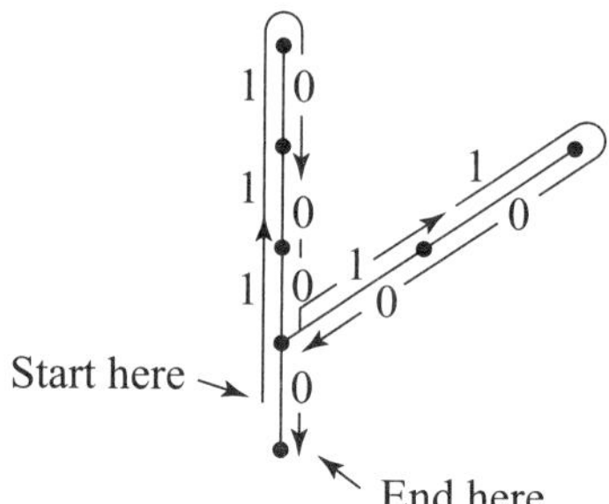

Figure 8.35 An Ordered Rooted Tree with Six Edges

Figure 8.35 shows the resulting tree; it is a rooted ordered tree with six edges (and seven vertices).

More generally, Bernhart's algorithm transforms a planted trivalent binary tree with n leaves into an ordered rooted tree with n edges.

On the other hand, consider an ordered rooted tree. In this case, every vertex of degree d can be expanded to an array of $d - 1$ vertices of degree three each. The resulting tree with horizontal and vertical edges is a planted trivalent binary tree.

For example, consider the ordered rooted tree in Figure 8.36. It can be transformed into the planted trivalent binary tree in Figure 8.37.

Thus there is a bijection between the two sets, so there are exactly C_n rooted trees with n edges (or $n + 1$ vertices). See Figures 8.32 and 8.38.

Accordingly, we have the following theorem.

Theorem 8.3 There are C_n ordered rooted trees with n edges.   ∎

An interesting observation: Notice that the binary number of the planted trivalent binary tree in Figure 8.34 is 1110001 1000. The same binary tree number can be obtained from Figure 8.36 by slightly altering the labeling. Suppose the bug starts crawling up the tree from the vertex above the root. Each time it crawls up an edge,

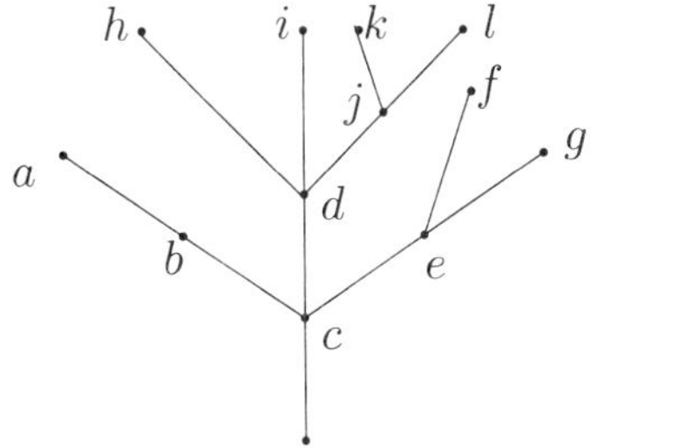

Figure 8.36 An Ordered Rooted Tree

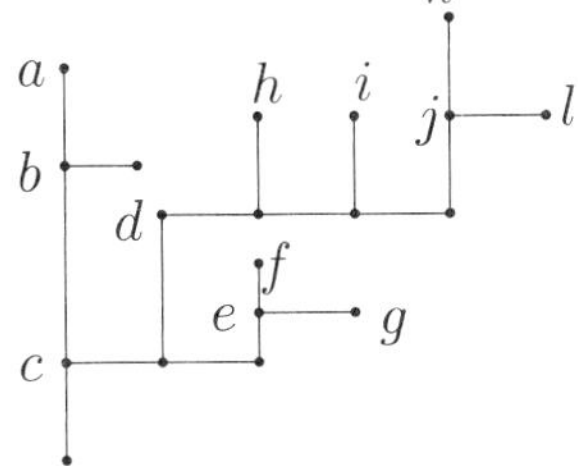

Figure 8.37 A Planted Trivalent Binary Tree

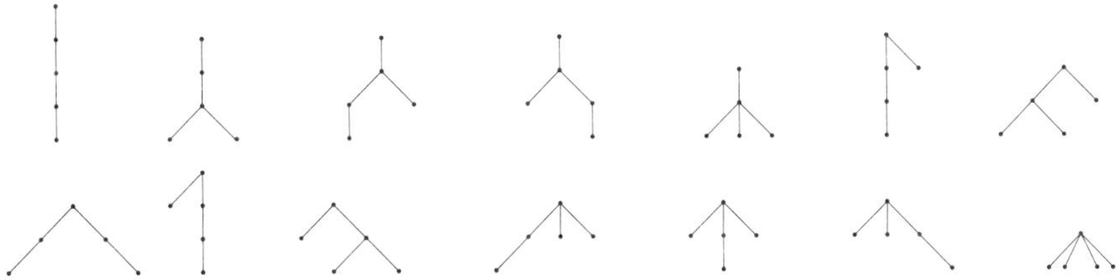

Figure 8.38 Ordered Rooted Trees with four edges

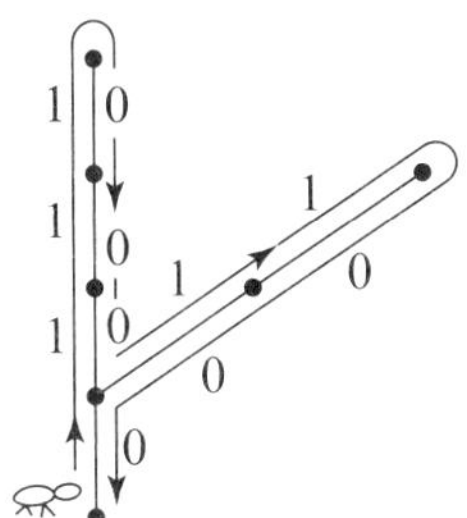

Figure 8.39 Binary Number 11100011000

it calls a 1, and every time it crawls down an edge, it calls a 0, until it returns to the ground. See Figure 8.39.

Number of Leaves in a Forest of Trees with n Edges

Since there are $C_n = \frac{1}{n+1}\binom{2n}{n}$ ordered rooted trees with n edges each, it follows that they contain a total of $(n+1)C_n = \binom{2n}{n}$ vertices. In 1999, L. W. Shapiro of Howard University, Washington, D.C., proved that exactly one-half of them are leaves.

For example, consider the fourteen trees with four edges each in Figure 8.38. They contain a total of $\binom{8}{4} = 70$ nodes, of which exactly 35 are leaves.

Ordered Rooted Trees with $n + 1$ Vertices

Next we investigate the number of ordered rooted trees with a given number of vertices.

Example 8.5 Find the number of ordered rooted trees with $n + 1$ vertices, where $n \geq 0$.

Solution   Figure 8.40 shows the possible ordered rooted trees for $0 \leq n \leq 3$.

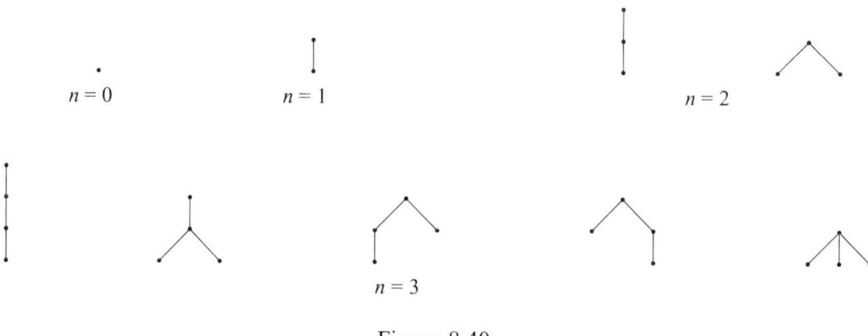

Figure 8.40

From these four cases, we predict that the desired number is C_n.

We confirm this conjecture by establishing a bijection between the set of well-formed sequences of n pairs and the set of ordered rooted trees with $n + 1$ vertices. To this end, consider the ordered rooted tree in Figure 8.41. Label each leaf with the correctly parenthesized sequence $w_i = ()$; see Figure 8.42.

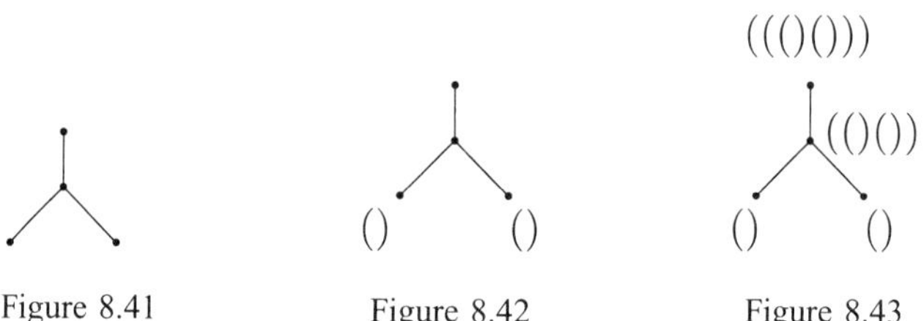

Figure 8.41   Figure 8.42   Figure 8.43

Now, label each internal vertex with children $w_1, w_2, \ldots, w_k$ with their parenthesized concatenation $(w_1 w_2 \ldots w_k)$, keeping the order of the children; see Figure 8.43.

Now drop the first left parenthesis and the last right parenthesis; in effect, this amounts to deleting the root of the tree and the edge(s) from it. This yields a correctly parenthesized sequence with n pairs. The sequence resulting from Figure 8.43 is $(()())$, a well-formed sequence with three pairs.

This process is clearly reversible. To see this, consider the well-formed sequence $((()(()))())$ with five pairs of parentheses. Using concatenation, we employ the sequence to construct the ordered rooted tree in Figure 8.44. Likewise, the sequence $((()()())(()))$ yields the ordered rooted tree in Figure 8.45.

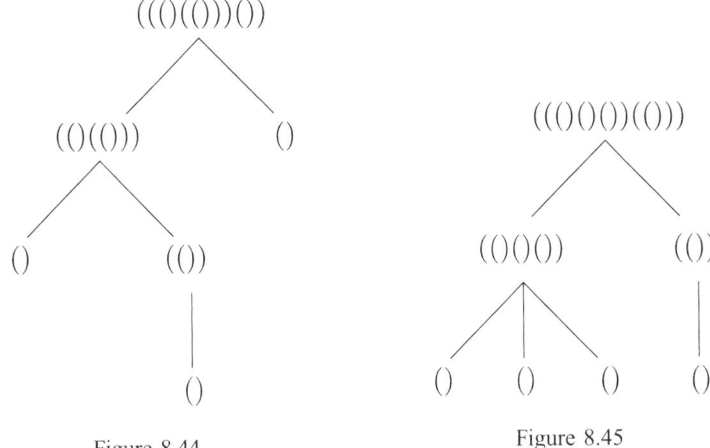

Figure 8.44

Figure 8.45

Thus, our algorithm establishes the desired bijection. Consequently, the number of ordered rooted trees with $n + 1$ vertices is C_n.

Because an ordered rooted tree with n edges has $n + 1$ vertices, it also follows by Theorem 8.3 that there are C_n ordered rooted trees with $n + 1$ vertices.

Example 6.10 Revisited

Using the *depth-first tree traversal algorithm*,[†] we can establish a one-to-one correspondence between the set of ordered trees with $n + 1$ vertices and the set of sequences of n 1s and $n - 1$s in Example 6.10. In such a tree traversal, we visit the root v_0. Let $v_1, v_2, \ldots, v_k$ be the vertices adjacent to v_0. Then traverse the subtree rooted at v_1 using the depth-first algorithm, followed by the subtrees rooted at $v_2, v_3, \ldots, v_k$ in that order.

For example, consider the ordered rooted tree rooted at A in Figure 8.46. Traverse the tree using the the depth-first algorithm. When an edge is traversed down, record a 1, and when it is traversed up, record a -1. This yields the sequence $11 - 111 - 1 - 1 - 11 - 1$, which is a valid sequence of five 1s and five -1s; see Figure 8.47.

Figure 8.48 shows the possible ordered rooted trees with $n + 1$ vertices, and the corresponding sequences n 1s and $n - 1$s, where $0 \le n \le 3$.

[†]   See R. L. Kruse et al., *Data Structures and Program Design in C*, Prentice-Hall, Englewood Cliffs, New Jersey, 1991.

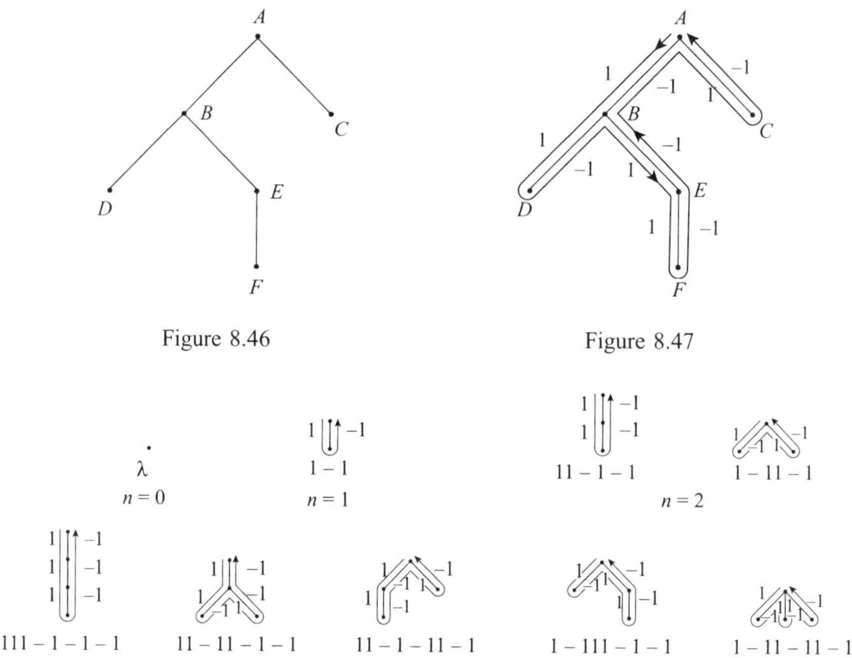

Figure 8.46

Figure 8.47

Figure 8.48

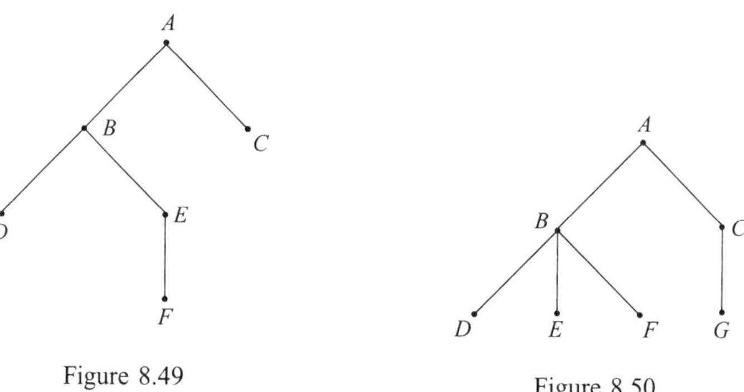

Figure 8.49

Figure 8.50

Example 6.28 Revisited

We now display a bijection between the set of ordered rooted trees with $n + 1$ vertices (see Example 8.5) and the set of n-tuples in Example 6.28. We accomplish this using the depth-first tree traversal algorithm. When a vertex is visited for the first time, record one less than the number of its descendants; ignore always the last vertex from the tree and each subtree.

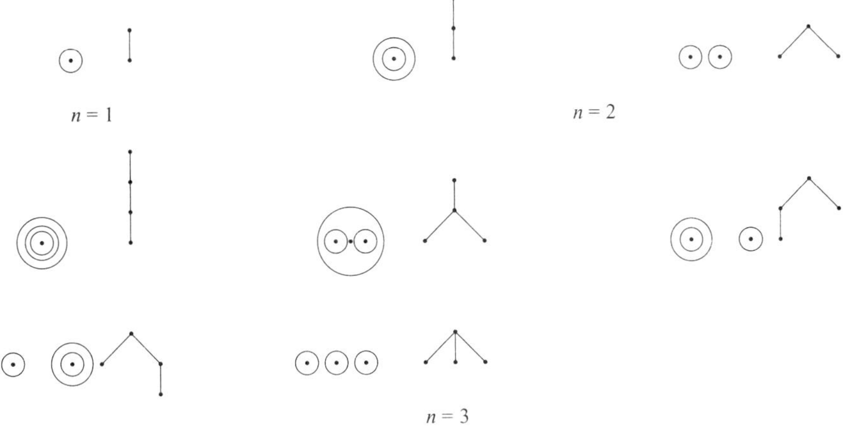

Figure 8.51

For example, consider the tree in Figure 8.49. We traverse the tree in the order A, B, D, E, F, and C; but ignore C. At the vertices A through F, we record $4 - 1, 0 - 1, 0 - 1$, and $0 - 1$, respectively, that is, $3, -1, -1$, and -1, respectively. Recall from Example 6.28 that $3 - 1 - 1 - 1$ is a valid 4-tuple; see Table 6.27.

Likewise, the tree in Figure 8.50 yields the 5-tuple $4 - 1 - 1 - 1 - 1$; you may confirm this.

The algorithm is clearly reversible, so the correspondence is a bijection.

Example 7.10 Revisited

We now exhibit a one-to-one correspondence between the set of n disjoint circles in Example 7.10 and the set of ordered rooted trees with $n + 1$ vertices. In the interest of brevity and clarity, we let Figure 8.51 speak for the bijection.

We ask you to display the correspondences between the set of circles corresponding to $n - 4$ and the set of ordered rooted trees with five vertices.

Next we pursue an interesting application of Example 8.5.

A Delightful Application

In Example 6.9, we conjectured that $2n$ people sitting at a round table can shake hands without their hands crossing in C_n different ways. We now confirm this by establishing a bijection between the set of handshakes and the set of ordered rooted trees with $n + 1$ vertices.

Suppose, for example, that there are $2n = 14$ people at a round table, and they shake their hands as in Figure 8.52. Using this configuration, we now construct an ordered rooted tree with eight vertices.

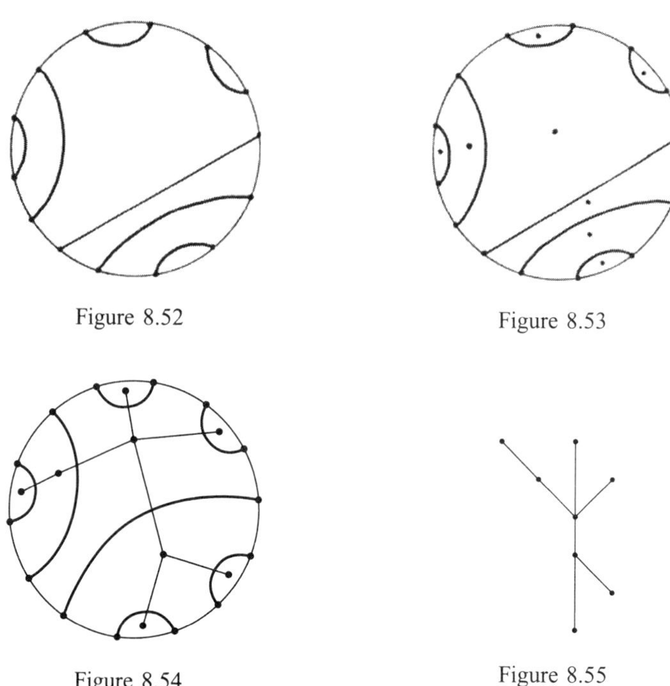

Figure 8.52

Figure 8.53

Figure 8.54

Figure 8.55

To this end, as we did in the case of the poygonal disection in Figure 8.21, we place a dot in each region; see Figure 8.53. We join every two dots if they are separated by exactly one arc or line segment; see Figure 8.54. The resulting graph is an ordered rooted tree with eight vertices; see Figure 8.55.

To see that this algorithm is reversible, consider the ordered rooted tree with nine vertices in Figure 8.56. Draw a circle around the tree. Place a dot on either side of every edge of the tree such that it lies on the circle also; see Figure 8.57. Join every two dots by an arc (or line segment) in such a way that the arc crosses an edge of the tree exactly once; see Figure 8.58. (The placement of the dots and the arcs or line segments determine the final appearance.) Dropping the tree results in the desired handshake configuration; it represents twenty people at a round table shaking hands without them crossing; see Figure 8.59.

Obviously, this technique can be extended to $2n$ people and an ordered rooted tree with $n + 1$ vertices. Thus, there is a bijection between the two sets, so there are C_n different ways $2n$ people at a round table can shake hands without their hands crossing.

The next example gives us yet another occurrence of Catalan numbers in the study of trees.

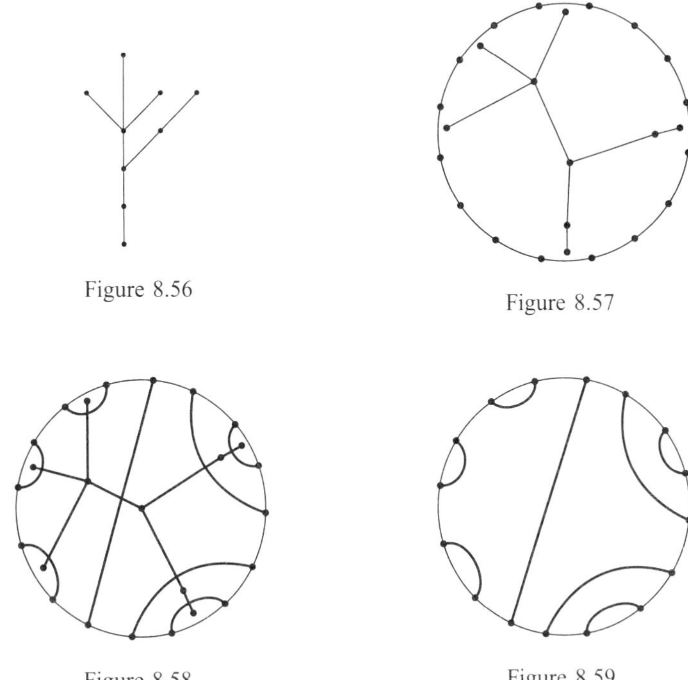

Figure 8.56

Figure 8.57

Figure 8.58

Figure 8.59

Example 8.6 Find the number of ordered rooted trees with $n+2$ vertices such that the length of the rightmost path of each subtree of the root is even, where $n \geq 0$.

Solution   Figure 8.60 shows the various possible such ordered rooted trees, where $0 \leq n \leq 3$. Once again, it appears that the desired answer is C_n. This is in fact the case.

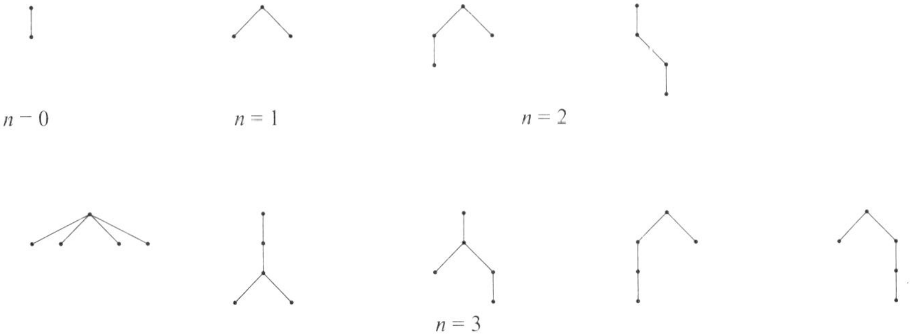

$n - 0$ $n = 1$ $n = 2$

$n = 3$

Figure 8.60 Ordered Rooted Trees with $n + 2$ Vertices

Next, we turn to ordered forests.

Ordered Forests

An *ordered forest* is a sequence of ordered rooted trees. For example, the ordered forest in Figure 8.61 consists of three ordered rooted trees, and a total of $4+5+6 = 15$ vertices. The tree rooted at A is the first tree in the forest, the tree rooted at E is the second, and the one rooted at J is the third.

We always draw the trees in an ordered forest in a such a way that their roots fall on an imaginary horizontal line, as in Figure 8.61.

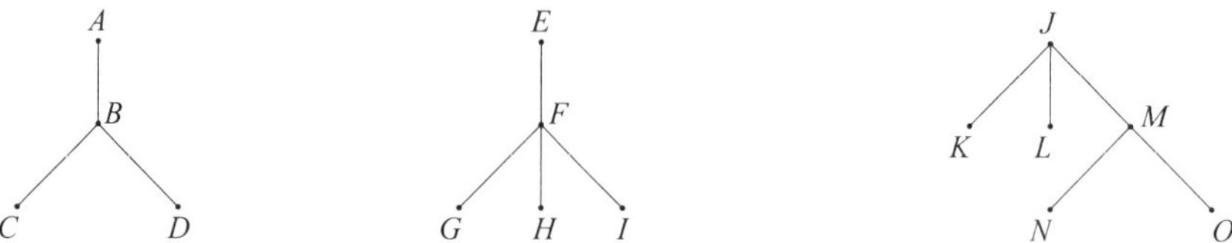

Figure 8.61 An Ordered Forest

Next, we would like to investigate the number of ordered forests with a total of n vertices. To this end, Figure 8.62 shows the various possible ordered forests with n vertices, where $0 \le n \le 4$.

It appears from Figure 8.62 that there are C_n ordered forests with n vertices. We confirm this by establishing a bijection between the set of ordered forests with n vertices and the set of binary trees with n vertices.

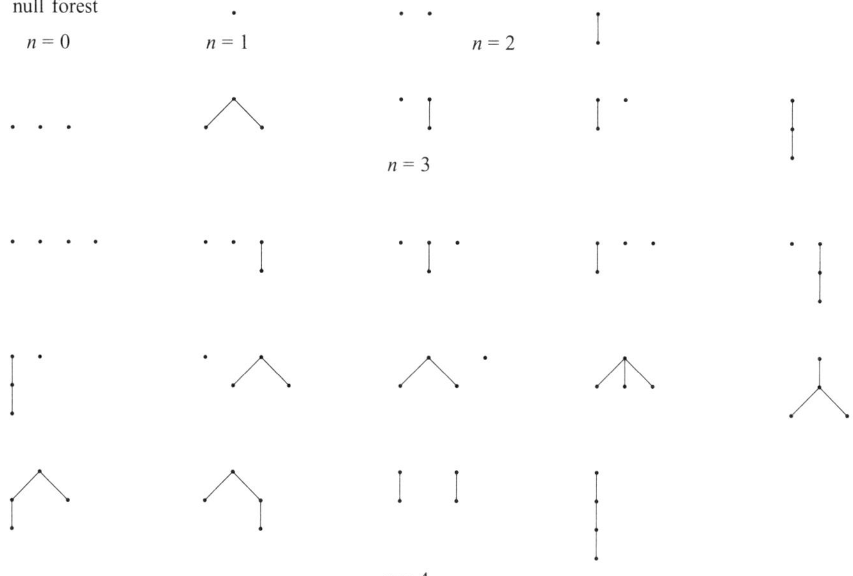

Figure 8.62 An Ordered Forest

To describe such a bijection, consider the forest in Figure 8.61. We combine the trees in the forest in a special way to construct a binary tree. The root *A* of the first tree becomes the root of the binary tree. The oldest child of a vertex in an ordered rooted tree becomes the left child of the vertex in the binary tree; its siblings, if any, become its descendants in the binary tree in that order. The next root of an ordered rooted tree becomes the right child of the previous root *A* (from the previous ordered rooted tree). Figure 8.63 shows the resulting binary tree.

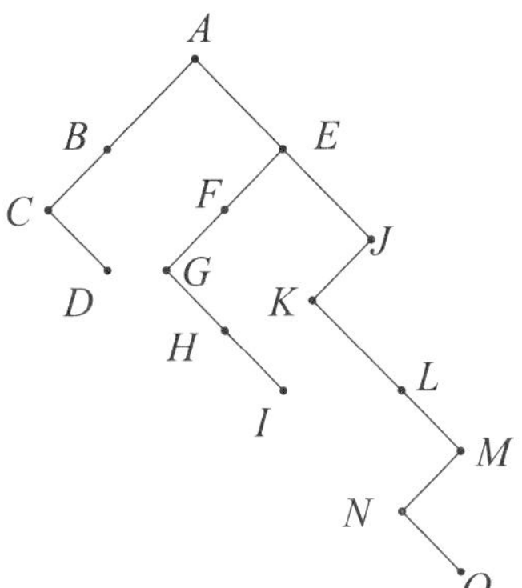

Figure 8.63 Corresponding Binary Tree

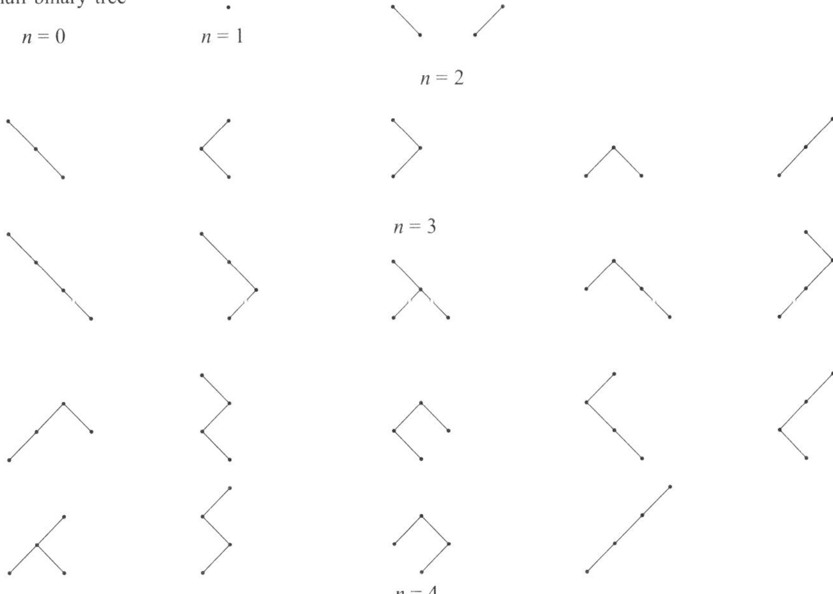

Figure 8.64 Binary Trees

Clearly, these steps can be reversed to construct the ordered forest from the binary tree, establishing the desired bijection. Because there are exactly C_n binary trees with n vertices (see Theorem 8.1), it follows that there are C_n ordered forests with n vertices.

Figure 8.64, for example, shows the binary trees corresponding to the ordered forests in Figure 8.62; you can confirm them.

9

Lattice Paths and Catalan Numbers

This chapter presents occurrences of Catalan numbers in the study of lattice paths in combinatorics. In the process, we shall revisit a number of examples from Chapters 6 and 7 and Catalan's parenthesization problem in Example 5.6.

Lattice Paths

The following example resembles very closely the rook's problem in Example 8.3. We employ it to derive yet another combinatorial formula for C_n.

Example 9.1 Consider a city with a square layout, as in Figure 9.1. The city is divided into blocks by n avenues and n streets. Avenues run east–west and streets north–south. A tourist at the lower left-hand corner (origin) would like to walk to

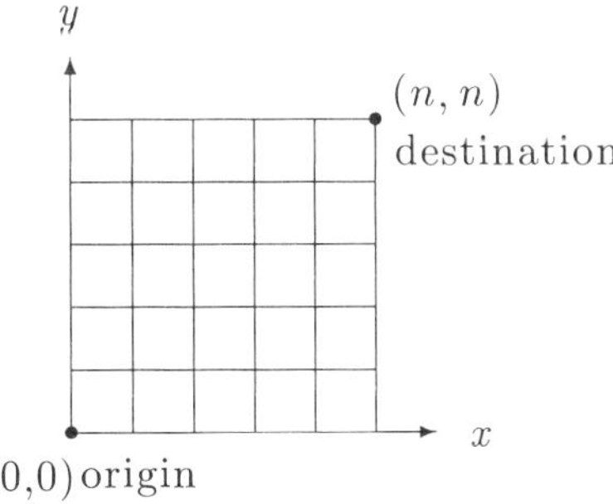

Figure 9.1 Lattice-Walking

the upper right-hand corner. How many different routes can he take? That is, find the number of possible lattice paths from the origin to the lattice point (n, n) on the cartesian plane such that from any lattice point (x, y), the tourist can walk one block east (E) or one block north (N), where $x, y \geq 0$.

Every path can be represented by a word made up of exactly n E's and n N's. So the total number[†] of possible paths equals $\frac{(2n)!}{n!n!} = \binom{2n}{n}$.

As in the rook's problem, we impose a restriction: *No path can cross the northeast diagonal.* Figure 9.2 shows the various possible lattice paths on an $n \times n$ grid, where $0 \leq n \leq 3$.

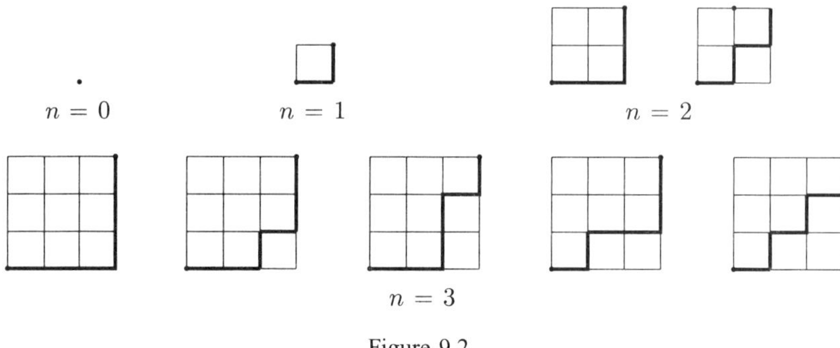

$$n = 0 \qquad n = 1 \qquad n = 2$$

$$n = 3$$

Figure 9.2

Figure 9.3 shows three such paths and their symbolic representations for a 5×5 grid; Figure 9.4 shows three invalid paths and their symbolic representations.

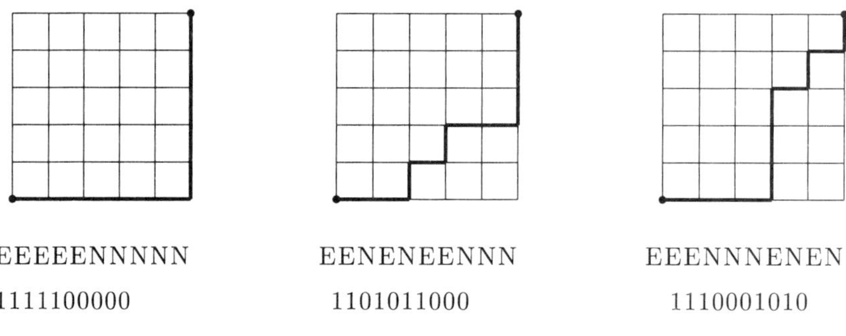

EEEEENNNNN   EENENEENNN   EEENNNENEN

1111100000   1101011000   1110001010

Figure 9.3 Valid Paths

We make three important observations:
- Every legal path must begin with an E and end in an N.
- Every legal string (that is, path) has the property that the number of E's is greater than or equal to the number of N's in each substring.

[†] This is a special case of the fact that there are $\binom{a+b}{a}$ paths from the origin to the lattice point (a, b).

- Replacing an E with a 1 and an N with a 0 results in a ten-bit word such that the number of 1s is greater than or equal to the number of 0s in each substring, when read from left to right. (Consequently, this problem is basically the same as Example 6.12).

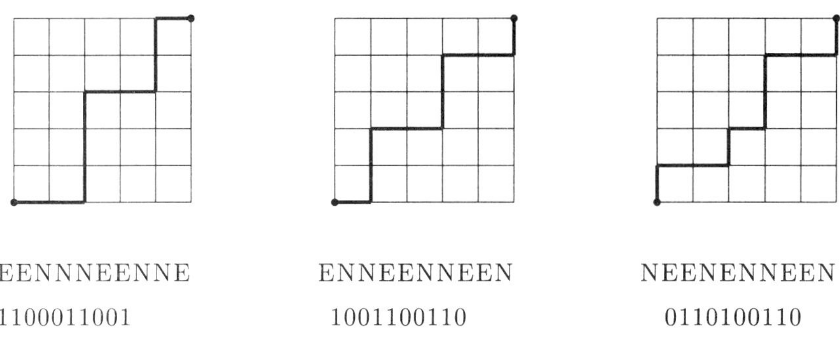

EENNNEENNE

1100011001

ENNEENNEEN

1001100110

NEENENNEEN

0110100110

Figure 9.4 Invalid Paths

Let us investigate closely an invalid path, say, EENNNEENNE in Figure 9.4. It crossed over the northeast diagonal with the third N (fifth move):

EENNN|EENNE

We make a transformation. Beginning with the next move (sixth move) E, swap E's and N's:

EENNN|EENNE ↔ EENNN|NNEEN

The resulting path, EENNN|NNEEN, consists of four E's and six N's; it takes us outside the grid and is still an invalid path; see Figure 9.5.

Figure 9.5 FENNNNNEEN   Figure 9.6 ENNNNEENNE

Consider another invalid path, ENNEENNEEN. It crossed the northeast diagonal with the second N (third move):

ENN|EENNEEN

Swap each E with an N beginning with the fourth move:

$$\text{ENN|EENNEEN} \leftrightarrow \text{ENN|NNEENNE}$$

The resulting path, ENN|NNEENNE consists of four E's and six N's; it does not take us to our destination, and hence is still an illegal path; see Figure 9.6.

The transformation exhibits a bijection between the set of invalid paths on a 5×5 grid and the set of invalid paths consisting of four E's and six N's. So, the number of invalid paths on a 5×5 grid is given by

$$\frac{10!}{4!6!} = \binom{10}{4}$$

Thus, the number of legal paths on a 5×5 grid is given by

$$\binom{10}{5} - \binom{10}{4} = 252 - 210 = 42$$

Obviously, this technique can be extended to an $n \times n$ grid. So the number of invalid paths equals the number of permutations of $2n$ letters (E's and N's), of which $n - 1$ are of one kind (E's) and $n + 1$ are of a second kind (N's), namely,

$$\frac{(2n)!}{(n+1)!(n-1)!} = \binom{2n}{n-1}$$

Thus, the number of legal paths on an $n \times n$ grid is given by

$$\binom{2n}{n} - \binom{2n}{n-1} = \frac{(2n)!}{n!n!} - \frac{(2n)!}{(n+1)!(n-1)!}$$

$$= \frac{(2n)!}{(n+1)n!}[(n+1) - n]$$

$$= \frac{(2n)!}{(n+1)n!}$$

$$= C_n \qquad\blacksquare$$

Examples 5.6 and 6.12 Revisited

Interestingly, there is a bijection between the set of well-formed sequences in Example 5.6 (or the set of binary words in Example 6.12) and the set of legal lattice paths in Example 9.1. The correspondence

$$(\leftrightarrow \text{E, \quad and \quad}) \leftrightarrow \text{N}$$

converts a well-formed sequence of parentheses into a valid lattice path and vice versa.

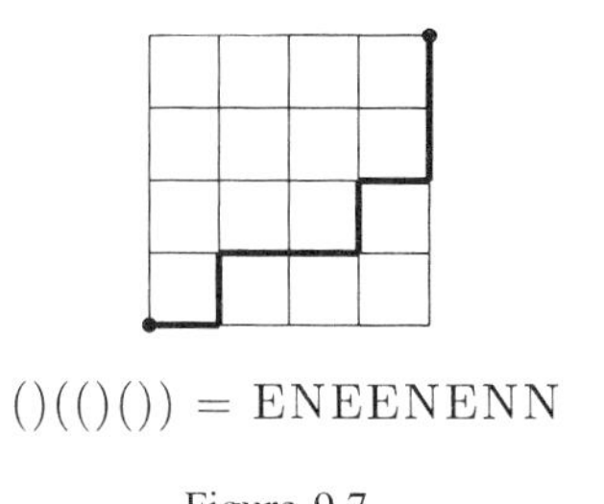

$()(()()) = \text{ENEENENN}$

Figure 9.7

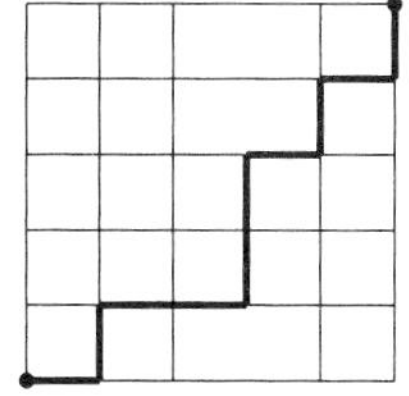

$\text{ENEENNENEN} = ()(())()()$

Figure 9.8

For instance, consider the valid sequence $()(()())$. Changing (to E and) to N results in the valid lattice path ENEENENN; see Figure 9.7.

On the other hand, consider the valid path ENEENNENEN. Switching E to (and N to) yields the valid sequence $()(())()()$; see Figure 9.8.

More generally, the correspondence ($\leftrightarrow$ E and) $\leftrightarrow$ N provides a bijection between the two sets, the set of well-formed sequences of parentheses and the set of valid lattice paths on an $n \times n$ grid.

Example 6.18 Revisited

Next, we exhibit a bijection between the set of lattice paths in Example 9.1 and the set of triangular arrays in Example 6.18. To this end, consider a lattice path on an $n \times n$ grid; see Figure 9.9. Flip it about the northeast diagonal $\overline{OA}$; see Figure 9.10. The reflected path is also obtained by converting E to N and N to E in the original path.

Fill the cells that lie above the path with 1s, and those that lie below the path and above the diagonal with 0s; see Figure 9.11. (With a little practice and imagination, we can directly go from Figure 9.9 to Figure 9.11.) Ignore the half-cells that lie above the diagonal.

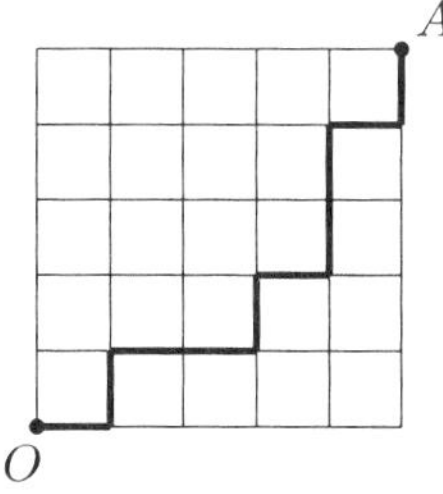

Figure 9.9 Valid Lattice Path

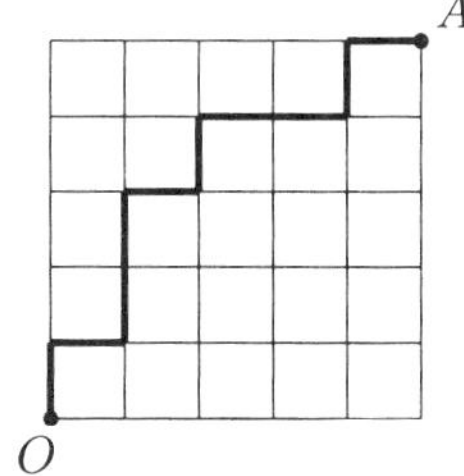

Figure 9.10 Reflected Path

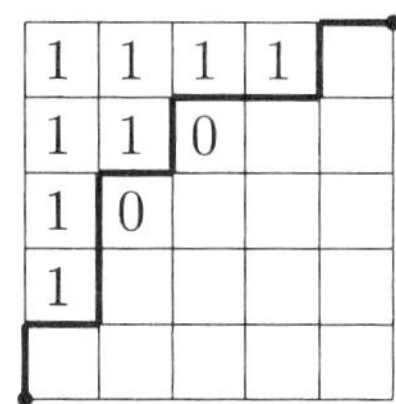

Figure 9.11

Since the new path goes up and to the right, it is guaranteed that in each row the 1s lie to the left of all 0s, and in each column the 1s lie above all 0s. In other words, the resulting array A of bits satisfies the conditions $a_{ij} \geq a_{i,j+1}$ and $a_{ij} \geq a_{i+1,j}$ for every i and j, where $1 \leq i, j \leq n - 1$. Since $i + j \leq n$ for every element in the array, it follows that $1 \leq j \leq n - i$. Thus, array A satisfies all required conditions. Consequently, a valid lattice path determines a unique triangular array.

For example, the triangular array corresponding to the lattice path in Figure 9.9 is

$$
\begin{matrix}
1\ 1\ 1\ 1 \\
1\ 1\ 0 \\
1\ 0 \\
1
\end{matrix}
$$

and the array corresponding to the lattice path in Figure 9.12 is

$$
\begin{matrix}
1\ 1\ 0 \\
1\ 0 \\
1
\end{matrix}
$$

(see the reflected path in Figure 9.13).

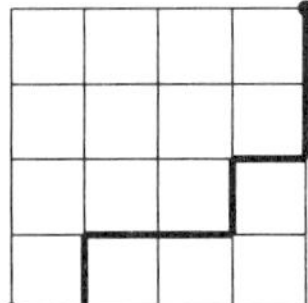

Figure 9.12
Valid Lattice
Path

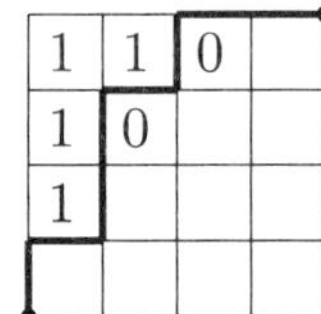

Figure 9.13
Reflected Path

It remains to show that there is a unique lattice path corresponding to every triangular array $A = (a_{ij})$, where $a_{ij} \geq a_{i,j+1}, a_{ij} \geq a_{i+1,j}, 1 \leq i \leq n - 1$, and $1 \leq j \leq n - i$. For example, see Figure 9.14. Place an $n \times n$ grid over the array; see Figure 9.15. Initially, draw a vertical unit bar in the lower left-hand corner. In each column, draw a horizontal unit bar below the last 1 in that column. In each row, draw a vertical unit bar on the right side of the last 1 in that row. If a row contains no 1s, place a vertical bar to the left of the leading 0; and if a column contains no 1s, place a unit horizontal bar above the top zero. Finally, add a horizontal bar on the top of the upper right-hand cell. The resulting path lies completely on or above the northeast diagonal; see Figure 9.16.

Reflect this path about the northeast diagonal. The reflected path does not rise above the diagonal; see Figure 9.17. So it is a valid lattice path.

For instance, the triangular array in Figure 9.14 yields the valid path EENNEE-NENN in Figure 9.17.

1 1 1 0	
1 1 0	
1 1	
0	

Figure 9.14   Figure 9.15   Figure 9.16   Figure 9.17

Thus the algorithm is reversible, establishing a bijection between the set of valid lattice paths and the set of triangular arrays. Consequently, there are exactly C_n distinct triangular arrays with the given characteristics.

Example 6.22 Revisited

In Example 6.22, we conjectured that there are C_n n-tuples $a_1 a_2 \ldots a_n$ of positive integers a_i such that $a_i \leq i$. We also established a one-to-one correspondence between the set of such n-tuples and the set of binary words in Example 6.12. We now establish a bijection between the set of such n-tuples and the set of lattice paths in Example 9.1.

The horizontal segments in a lattice path determine the vertical segments in it; see Figure 9.18. So we ignore the vertical segments. Label each horizontal segment at height k above the x-axis with $k + 1$; see Figure 9.19. This yields an n-tuple with the desired property.

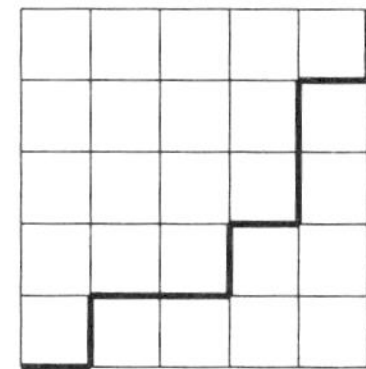

Figure 9.18

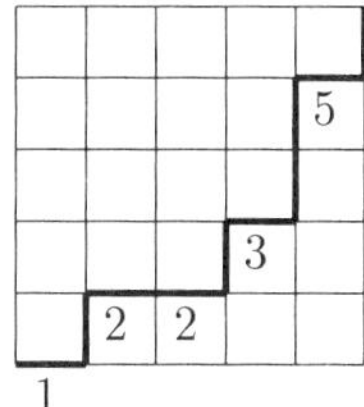

Figure 9.19

Clearly, this algorithm works in the reverse direction as well, thereby establishing the desired bijection.

For example, the lattice path ENEENENNEN in Figure 9.18 determines the 5-tuple 12235 in Figure 9.19 and vice versa. See Figure 9.20 also.

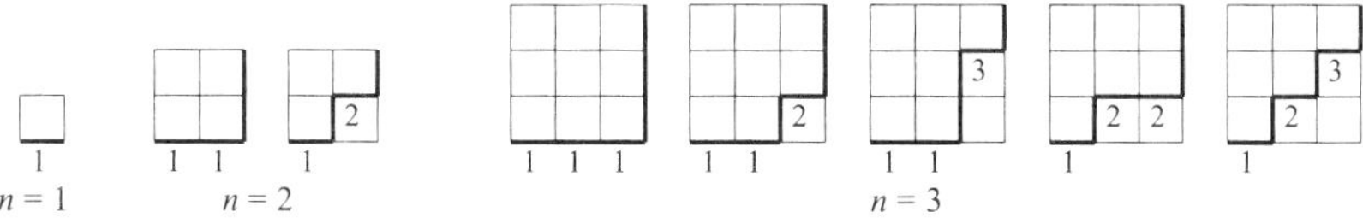

Figure 9.20

With a little practice, the n-tuple can be read directly from the path name. First, assign a 1 to the first E. Then assign a label to each remaining E by incrementing the previous label by the number of N's passed over to reach that E:

$$
\begin{array}{ccccccccccc}
E & N & E & E & N & E & N & N & E & N \\
1 & & 2 & 2 & & 3 & & & 5 &
\end{array}
$$

Clearly, this shortcut is reversible.

Example 6.22 Revisited: An Alternate Bijection

A bijection between the set of n-tuples $a_1 a_2 \dots a_n$ of positive integers $a_i \leq i$ in Example 6.22 and the set of legal lattice paths in Example 9.1 can be established in a slightly different way also. To this end, consider any valid lattice path on an $n \times n$ chessboard. Let b_i denote the area bounded by $x = i - 1, x = i$, the lattice path, and the x-axis, where $i \leq i \leq n$. Let $a_i = 1 + b_i$. The resulting n-tuple $a_1 a_2 \dots a_n$ has the property that $1 \leq a_i \leq i$. Furthermore, this correspondence sets up the desired bijection.

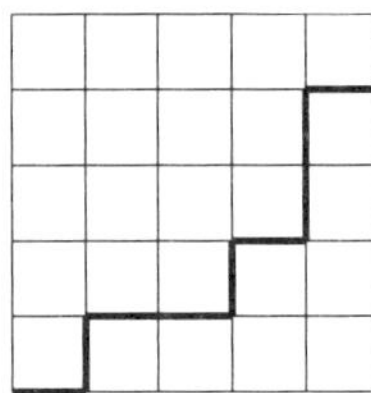

Figure 9.21

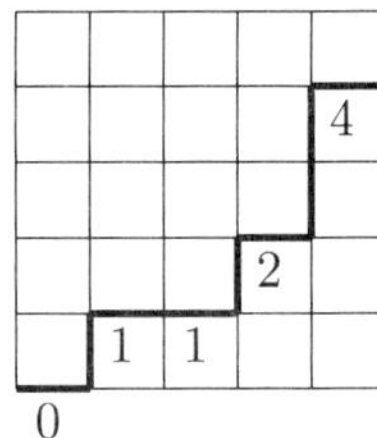

Figure 9.22

For example, consider the path in Figure 9.21. Find the area between the x-axis, each horizontal line segment, and the vertical lines $x = i - 1$ and $x = i$; they are 0, 1, 1, 2, and 4; see Figure 9.22. Add 1 to each. This results in the 5-tuple 12235, which has the property that $1 \leq a_i \leq i$. Clearly, we can go in the reverse direction as well.

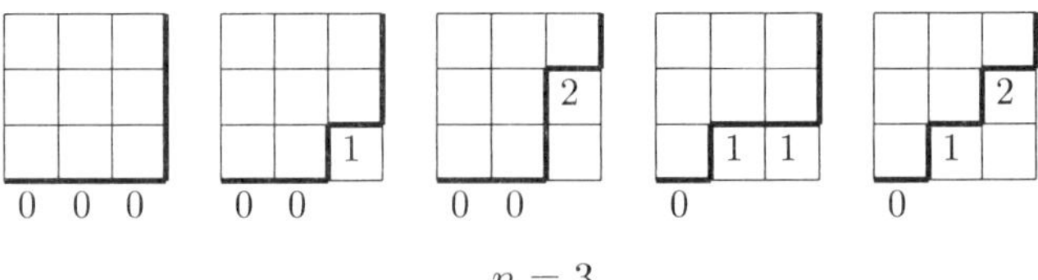

$$n = 3$$

Figure 9.23

Figure 9.23 shows the possible lattice paths on a 3×3 chessboard, the corresponding areas, and the resulting 3-tuples.

Example 6.20 Revisited

The algorithm also establishes a one-to-one correspondence between the set of legal lattice paths in Example 9.1 and the set of n-tuples $a_1 a_2 \ldots a_n$ of positive integers a_i in Example 6.20, where $a_{i+1} \leq a_i \leq n - i + 1$. This time, read the labels of the horizontal segments in the reverse order; see Figure 9.20.

Example 6.21 Revisited

The same algorithm also has an interesting byproduct. It provides a bijection between the set of legal lattice paths in Example 9.1 and the set of $(n - 1)$-tuples $a_1 a_2 \ldots a_{n-1}$ of nonnegative integers a_i in Example 6.21, where $a_{i+1} \leq a_i \leq n - i$. To see this, consider a valid lattice path on an $n \times n$ grid; see Figure 9.24.

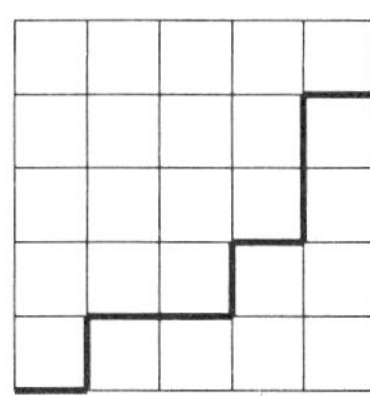

Figure 9.24

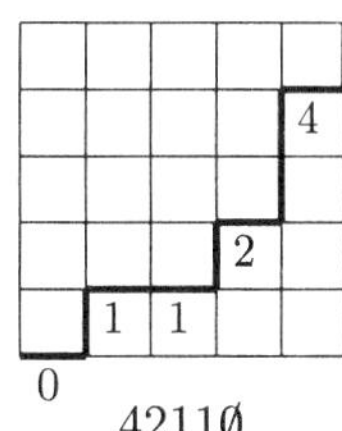

Figure 9.25

As before, ignore the vertical segments. Read the y-coordinates of the horizontal segments from right to left, ignoring the leftmost y-coordinate, which is always zero.

For instance, the path ENEENENNEN in Figure 9.24 yields the 4-tuple 4211; see Figure 9.25. See Figure 9.26 also.

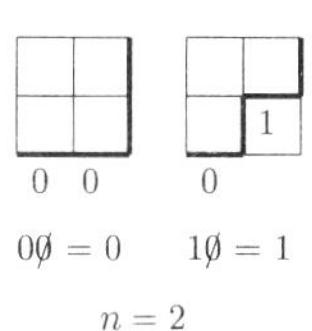
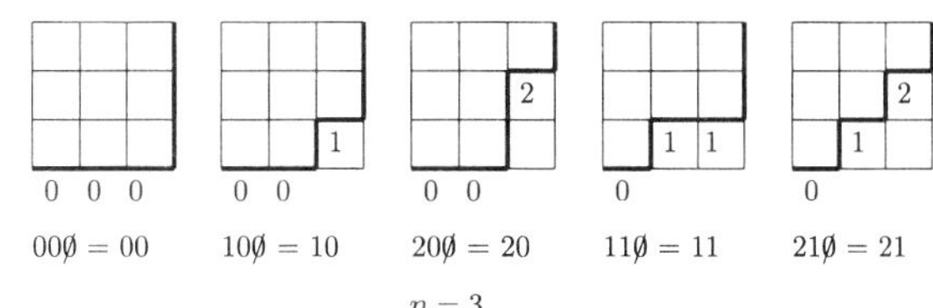

Figure 9.26

Example 6.23 Revisited

The above lattice-path n-tuple algorithm can be applied to Example 6.23 with a modest modifications. As in Figure 9.19, assign labels to the horizontal line segments, read them from left to right, and then append $n + 1$ to the resulting n-tuple; see Figure 9.27.

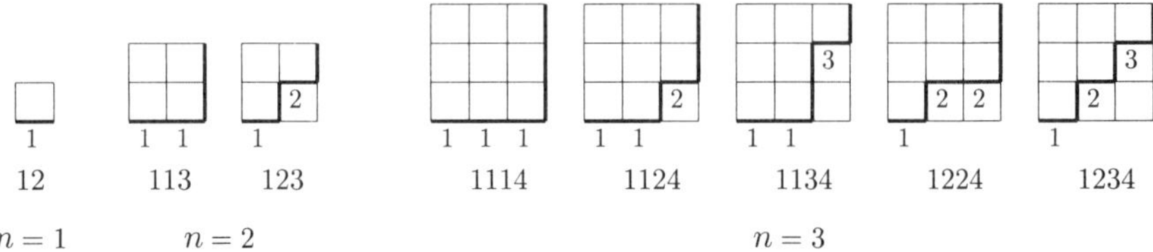

Figure 9.27

Example 6.24 Revisited

Again, we return to the lattice-path n-tuple algorithm and show how it can be slightly modified to recover the n-tuples in Example 6.24 from the lattice paths and vice versa. To this end, consider a valid lattice path on an $n \times n$ grid; see Figure 9.28. As before, assign labels to the horizontal line segments using one more than their heights above the x-axis; see Figure 9.29.

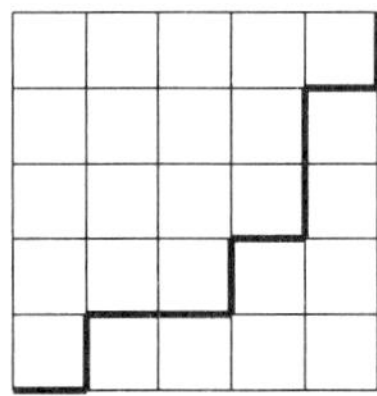

Figure 9.28

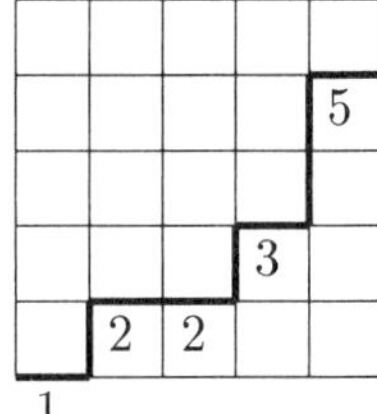

Figure 9.29

Use the labels to construct the corresponding n-tuple as follows: Count the number of 1s, 2s, 3s, and so on among the labels. Use the counts to form the n-tuple $a_1 a_2 \ldots a_n$, where a_i denotes the number of labels i, where $1 \le i \le n$.

For instance, consider the labels 1, 2, 2, 3, and 5 in Figure 9.29. Then $a_1 = 1, a_2 = 2, a_3 = 1, a_4 = 0$, and $a_5 = 1$. The corresponding 5-tuple is 12101. Notice that it has the desired properties:

- Every component is nonnegative.
- The sum of every partial sum with k summands is $\ge k$.
- The total sum is 5.

You may confirm that the lattice paths in Figure 9.20 yield the following n-tuples, where $1 \leq n \leq 3$:

$$
\begin{array}{llllll}
n = 1 & 1 \\
n = 2 & 20 & 11 \\
n = 3 & 300 & 210 & 201 & 120 & 111
\end{array}
$$

Examples 7.7 and 6.2 Revisited

Figure 9.30 illustrates an obvious correspondence between the set of binary words in Example 7.7 (and hence the set of mountain ranges in Example 6.2) and the set of lattice paths in Example 9.1.

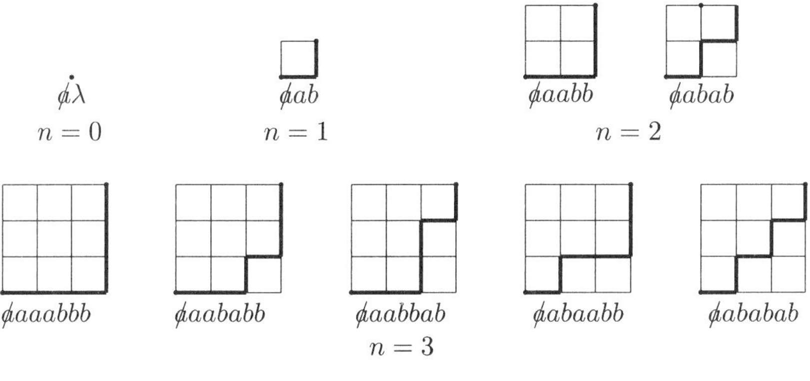

Figure 9.30

The lattice path problem in Example 9.1 can be reworded, as the next example shows.

Example 9.2 There are $2n$ adults $a_1, a_2, \ldots, a_{2n}$ standing at a ticket counter of a carnival. Exactly n of them have a \$10 bill each and each of the other n people has a \$5 bill. Each ticket costs \$5. Initially, the cashier has no change. Find the number of $2n$-tuples $a_1 a_2 \cdots a_{2n}$ such that the cashier will have sufficient change for person a_i, where $2 \leq i \leq 2n$.

To see that this example is basically the same as the lattice-walking problem, let us denote a person with a \$10 bill by a T (same as E in Example 9.1) and one with a \$5 bill with an F (same as N in Example 9.1). Since the cashier has no change initially, every $2n$-tuple must begin with a T. In addition, since the cashier must have enough change for each person, the number of T's must always be greater than or equal to the number of F's in each sequence.

For instance, TTFTFTTFFF is a valid sequence (see Figure 9.3), whereas TFFTTFFTTF is an invalid sequence (see Figure 9.4).

It now follows that Examples 9.1 and 9.2 are the same except for the wording.

Example 6.10 Revisited

Suppose we change T into a 1 and F into a -1 in Example 9.2 (that is, E into 1 and N into -1 in Example 9.1). Then 11-11-111-1-1-1 is a valid solution (see Figure 9.3), whereas 11-1-1-111-1-11 is an invalid solution (see Figure 9.4). Using the labels ± 1, the problem can be restated as follows: Find the number of $2n$-tuples $a_1 a_2 \cdots a_{2n}$ consisting of n 1s and n -1s such that every partial sum is nonnegative.

Consequently, Examples 9.1 and 9.2 are exactly the same as Example 6.10 or 6.11. Thus the answer in each case is

$$\binom{2n}{n} - \binom{2n}{n-1} = C_n$$

Example 6.17 Revisited

We now show that there is a bijection between the set of $2 \times n$ arrays in Example 6.17 and the set of $2n$-tuples in Example 9.2. Consider, for instance, the 2×4 array

$$1\ 3\ 4\ 7$$
$$2\ 5\ 6\ 8$$

(see Table 6.10). To find the corresponding 8-tuple, place a T in positions 1, 3, 4, and 7; and an F in positions 2, 5, 6, and 8. This yields the 8-tuple TFTTFFTF, which is clearly a valid solution in Example 9.2.

Conversely, consider the valid solution TTFTFTTFFF in Example 9.2. We just reverse the algorithm to find the corresponding 2×5 array: It has a T in positions 1, 2, 4, 6, and 7; these positions form the top row. It has an F positions 3, 5, 8, 9, and 10, and they form the second row. The resulting 2×5 array is

$$1\ 2\ 4\ 6\ 7$$
$$3\ 5\ 8\ 9\ 10$$

which is clearly valid.

Obviously, this algorithm works in the general case and establishes the desired one-to-one correspondence.

The next example investigates a special class of lattice paths in the first quadrant on the cartesian plane.

Example 9.3 Find the number of (unordered) pairs of lattice paths of length $n+1$ from the origin that have the same endpoint and that meet only at the endpoint. The paths can go lattice point to lattice point either north or east, one unit at a time.

Solution   Figure 9.31 shows the various pairs of paths for $0 \leq n \leq 3$. From these experiments, we conjecture that the answer is once again C_n, which is in fact true.

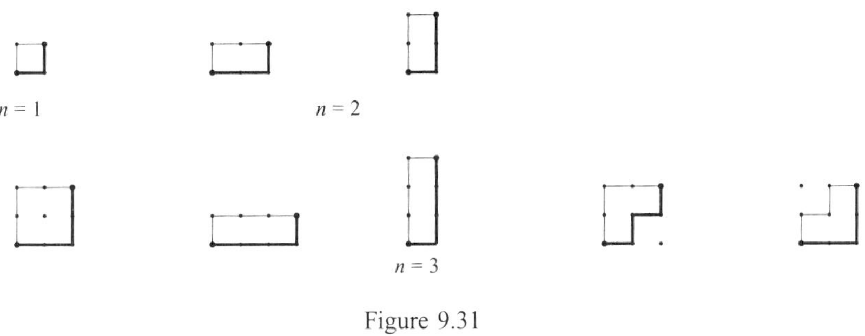

Figure 9.31

The following example closely resembles this work.

Example 9.4 Find the number of (unordered) pairs of lattice paths with $n-1$ steps, originating at the origin and ending at the same point, using easterly or northerly unit steps such that one path does not rise above the other.

Solution   Figure 9.32 shows the possible pairs of such lattice paths, where $1 \leq n \leq 4$. The number of such paths with $n - 1$ steps is C_n.

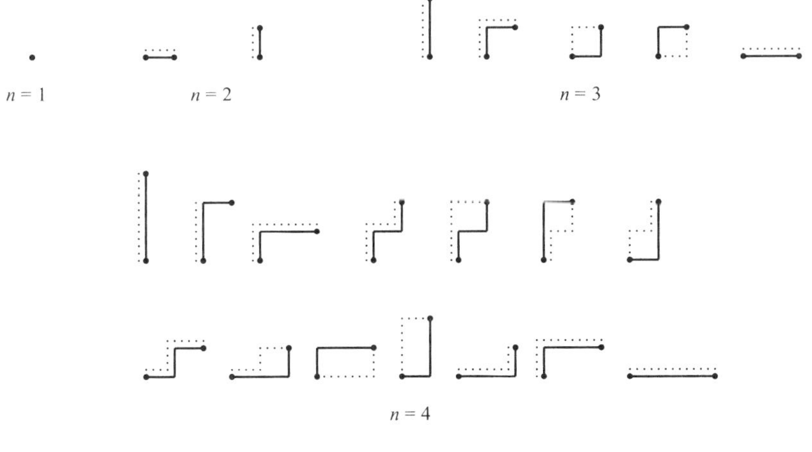

Figure 9.32

The following example was developed by Deutsch.

Example 9.5 Find the number of lattice paths from the origin to $(n - 1, n - 1)$ with unit steps $(1,0)$ 0γ $(0,1)$, 0γ or a diagonal step $(1,1)$ such that:
- No path falls below the line $y = x$; and
- The diagonal steps appear only on the line $y = x$.

Solution   Figure 9.33 shows the possible paths, where $1 \leq n \leq 4$.

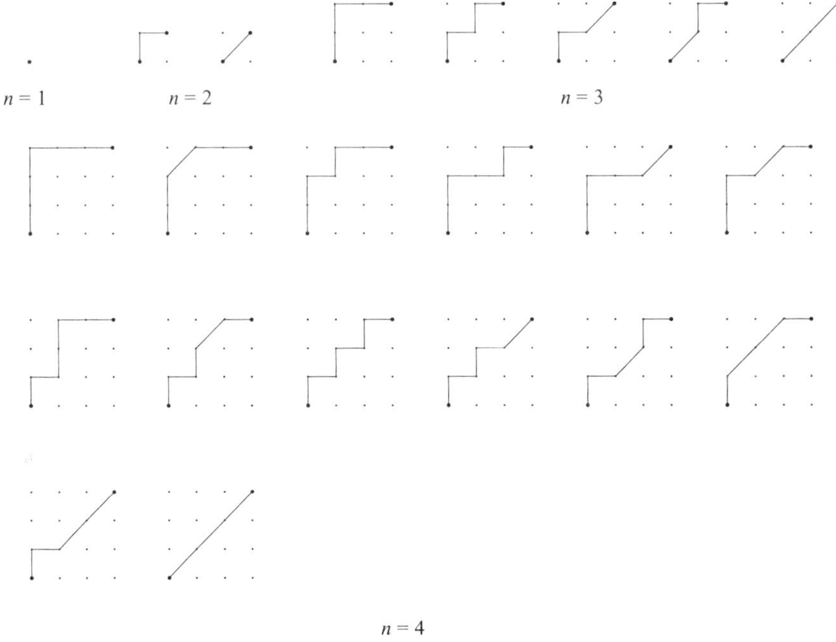

Figure 9.33   ∎

Example 6.15 Revisited

Interestingly, there is an obvious bijection between the set of lattice paths in Example 9.5 and the set of sequences in Example 6.15. To see this, we replace steps $(0,1)$ and $(1,1)$ by 1, and step $(1,0)$ by -1, and vice versa.

For example, using this algorithm, the lattice paths in Figure 9.34 yield the sequences $1 - 11$ and $1 - 11 - 11$, respectively. Reversing the algorithm, the sequences $11 - 1$ and $1 - 111 - 1$ yield the lattice paths in Figure 9.35. You may confirm both cases.

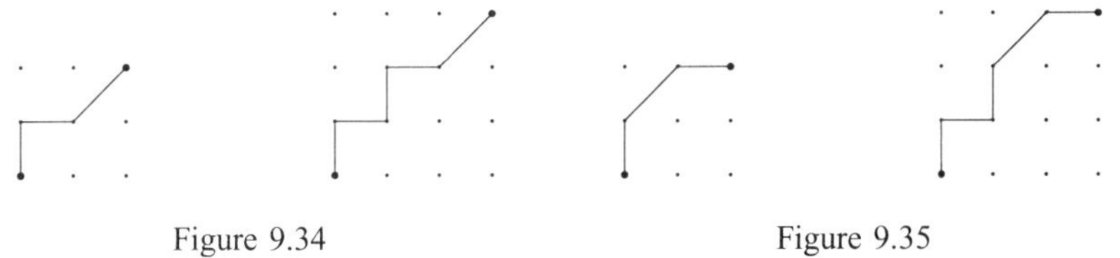

Figure 9.34   Figure 9.35

Example 6.6 Revisited

It follows from the preceding discussion that there exists a bijection between the set of lattice paths in Example 9.4 and the set of incomplete Dyck paths in Example 6.6 as well. Replace steps $(0,1)$ and $(1,1)$ with an upstep $(1,1)$, and each step $(1,0)$ with a downstep $(1,-1)$ and vice versa.

For example, the lattice paths in Figure 9.34 yield the incomplete Dyck paths in Figure 9.36 and the incomplete Dyck path in Figure 9.37 yields the lattice path in Figure 9.38; see Figures 6.8 and 9.32.

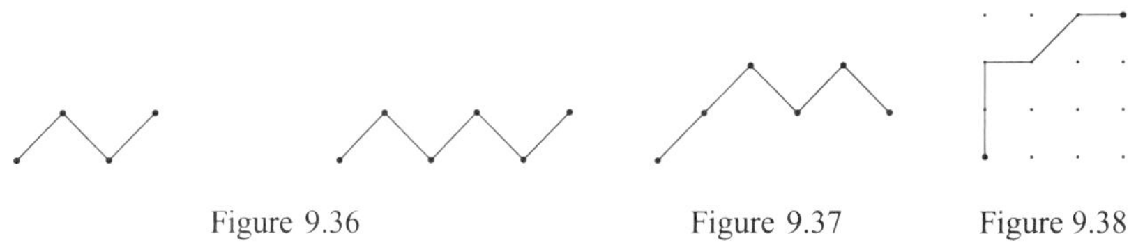

Figure 9.36   Figure 9.37   Figure 9.38

Once again, we return to Catalan's parenthesization problem.

Singmaster's Method for the Parenthesization Problem

In 1978, D. Singmaster of South Bank University, London, investigated the parenthesization problem using postfix (reversed Polish) expressions and cyclic shifts of such expressions. In the process, he developed yet another combinatorial formula for C_n.

In a postfix expression, a binary operator follows its two operands. For example, $(a + b) * c = ab + c*$ and $(a * b) - (c \uparrow d) = ab * cd \uparrow -$. Every algebraic expression can be converted into a postfix expression and vice versa, as in the case of prefix expressions.

We would like to find the number of ways of multiplying (or parenthesizing) $n + 1$ symbols (operands), where $n \geq 1$. Notice that it involves n multiplications (or operators). Let S denote an arbitrary symbol (operand) and X an operator.

Recall, for instance, that there are two different ways of multiplying three items. Using the reversed Polish notation, these two expressions are given by the expressions SSSXX and SSXSX, which are both valid. Notice that SXXSSX is not a valid postfix expression.

Two interesting and useful observations:

- A postfix expression is valid only if the number of S's is greater than the number of X's in each substring, when scanned from left to right.
- The number of ways of multiplying $n + 1$ symbols equals the number of valid postfix expressions consisting of $n + 1$ symbols.

There are $n + 1$ S's and n X's, a total of $2n + 1$ symbols, of which $n + 1$ are of one kind and n are of a second kind. There are $\binom{2n+1}{n}$ such expressions.

But not all of them are valid postfix expressions. For instance, let $n = 3$. Then, for example, SXSSXSX is not a valid postfix expression, but XSXSXSS is valid.

Fortunately, all is not lost. Notice that the valid expresion XSXSXSS is a cyclic permutation of the elements of the string SXSSXSX; it is obtained by cylically shifting the elements in XSXSXSS one letter at a time in the counterclockwise direction:

$$\text{SXSSXSX} \;\to\; \text{XSXSSXS} \;\to\; \text{SXSXSSX} \;\to\; \text{XSXSXSS}$$
$$\uparrow \qquad\qquad \uparrow \qquad\qquad \uparrow \qquad\qquad \uparrow$$
$$\text{invalid} \qquad \text{invalid} \qquad \text{invalid} \qquad \text{valid}$$

Suppose we continue to permute cyclically all elements. Then we get the cyclic chain of words in Figure 9.39.

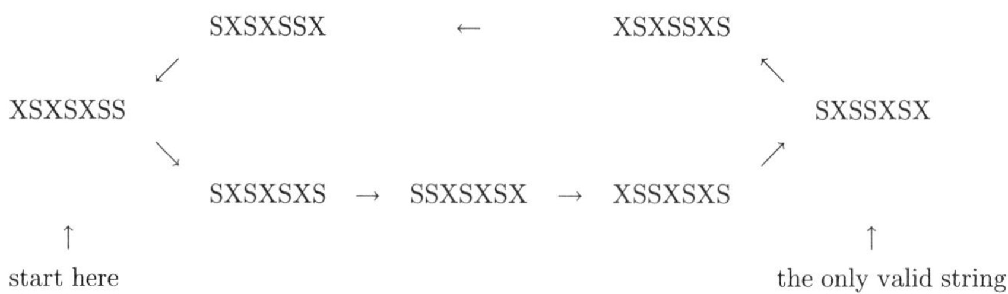

Figure 9.39

An interesting observation: Exactly one of the seven possible cyclic permutations is a valid postfix expression.

More generally, consider any string of $n + 1$ S's and n X's. Is it true that the $2n + 1$ cyclic shifts yield exactly one valid postfix expression? The next theorem provides the answer. Before we present it, we make two additional observations, which play an important role in the development of the proof.

Consider, for instance, the invalid expression SXSSXSX. Suppose we delete the substring SX from the middle: SXS-SX, where we have left a dash to remind us of the deletion point. Cyclically shifting the elements to the right in a cyclic fashion yields the following expressions:

$$\text{SXS–SX} \quad \rightarrow \quad \text{XSXS–S} \quad \rightarrow \quad \text{SXSXS–} \quad \rightarrow \quad \text{–SSXSX}$$

↑
valid

Notice that -SSXSX is a valid postfix expression. Reinserting the pair SX at the insertion point yields a postfix expression consisting of four S's and three X's: SXSSXSX.

On the other hand, suppose we delete the first pair SX: -SSXSX. As before, continuing to shift the elements to the right yields the following expressions:

$$\text{-SSXSX} \quad \rightarrow \quad \text{X–SSXS} \quad \rightarrow \quad \text{SX–SSX} \quad \rightarrow \quad \text{XSX–SS} \quad \rightarrow \quad \text{SXSX–S} \quad \rightarrow \quad \text{SSXSX–}$$

↑
valid

SSXSX- is a valid postfix expression. Now reinsert the pair SX at the deletion point. This results in a valid postfix expression of length seven: SSXSXSX.

We now present the theorem, which will be proved by induction.

Theorem 9.1 (*Singmaster, 1978*) Exactly one of the cyclic permutations of any string of $n + 1$ S's and n X's is a valid postfix expresssion.

Proof (*by induction*)   Suppose $n = 1$. Then we have two S's and one X. There are three possible strings of length three: XSS, SXS, and SSX; each is a cyclic permutation of the preceding string. Exactly one of them is a valid postfix expression, namely, SSX.

Assume that the result is true for an arbitrary integer $k \geq 1$. Let A be an arbitrary string of length $2(k + 1) + 1 = 2k + 3$, consisting of $k + 2$ S's and $k + 1$ X's. Then A must contain the substring SX of length two. Deleting it from the string A yields a substring B of $k + 1$ S's and k X's. By our inductive hypothesis, B has a unique cyclic permutation that is a valid postfix expression. Now insert SX in the substring B, where it was deleted, if the insertion point is not at the beginning; if the insertion point is at the beginning of B, then insert it at the end of B. The resulting string is a valid postfix expression.

Suppose A has two distinct, valid postfix shifts. Then the deletion of the pair SX from each will produce two distinct, valid postfix shifts of length $2k + 1$. This contradicts the inductive hypothesis, so A has a unique valid postfix cyclic permutation.

Thus, by induction, exactly one of the cyclic permutations of any string of $n + 1$ S's and n X's is a valid postfix expresssion.   ∎

Number of Valid Postfix Expressions

Singmaster's theorem enables us to derive an explicit formula for the number of valid postfix expressions consisting of $n + 1$ S's and n X's, as follows:

$$\text{Total number of strings} = (2n + 1)(\text{number of valid postfix expressions})$$

$$\binom{2n + 1}{n} = (2n + 1)(\text{number of valid postfix expressions}).$$

Therefore,

$$\text{Number of valid postfix expressions} = \frac{1}{2n + 1}\binom{2n + 1}{n}$$

$$= \frac{1}{2n + 1} \cdot \frac{2n + 1}{n + 1} \cdot \binom{2n}{n}$$

$$= \frac{1}{n + 1}\binom{2n}{n}$$

$$= C_n.$$

Next we present a geometric-combinatorial solution to the parenthesization problem. This hybrid device was developed in 1887 by French mathematician Désiré André (1840–1917) in his elegant solution to the ballot problem in Example 7.5.

André's Solution to the Parenthesization Problem

Once again, consider a string of $n + 1$ S's and n X's, where S denotes an arbitrary symbol (operand) and X denotes multiplication (operator), and $n \geq 1$. Notice that the number of different ways of parenthesizing (multiplying) the symbols equals the number of different postfix expressions that can be formed. Each postfix expression is a string of length $2n + 1$ consisting of $n + 1$ S's and n X's.

Each postfix expression can be represented graphically, where S denotes an upstroke and X a downstroke (see Example 6.2). For example, the string SSXSXSX can be represented by the mountain range in Figure 9.40.

Notice that every valid string must begin with an S. So the first move is always from the origin to the lattice point $(1,1)$. Thus we are left with n S's (upstrokes) and n X's (downstrokes). Because the number of S's is always greater than the number of X's in each substring, the graph never touches or crosses the x-axis; further, it must end at the lattice point $(2n + 1, 1)$; see Figure 9.40.

Thus the number of valid postfix expressions with $n + 1$ S's and n X's equals the number of paths of length $2n$ from the point $(1,1)$ to the point $(2n + 1, 1)$ that

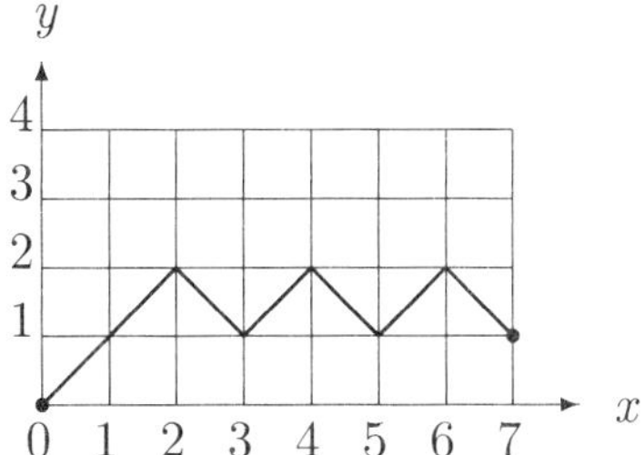

Figure 9.40 Graph of the Postfix Expression SSXSXSX

do not touch (or cross) the horizontal axis. Each such path can be represented by a string of length $2n$, with n characters of one kind (S's) and n characters of a second kind (X's). The number of such words (paths) is given by

$$\frac{(2n)!}{n!n!} = \binom{2n}{n}$$

Unfortunately, however, many of those paths from $(1,1)$ to $(2n+1, 1)$ meet the x-axis; they are not acceptable and hence need to be excluded from this count. See Figure 9.41.

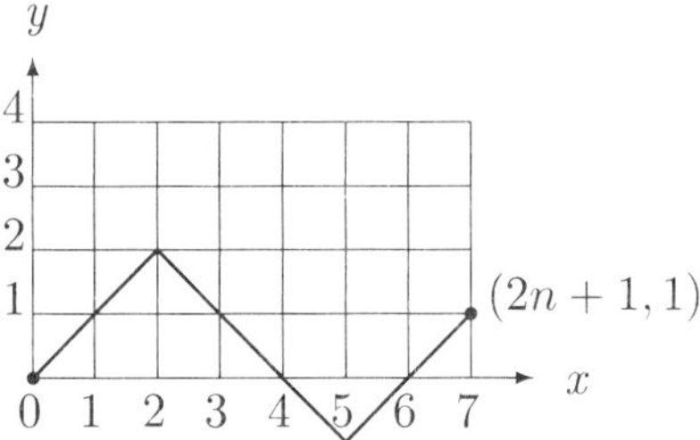

Figure 9.41 Graph Crosses the x-axis

To this end, we will exhibit a bijection between the set A of paths from $(1,1)$ to $(2n+1, 1)$ that meet the x-axis and the set B of paths from $(1,1)$ to $(2n+1, -1)$. The technique we employ is very basic: As soon as the graph meets the x-axis, reflect the rest of the graph about the axis; the resulting graph ends at $(2n+1, -1)$. See Figure 9.42.

To confirm that this procedure is reversible, consider a path from $(1,1)$ to $(2n+1, -1)$. It must cross the x-axis at some point P (see Figure 9.42). Reflect about the horizontal axis the rest of the graph lying to the right of P. This results in a graph that still meets the x-axis but ends at $(2n+1, 1)$.

Thus, there exists a one-to-one correspondence between the sets A and B, so they have the same cardinality.

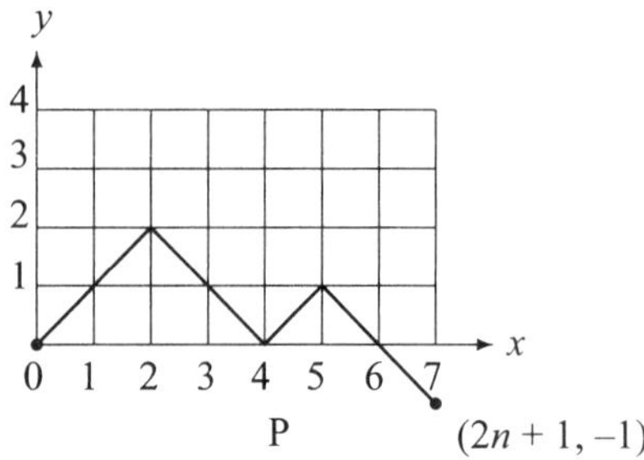

Figure 9.42 André's Reflection Method

Now, every path from $(1,1)$ to $(2n + 1, -1)$ consisits of $n - 1$ upstrokes (S's) and $n + 1$ downstrokes (X's). There are

$$\frac{(2n)!}{(n - 1)!(n + 1)!} = \binom{2n}{n}$$

such paths.

Thus, the number of paths from $(1,1)$ to $(2n + 1, 1)$ that do not meet the x-axis is given by

$$\binom{2n}{n} - \binom{2n}{n - 1} = \frac{(2n)!}{n!n!} - \frac{(2n)!}{(n - 1)!(n + 1)!}$$

$$= \frac{(2n)!}{n!(n + 1)!} \cdot [(n + 1) - n]$$

$$= \frac{(2n)!}{n!(n + 1)!}$$

$$= \frac{1}{n + 1}\binom{2n}{n}$$

$$= C_n$$

Thus, there are exactly C_n lattice paths from the origin to the lattice point $(2n+1, 1)$ without touching or crossing the x-axis. (This is *André's formula*.)   ∎

Rodrigues's Derivation of C_n

In 1838, French mathematician B. O. Rodrigues (1794–1851) developed the explicit formula for C_n using a still different technique. To pursue his derivation, let R_n denote the number of ways of multiplying $n + 1$ elements, where order matters. For example, there are two different ways of multiplying two elements a and b, namely, $a \cdot b$ and $b \cdot a$; so $R_1 = 2$.

Thus R_n counts the number of ways of multiplying $n + 1$ items, where order is important and C_n counts the number of ways, where the order of the elements is fixed. Since $n + 1$ items can be arranged in $(n + 1)!$ ways, this implies that $(n + 1)! C_n = R_n$, where $n \geq 0$.

We will find a recurrence relation for R_n. To this end, consider $n - 1$ symbols $S_1, S_2, \cdots, S_{n-1}$. Consider the product $S_i \cdot S_j$. There are four ways a new symbol S_n can be inserted into this product; they are

$$(S_n \cdot S_j) \cdot S_j, \;\; (S_i \cdot S_n) \cdot S_j, \;\; S_i \cdot (S_n \cdot S_j), \;\; \text{and} \;\; S_i \cdot (S_j \cdot S_n)$$

S_n can also be inserted on the left or right of each product P with $n - 1$ symbols, namely, $S_n \cdot P$ and $P \cdot S_n$.

Thus the total number R_n of ways of multiplying $n + 1$ elements, taking order into consideration, is given by

$$R_n = [4(n - 1) + 2]R_{n-1}$$

$$= (4n - 2)R_{n-1}$$

Using iteration, this yields

$$R_n = (4n - 2) \cdots 10 \cdot 6 \cdot 2 \cdot R_0$$

where we define $R_0 = 1$. So

$$R_n = 2 \cdot 6 \cdot 10 \cdots (4n - 2)$$

Thus

$$C_n = \frac{R_n}{(n + 1)!}$$

$$= \frac{2 \cdot 6 \cdot 10 \cdots (4n - 2)}{(n + 1)!}$$

$$= \frac{1 \cdot 3 \cdot 5 \cdots (2n - 1)}{(n + 1)!} \cdot 2^n$$

$$= \frac{1 \cdot 3 \cdot 5 \cdots (2n-1)}{(n+1)!n!} \cdot n! \cdot 2^n$$

$$= \frac{(2n)!}{(n+1)!n!}$$

$$= \frac{1}{n+1}\binom{2n}{n}$$

where $n \geq 0$. ∎

The next chapter presents some additional occurrences of Catalan numbers.

10

Partitions and Catalan Numbers

In the preceding chapters, we studied occurrences of Catalan numbers in various contexts. In this chapter, we will study their occurrences in the theory of partitioning in combinatorics. We begin with a brief introduction to partitioning.

Partitions

A family of subsets S_i of a set S is a *partition* of S if:

- Each S_i is nonempty;
- The subsets are pairwise disjoint; that is, $S_i \cap S_j = \emptyset$ if $i \neq j$; and
- The union of the subsets S_i is S; that is, $\bigcup_{i \in I} S_i = S$, where I denotes an index set.

Each subset S_i is a *block* of the partition.

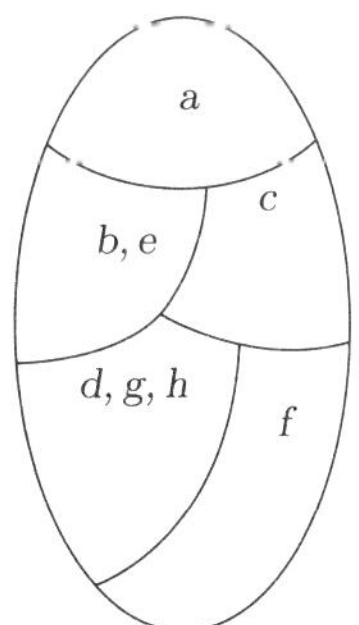

Figure 10.1 A Partition of S

For example, consider the set $S = \{a, b, c, d, e, f, g, h\}$. Then $P = \{\{a\}, \{b, e\},$ $\{c\}, \{d, g, h\}, \{f\}\}$ is a partition of of S. The subsets $B_1 = \{a\}, B_2 = \{b, e\}, B_3 = \{c\},$ $B_4 = \{d, g, h\}$, and $B_5 = \{f\}$ are the blocks of the partitioning; see Figure 10.1. In the interest of brevity, we denote the partition P as $a - be - c - dgh - f$.

We turn to noncrossing partitions, which originated in 1948 in the study of *planar rhyme schemes* by H. W. Becker of Omaha, Nebraska.

Noncrossing Partitions

A *noncrossing partition* π of the set $S = \{1, 2, \ldots, n\}$ is a partition $\{B_1, B_2, \ldots, B_k\}$ of S such that if $a < b < c < d$ and $a, c \in B_i$, and $b, d \in B_j$, then $i = j$.

For example, let $n = 8$. Then $\sigma = 125 - 34 - 68 - 7$ is a noncrossing partition of S. But $125 - 37 - 4 - 68$ is crossing, since $3 < 6 < 7 < 8$, but the blocks $\{3, 7\}$ and $\{6, 8\}$ are not equal.

Interestingly, noncrossing partitions of the set $S = \{1, 2, \ldots, n\}$ can be described geometrically also. Such a characterization is due to Robert Steinberg* of the University of California at Los Angeles. To this end, consider a circle with n points on it, labeled 1 through n in order. A *noncrossing partition* of S is a partition such that the convex hulls of the blocks are pairwise disjoint.

For example, again we let $n = 8$. Then σ is a noncrossing partition of $S = \{1, 2, \ldots, 8\}$ (see Figure 10.2), whereas τ is a crossing partition (see Figure 10.3).

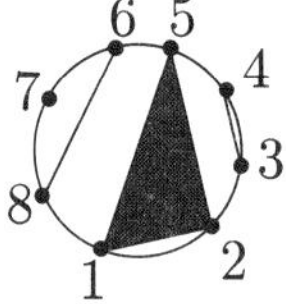

$125 - 34 - 68 - 7$

Figure 10.2 A Noncrossing Partition of S

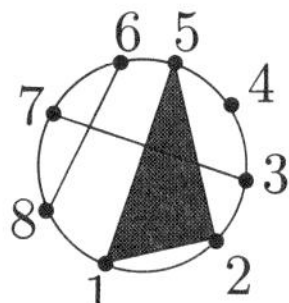

Figure 10.3 Not a Noncrossing Partition of S

We now present three examples to demonstrate the occurrence of Catalan numbers in the theory of noncrossing partitioning.

---

* Steinberg, a member of the National Academy of Sciences, won the Leroy P. Steele Prize in 1985.

Example 10.1 Find the number of noncrossing partitions of the set $\{1, 2, \ldots, n\}$.

Solution   Figure 10.4 shows the geometric representations of the possible noncrossing partitions of the set for $1 \leq n \leq 4$. Table 10.1 summarizes the findings.

Table 10.1

n	Noncrossing Partitions					Count
1	1					1
2	$1 - 2$	12				2
3	$1 - 2 - 3$	$12 - 3$	$13 - 2$	$1 - 23$	123	5
4	$1 - 2 - 3 - 4$	$12 - 3 - 4$	$13 - 2 - 4$	$14 - 2 - 3$	$1 - 23 - 4$	14
	$1 - 24 - 3$	$1 - 2 - 34$	$12 - 34$	$14 - 23$	$123 - 4$	
	$124 - 3$	$134 - 2$	$1 - 234$	1234		

Example 10.2 Find the number of noncrossing partitions of some set $\{1, 2, \ldots, k\}$ into $n + 1$ blocks such that every two distinct elements of the same block differ by at least three.

Solution   Figure 10.5 show the various noncrossing partitions with the desired property, where $0 \leq n \leq 4$. Table 10.2 summarizes the data.

Table 10.2

n	Noncrossing Partitions			Count
0	1			1
1	$1 - 2$			1
2	$1 - 2 - 3$	$14 - 2 - 3$		2
3	$1 - 2 - 3 - 4$	$14 - 2 - 3 - 5$	$15 - 2 - 3 - 4$	5
	$1 \quad 25 \quad 3 \quad 4$	$16 \quad 25 \quad 3 \quad 4$		
4	$1 - 2 - 3 - 4 - 5$	$14 - 2 - 3 - 5 - 6$	$15 - 2 - 3 - 4 - 6$	14
	$16 - 2 - 3 - 4 - 5$	$1 - 25 - 3 - 4 - 6$	$1 - 26 - 3 - 4 - 5$	
	$1 - 2 - 36 - 4 - 5$	$16 - 25 - 3 - 4 - 7$	$17 - 26 - 3 - 4 - 5$	
	$17 - 2 - 36 - 4 - 5$	$17 - 25 - 3 - 4 - 6$	$1 - 27 - 36 - 4 - 5$	
	$147 - 2 - 3 - 5 - 6$	$18 - 27 - 36 - 4 - 5$		

In 1996, using generating functions, Martin Klazar of Charles University, Czech Republic, established that the number of such noncrossing partitions is indeed C_n.

1
1
n = 1
2
2
1
1 − 2
1
12
n = 2
3
2
1
123
3
2
1
12 − 3
3
2
1
13 − 2
3
2
1
1 − 23
3
2
1
1 − 2 − 3
n = 3
3
4
2
1
1 − 2 − 3 − 4
3
4
2
1
12 − 3 − 4
3
4
2
1
13 − 2 − 4
3
4
2
1
14 − 2 − 3
3
4
2
1
1 − 23 − 4
3
4
2
1
1 − 24 − 3
3
4
2
1
1 − 2 − 34
3
4
2
1
12 − 34
3
4
2
1
14 − 23
3
4
2
1
123 − 4
3
4
2
1
124 − 3
3
4
2
1
134 − 2
3
4
2
1
1 − 234
3
4
2
1
1234
n = 4

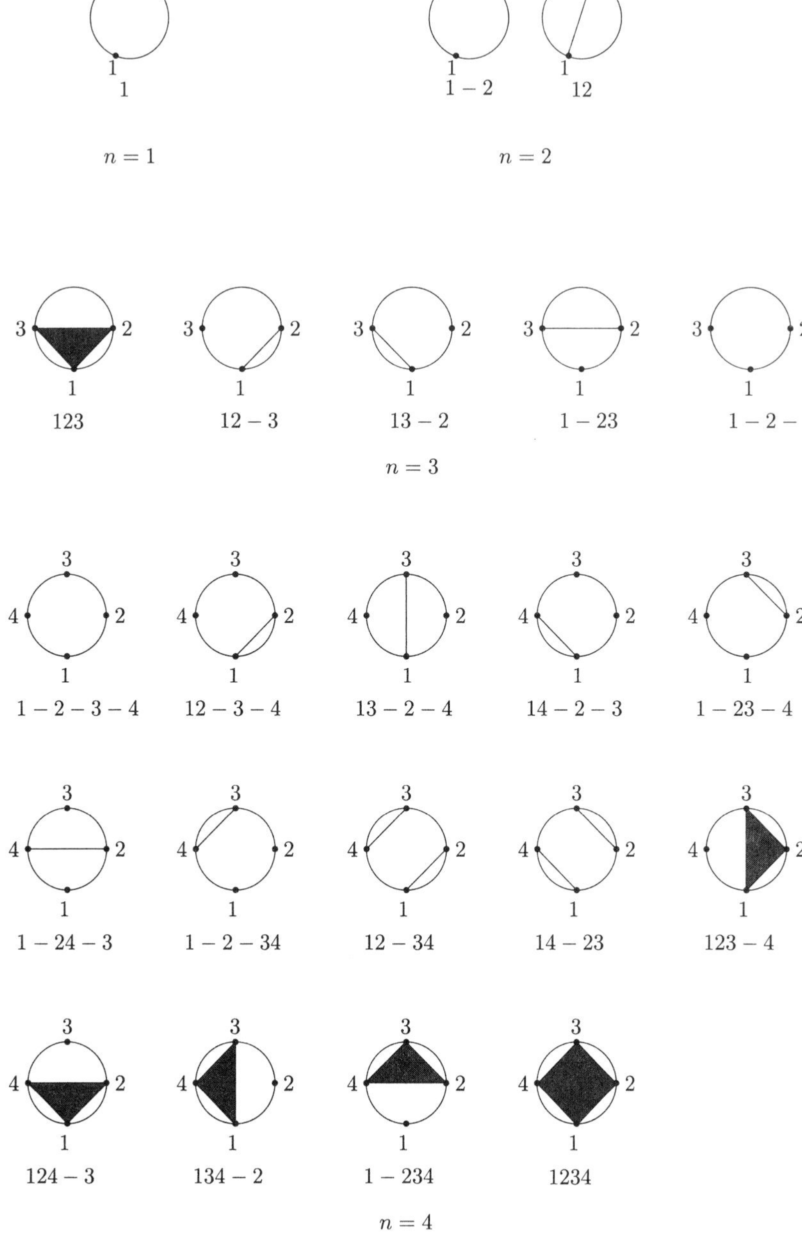

Figure 10.4

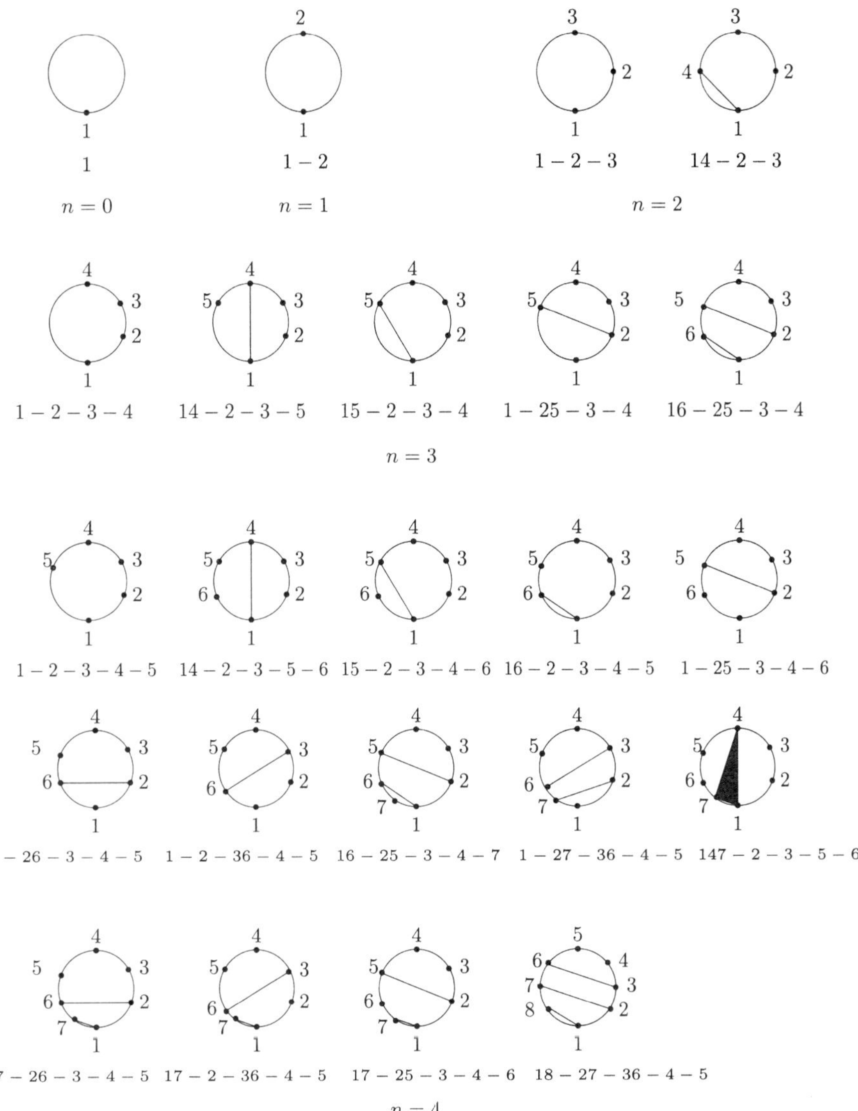

Figure 10.5

Example 10.3 Find the number of noncrossing partitions of the set $\{1, 2, \ldots, 2n + 1\}$ into $n + 1$ blocks such that no block contains two consecutive integers.

Solution Figure 10.6 shows the various possible noncrossing partitions for $0 \leq n \leq 4$. The resulting data are summarized in Table 10.3.

In 1972, R. C. Mullin of the University of Waterloo, Ontario, Canada, and R. G. Stanton of the University of Manitoba, Winnipeg, Canada, established a bijection between the set of such noncrossing partitions and the set of ordered rooted trees with $n + 1$ vertices in Example 8.5.

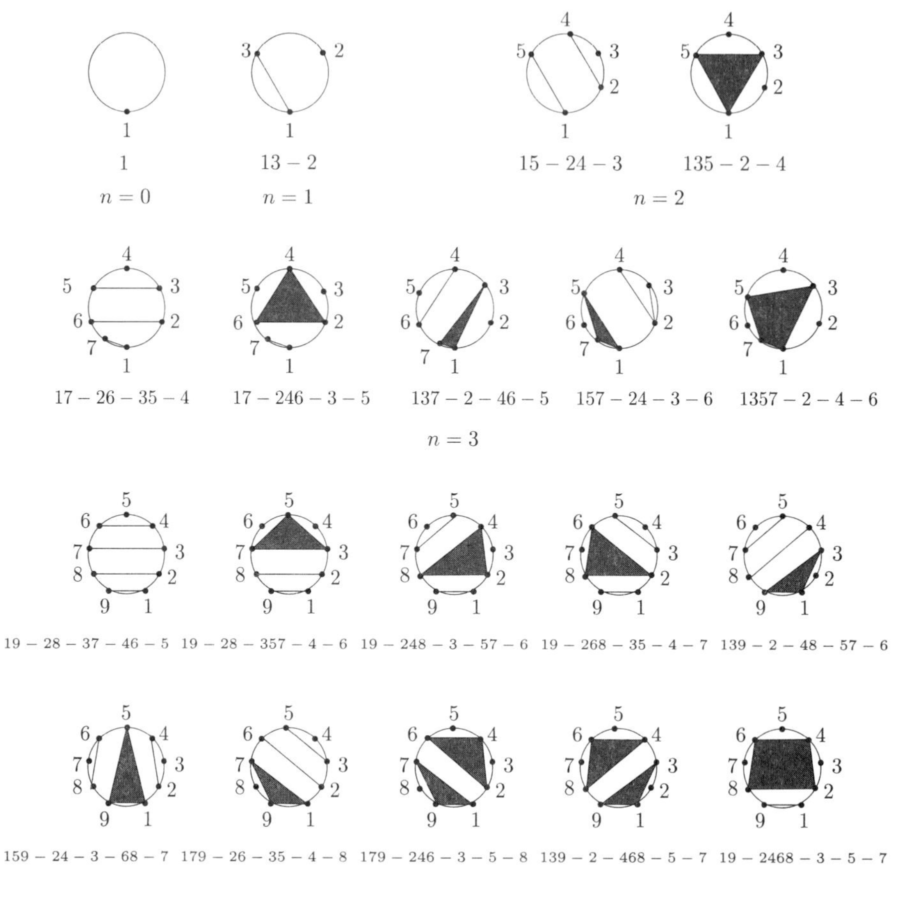

Figure 10.6

Table 10.3

n	Noncrossing Partitions			Count
0	1			1
1	$13 - 2$			1
2	$15 - 24 - 3$	$135 - 2 - 4$		2
3	$17 - 26 - 35 - 4$	$17 - 246 - 3 - 5$	$137 - 2 - 46 - 5$	5
	$157 - 24 - 3 - 6$	$1357 - 2 - 4 - 6$		
4	$19 - 28 - 37 - 46 - 5$	$19 - 28 - 357 - 4 - 6$	$19 - 248 - 3 - 57 - 6$	14
	$19 - 268 - 35 - 4 - 7$	$139 - 2 - 48 - 57 - 6$	$159 - 24 - 3 - 68 - 7$	
	$179 - 26 - 35 - 4 - 8$	$179 - 246 - 3 - 5 - 8$	$139 - 2 - 468 - 5 - 7$	
	$19 - 2468 - 3 - 5 - 7$	$1579 - 24 - 3 - 6 - 8$	$1379 - 2 - 46 - 5 - 8$	
	$1359 - 2 - 4 - 68 - 7$	$13579 - 2 - 4 - 6 - 8$		

11

Algebra, Sports, and Catalan Numbers

This chapter presents some interesting occurrences of Catalan numbers in abstract algebra, linear algebra, and sports. In addition, it presents three examples where the answers seem to be Catalan numbers but are in fact otherwise.

We begin with two applications of Catalan numbers to the theory of groups in abstract algebra.

Groups and Catalan Numbers

The next two examples deal with the additive group[†] Z_m of integers modulo m. It is based on a problem[*] proposed in the *American Mathematical Monthly* by Ernest T. Parker (1926–1991) of the University of Illinois at Urbana-Champaign, later reworded by John L. Selfridge of Northern Illinois University at DeKalb.

Example 11.1 Find the number of n-element multisets[‡] $\{a_1, a_2, \ldots, a_n\}$ of elements $a_i \in \mathbf{Z_{n+1}}$ such that $a_1 + a_2 + \ldots + a_n = 0$, the additive identity of $\mathbf{Z_{n+1}}$.

Solution In the interest of brevity, we denote the multiset $\{a_1, a_2, \ldots, a_n\}$ as the n-tuple $a_1 a_2 \ldots a_n$. Table 11.1 shows the various possible n-element multisets, where $0 \le n \le 4$.

[†] For a discussion of group theory, see J. B. Fraleigh, *A First Course in Abstract Algebra*, 7th ed., Addison-Wesley, Reading, Massachusetts, 2003.

[*] R. K. Guy, "Parker's Permutation Problem Involves the Catalan Numbers," *American Mathematical Monthly* 100 (March 1993), 287–289.

[‡] Recall that a *multiset* is a set with repeated elements. For example, $\{2, 3, 3, 5, 5, 5\}$ is a multiset.

Table 11.1

n	$\mathbf{Z_{n+1}}$		n-element Multisets				Count
1	0,1						1
2	0,1,2	00	12				2
3	0,1,2,3	000	013	022	112	233	5
		0000	0014	0023	0113	0122	
4	0,1,2,3,4	0244	0334	1112	1144	1234	14
		1333	2224	2233	3444		

Using primitive roots and cyclotomic polynomials, Ira M. Gessel of Brandies University, Waltham, Massachusetts, established in 1993 that the number of such multisets is C_n. ■

The next example, closely related to this example, is due to Sergey Fomin of the University of Michigan at Ann Arbor.

Example 11.2 Find the number of $(n + 1)$-element multisets $a_1 a_2 \ldots a_{n+1}$ over the additive group $\mathbf{Z_n} = \{0, 1, 2, \ldots, n - 1\}$ of integers modulo n such that $a_1 + a_2 + \cdots + a_{n+1} = 0$, the additive identity of $\mathbf{Z_n}$.

Solution   Table 11.2 lists the possible multisets for $1 \le n \le 4$.

Table 11.2

n	$\mathbf{Z_n}$		$(n + 1)$-element Multisets				Count
1	0	00					1
2	0,1	000	011				2
3	0,1,2	0000	0012	0111	0222	1122	5
		00000	00013	00022	00112	00233	
4	0,1,2,3	01111	01133	01223	02222	03333	14
		11123	11222	12333	22233		

■

Next, we present two interesting applications of Catalan numbers to the theory of rings. Both are consequences of André's theorem.

Rings, Ideals, and Catalan Numbers

There are exactly C_n lattice paths from the origin to the lattice point $(2n, 0)$ that may touch the x-axis. Using this fact, in 1975 L. W. Shapiro established two interesting results about two-sided ideals in the ring R_n of $n \times n$ upper triangular matrices over a field F.

- The number of ideals in the ring is C_{n+1}.
- The number of nilpotent ideals in the ring is C_n.

For example, let $n = 2$ and let F be the binary field Z_2. Then

$$T_2 = \left\{ \begin{pmatrix} a & b \\ 0 & c \end{pmatrix} \mid a, b, c \in Z_2 \right\}$$

It has exactly $5 = C_3$ ideals:

$$I_1 = \left\{ \begin{pmatrix} 0 & 0 \\ 0 & 0 \end{pmatrix} \right\}, \quad I_2 = \left\{ \begin{pmatrix} a & 0 \\ 0 & 0 \end{pmatrix} \right\}, \quad I_3 = \left\{ \begin{pmatrix} 0 & b \\ 0 & 0 \end{pmatrix} \right\}$$

$$I_4 = \left\{ \begin{pmatrix} 0 & 0 \\ 0 & c \end{pmatrix} \right\}, \quad I_5 = \left\{ \begin{pmatrix} a & b \\ 0 & c \end{pmatrix} \right\}$$

The ring R_2 has exactly $C_2 = 2$ nilpotent ideals: I_1 and I_3.

We turn now to three classes of matrices.

Positive Definite Matrices and Catalan Numbers

In 1980, F. T. Leighton and M. Newman of the National Bureau of Standards, Washington, D.C., investigated a class of $n \times n$ integral matrices that are positive definite and tridiagonal, whose sub- and superdiagonals consist of only ones, and whose determinants are one. They showed that the class S_n of such matrices is finite and there are exactly C_n such matrices. Four years later, they were studied by L. W. Shapiro of Howard University, who called them *suitable matrices*.

Such matrices are not commonly known, so we need to define certain terms. To begin with, a symmetric matrix $\mathbf{A} = (a_{ij})_{n \times n}$ is *positive definite* if $\mathbf{X}^T \mathbf{A} \mathbf{X} > \mathbf{0}$ for all $\mathbf{X} \neq \mathbf{0}$.

Principal submatrices of $\mathbf{A}$, which are submatrices consisting of the first r rows and r columns, provide a useful criterion to determine the positive definiteness of $\mathbf{A}$, where $1 \leq r \leq n$. A symmetric matrix is positive definite if every principal submatrix is positive.

For example, let

$$\mathbf{M} = \begin{bmatrix} 2 & 3 & -1 \\ 3 & 5 & -2 \\ -1 & -2 & 7 \end{bmatrix}$$

It is positive definite, since $|2| = 2$, $\begin{vmatrix} 2 & 3 \\ 3 & 5 \end{vmatrix} = 1$, and $|\mathbf{M}| = 6$ are all positive numbers.

For the curious-minded, we add an interesting property of positive definite matrices: A symmetric matrix is positive definite if and only if its eigenvalues are positive.

For example, the eigenvalues of the above 3×3 matrix $\mathbf{M}$ are given by the characteristic equation

$$\mathbf{M} - \lambda \mathbf{I} = \begin{bmatrix} 2-\lambda & 3 & -1 \\ 3 & 5-\lambda & -2 \\ -1 & -2 & 7-\lambda \end{bmatrix} = 0$$

That is,

$$\lambda^3 - 14\lambda^2 + 45\lambda - 6 = 0$$

The solutions of this equation are clearly positive, so the eigenvalues of the matrix $\mathbf{M}$ are all positive, as expected.

The matrix $\mathbf{A}$ is *tridiagonal* if $a_{ij} = 0$ if $j > i+1$ and if $j < i-1$. For example, the matrix

$$\begin{bmatrix} a & b & 0 & 0 & 0 \\ c & d & e & 0 & 0 \\ 0 & f & g & h & 0 \\ 0 & 0 & i & j & k \\ 0 & 0 & 0 & l & m \end{bmatrix}$$

is tridiagonal. The diagonal below and the diagonal above the main diagonal are the *sub-* and *superdiagonals* of the matrix.

Thus, a square, integral, tridiagonal matrix whose sub- and superdiagonals consist of 1s has the form

$$\begin{bmatrix} a_1 & 1 & 0 & 0 & \cdots & 0 \\ 1 & a_2 & 1 & 0 & \cdots & 0 \\ 0 & 1 & a_3 & 1 & \cdots & 0 \\ & & & \vdots & & \\ & & & \cdots & a_{n-1} & 1 \\ 0 & 0 & 0 & \cdots & 1 & a_n \end{bmatrix}$$

Clearly, this matrix yields the n-tuple $(a_1, a_2, \cdots, a_n)$ and vice versa. Consequently, knowing the n-tuple $(a_1, a_2, \cdots, a_n)$, we can find the corresponding suitable matrix $\mathbf{A}$. Let B_n denote the set of corresponding n-tuples.

Suppose **A** is suitable. Leighton and Newman showed that $a_k = 1$ for some k. They also showed that:

$(a_1, a_2, \cdots, a_n) \in B_n$

$$\Leftrightarrow (a_1, a_2, \cdots, a_k + 1, 1, a_{k+1} + 1, \cdots, a_n) \in B_{n+1} \qquad (11.1)$$

$$\Leftrightarrow (1, a_1 + 1, a_2, \cdots, a_n) \in B_{n+1} \qquad (11.2)$$

$$\Leftrightarrow (a_1, a_2, \cdots, a_{n-1}, a_n + 1, 1) \in B_{n+1}, \qquad (11.3)$$

where $\Leftrightarrow$ means *if and only if*.

These three properties enable us to build up B_{n+1} from B_n. Consequently, we do not need to compute $|A|$ to determine whether A is positive definite. This saves us a lot of computational time.

For example, when $n = 1$, $A = [a_1]$; so $|A| = a_1 = 1$. The corresponding 1-tuple is (1): $B_1 = \{(1)\}$. Using B_1, the elements of B_2 can be constructed as follows: Using condition (11.1), the 1-tuple (1) yields (2,1); using condition (11.2), it yields (1,2). Thus, $B_2 = \{(2, 1), (1, 2)\}$. Similarly, B_3 can be constructed from B_2, B_4 from B_3, and so on.

Table 11.3 lists as permuations the n-tuples corresponding to $n \times n$ suitable matrices, where $1 \leq n \leq 4$.

Table 11.3

n	n-tuples					Count
1	1					1
2	21	12				2
3	131	312	221	122	213	5
	1231	2141	1412	1321	4122	
4	3213	3131	2312	2221	1222	14
	2132	1313	3123	2214		

Euler's Triangulation Problem Revisited

Shapiro exhibited a bijection between the set of triangulations of a regular $(n+2)$-gon (see Example 5.1) and the set S_n of $n \times n$ suitable matrices. To see this, consider, for example, the various triangulations of a regular pentagon in Figure 11.1. Ignoring the base edge, enter the degrees of the remaining vertices of the graphs in the clockwise direcion: 233, 332, 324, 242, and 422. Subtract one from each entry. This results in the triplets 122, 221, 213, 131, and 311. Interestingly, they are the same 3-tuples we obtained earlier; see Table 11.3.

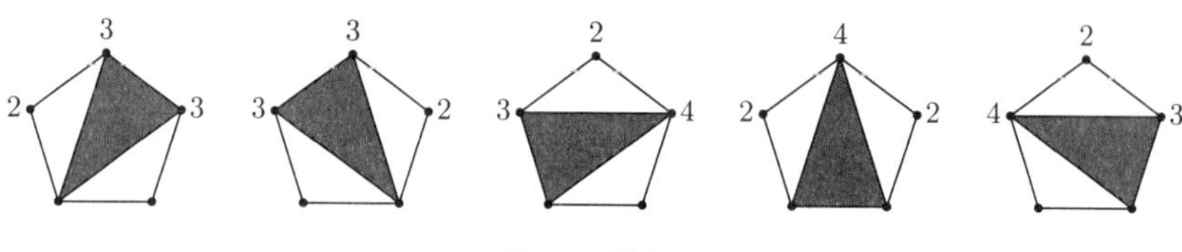

Figure 11.1

Figure 11.2 shows the triangulations of the regular $(n + 2)$-gons, the degrees of the nonbase vertices, and the resulting n-tuples, where $n = 1, 2,$ or 4.

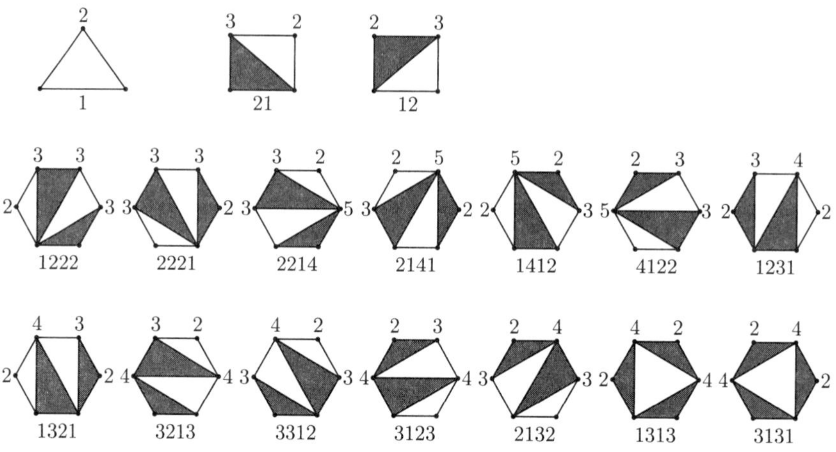

Figure 11.2

This algorithm is certainly reversible. To see this, consider the 4-tuple 2312. Add one to each element: 3423. Now, draw a hexagon; see Figure 11.3. Ignore its base and enter the degrees 3, 4, 2, and 3 of the remaining vertices in the clockwise direction; see Figure 11.4. Draw nonintersecting diagonals in such a way that the degrees of the four vertices match; see Figure 11.5. The resulting configuration is a triangulation of the hexagon.

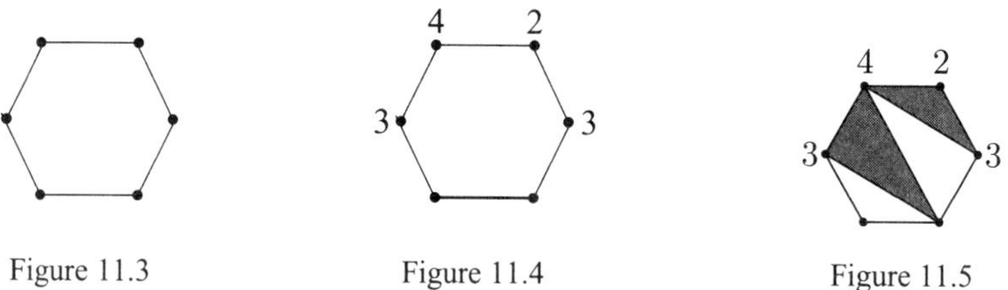

Figure 11.3   Figure 11.4   Figure 11.5

Catalan's Parenthesization Problem Revisited

It follows from earlier discussions that there exists a bijection between the set of correctly parenthesized expressions with $n + 1$ operands and the set of $n \times n$ suitable matrices. Nonetheless, we illustrate a straightforward bijection between the two sets, discovered by Shapiro.

Consider, for example, the correctly parenthesized expression $((a(bc))(de))$ with five operands. Count the number of parentheses, left and right, between any two consecutive operands. Add one to each count. The resulting 4-tuple belongs to B_4:

$$\begin{array}{rl} \text{Expression:} & ((a(bc))(de)) \\ \text{Count parentheses:} & 1\ 0\ 3\ 0 \\ \text{Add 1 to each:} & 2\ 1\ 4\ 1 \end{array}$$

Notice that $2141 \in B_4$; see Table 11.3.

Likewise, the expression $((a(b(cd)(ef)))(gh))$ yields the 7-tuple 2213151 in B_7:

$$\begin{array}{rl} \text{Expression:} & ((a(b(cd)(ef)))(gh)) \\ \text{Count parentheses:} & 1\ 10\ \ 20\ \ 4\ \ 0 \\ \text{Add 1 to each:} & 2\ 21\ \ 31\ \ 5\ \ 1 \end{array}$$

Clearly, this procedure is reversible. For example, the 4-tuple 2312 in B_4 can be used to generate a correctly parenthesized expression with five operands:

$$\begin{array}{rl} \text{Given 4-tuple:} & 2\ \ 3\ \ 1\ \ 2 \\ \text{Subtract one from each:} & 1\ \ 2\ \ 0\ \ 1 \\ \text{Place five operands:} & a\ \ b\ \ cd\ e \\ \text{Insert parentheses:} & (a\,(b((cd)e))) \end{array}$$

The resulting expression $(a(b((cd)e)))$ is correctly parenthesized.
The following example deals with an interesting class of matrices.

Example 11.3 Let $f(n)$ denote the number of nonnegative integral matrices $(m_{ij})_{n \times n}$ such that:

- $m_{ij} = 0$ unless $i = n, i = j,$ or $i = j - 1$; and
- The kth row and column sums are each k, where $1 \le k \le n$.

Find a formula for $f(n)$.

Solution   Table 11.4 lists the various such matrices for $1 \le n \le 4$.

Using the data in Table 11.4, we conjecture that $f(n) = C_n$. Indeed, this is the case.[†]

[†] See R. P. Stanley, *Catalan Addendum*.

Table 11.4

$$[1]\qquad \begin{bmatrix} 0 & 1 \\ 1 & 1 \end{bmatrix}\qquad \begin{bmatrix} 1 & 0 \\ 0 & 2 \end{bmatrix}$$

$$n = 1 \qquad\qquad\qquad n = 2$$

$$\begin{bmatrix} 0 & 1 & 0 \\ 0 & 1 & 1 \\ 1 & 0 & 2 \end{bmatrix}\quad \begin{bmatrix} 0 & 1 & 0 \\ 0 & 0 & 2 \\ 1 & 1 & 1 \end{bmatrix}\quad \begin{bmatrix} 1 & 0 & 0 \\ 0 & 0 & 2 \\ 0 & 2 & 1 \end{bmatrix}\quad \begin{bmatrix} 1 & 0 & 0 \\ 0 & 1 & 1 \\ 0 & 1 & 2 \end{bmatrix}\quad \begin{bmatrix} 1 & 0 & 0 \\ 0 & 2 & 0 \\ 0 & 0 & 3 \end{bmatrix}$$

$$n = 3$$

$$\begin{bmatrix} 0 & 1 & 0 & 0 \\ 0 & 0 & 2 & 0 \\ 0 & 0 & 0 & 3 \\ 1 & 1 & 1 & 1 \end{bmatrix}\begin{bmatrix} 1 & 0 & 0 & 0 \\ 0 & 0 & 2 & 0 \\ 0 & 0 & 0 & 3 \\ 0 & 2 & 1 & 1 \end{bmatrix}\begin{bmatrix} 0 & 1 & 0 & 0 \\ 1 & 1 & 1 & 0 \\ 0 & 0 & 0 & 3 \\ 1 & 0 & 2 & 1 \end{bmatrix}\begin{bmatrix} 1 & 0 & 0 & 0 \\ 0 & 1 & 1 & 0 \\ 0 & 0 & 0 & 3 \\ 0 & 1 & 2 & 1 \end{bmatrix}\begin{bmatrix} 1 & 0 & 0 & 0 \\ 0 & 2 & 0 & 0 \\ 0 & 0 & 0 & 3 \\ 0 & 0 & 3 & 1 \end{bmatrix}\begin{bmatrix} 0 & 1 & 0 & 0 \\ 0 & 0 & 2 & 0 \\ 0 & 0 & 1 & 2 \\ 1 & 1 & 0 & 2 \end{bmatrix}$$

$$\begin{bmatrix} 1 & 0 & 0 & 0 \\ 0 & 0 & 2 & 0 \\ 0 & 0 & 1 & 2 \\ 0 & 2 & 0 & 2 \end{bmatrix}\begin{bmatrix} 0 & 1 & 0 & 0 \\ 0 & 1 & 1 & 0 \\ 0 & 0 & 1 & 2 \\ 1 & 0 & 1 & 2 \end{bmatrix}\begin{bmatrix} 1 & 0 & 0 & 0 \\ 0 & 1 & 1 & 0 \\ 0 & 0 & 1 & 2 \\ 0 & 1 & 1 & 2 \end{bmatrix}\begin{bmatrix} 1 & 0 & 0 & 0 \\ 0 & 2 & 0 & 0 \\ 0 & 0 & 1 & 2 \\ 0 & 0 & 2 & 2 \end{bmatrix}\begin{bmatrix} 0 & 1 & 0 & 0 \\ 0 & 1 & 1 & 0 \\ 0 & 0 & 2 & 1 \\ 1 & 0 & 0 & 3 \end{bmatrix}\begin{bmatrix} 1 & 0 & 0 & 0 \\ 0 & 1 & 1 & 0 \\ 0 & 0 & 2 & 1 \\ 0 & 1 & 0 & 3 \end{bmatrix}$$

$$\begin{bmatrix} 1 & 0 & 0 & 0 \\ 0 & 2 & 0 & 0 \\ 0 & 0 & 2 & 1 \\ 0 & 0 & 1 & 3 \end{bmatrix}\begin{bmatrix} 1 & 0 & 0 & 0 \\ 0 & 2 & 0 & 0 \\ 0 & 0 & 3 & 6 \\ 0 & 0 & 0 & 4 \end{bmatrix}$$

$$n = 4$$

$$\begin{bmatrix} 0 & 1 & 0 & 0 & 0 \\ 0 & 0 & 2 & 0 & 0 \\ 0 & 0 & 0 & 3 & 0 \\ 0 & 0 & 0 & 0 & 4 \\ 1 & 1 & 1 & 1 & 1 \end{bmatrix}\begin{bmatrix} 1 & 0 & 0 & 0 & 0 \\ 0 & 0 & 2 & 0 & 0 \\ 0 & 0 & 0 & 3 & 0 \\ 0 & 0 & 0 & 0 & 4 \\ 0 & 2 & 1 & 1 & 1 \end{bmatrix}\begin{bmatrix} 0 & 1 & 0 & 0 & 0 \\ 0 & 1 & 1 & 0 & 0 \\ 0 & 0 & 0 & 3 & 0 \\ 0 & 0 & 0 & 0 & 4 \\ 1 & 0 & 2 & 1 & 1 \end{bmatrix}\begin{bmatrix} 1 & 0 & 0 & 0 & 0 \\ 0 & 1 & 1 & 0 & 0 \\ 0 & 0 & 0 & 3 & 0 \\ 0 & 0 & 0 & 0 & 4 \\ 0 & 1 & 2 & 1 & 1 \end{bmatrix}\begin{bmatrix} 1 & 0 & 0 & 0 & 0 \\ 0 & 2 & 0 & 0 & 0 \\ 0 & 0 & 0 & 3 & 0 \\ 0 & 0 & 0 & 0 & 4 \\ 0 & 0 & 3 & 1 & 1 \end{bmatrix}$$

$$
\begin{bmatrix} 0&0&0&4&1 \\ 0&0&2&0&2 \\ 0&0&1&0&2 \\ 0&2&0&0&0 \\ 1&0&0&0&0 \end{bmatrix}
\quad
\begin{bmatrix} 0&0&0&3&2 \\ 0&0&3&1&0 \\ 0&2&0&0&1 \\ 1&0&0&0&1 \\ 0&0&0&0&1 \end{bmatrix}
\quad
\begin{bmatrix} 0&0&0&3&2 \\ 0&0&2&1&1 \\ 0&2&1&0&0 \\ 1&0&0&0&1 \\ 0&0&0&0&1 \end{bmatrix}
\quad
\begin{bmatrix} 0&0&0&3&2 \\ 0&0&1&1&2 \\ 0&1&2&0&0 \\ 1&1&0&0&0 \\ 0&0&0&0&1 \end{bmatrix}
\quad
\begin{bmatrix} 0&0&0&2&3 \\ 0&0&2&2&0 \\ 0&2&1&0&0 \\ 0&0&0&0&2 \\ 1&0&0&0&0 \end{bmatrix}
$$

$$
\begin{bmatrix} 0&0&0&4&1 \\ 0&0&2&0&2 \\ 0&1&1&0&1 \\ 0&1&0&0&1 \\ 1&0&0&0&0 \end{bmatrix}
\quad
\begin{bmatrix} 0&0&0&4&1 \\ 0&0&0&0&4 \\ 0&0&3&0&0 \\ 0&2&0&0&0 \\ 1&0&0&0&0 \end{bmatrix}
\quad
\begin{bmatrix} 0&0&0&3&2 \\ 0&0&3&1&0 \\ 0&0&0&0&3 \\ 0&2&0&0&0 \\ 1&0&0&0&0 \end{bmatrix}
\quad
\begin{bmatrix} 0&0&0&3&2 \\ 0&0&2&1&1 \\ 0&0&1&0&2 \\ 0&2&0&0&0 \\ 1&0&0&0&0 \end{bmatrix}
\quad
\begin{bmatrix} 0&0&0&2&3 \\ 0&0&2&2&0 \\ 0&2&1&0&0 \\ 1&0&0&0&1 \\ 0&0&0&0&1 \end{bmatrix}
$$

$$
\begin{bmatrix} 0&0&0&4&1 \\ 0&0&2&0&2 \\ 0&1&1&0&1 \\ 1&1&0&0&0 \\ 0&0&0&0&1 \end{bmatrix}
\quad
\begin{bmatrix} 0&0&0&4&1 \\ 0&0&1&0&3 \\ 0&0&2&0&1 \\ 0&2&0&0&0 \\ 1&0&0&0&0 \end{bmatrix}
\quad
\begin{bmatrix} 0&0&0&3&2 \\ 0&0&3&1&0 \\ 0&1&0&0&2 \\ 0&1&0&0&1 \\ 1&0&0&0&0 \end{bmatrix}
\quad
\begin{bmatrix} 0&0&0&3&2 \\ 0&0&2&1&1 \\ 0&1&1&0&1 \\ 0&1&0&0&1 \\ 1&0&0&0&0 \end{bmatrix}
\quad
\begin{bmatrix} 0&0&0&3&2 \\ 0&0&0&1&3 \\ 0&0&3&0&0 \\ 0&2&0&0&0 \\ 1&0&0&0&0 \end{bmatrix}
$$

$$
\begin{bmatrix} 0&0&0&4&1 \\ 0&0&2&0&2 \\ 0&2&1&0&0 \\ 0&0&0&0&2 \\ 1&0&0&0&0 \end{bmatrix}
\quad
\begin{bmatrix} 0&0&0&4&1 \\ 0&0&1&0&3 \\ 0&1&2&0&0 \\ 0&1&0&0&1 \\ 1&0&0&0&0 \end{bmatrix}
\quad
\begin{bmatrix} 0&0&0&3&2 \\ 0&0&3&1&0 \\ 0&1&0&0&2 \\ 1&1&0&0&0 \\ 0&0&0&0&1 \end{bmatrix}
\quad
\begin{bmatrix} 0&0&0&3&2 \\ 0&0&2&1&1 \\ 0&1&1&0&1 \\ 1&1&0&0&0 \\ 0&0&0&0&1 \end{bmatrix}
\quad
\begin{bmatrix} 0&0&0&3&2 \\ 0&0&1&1&2 \\ 0&0&2&0&1 \\ 0&2&0&0&0 \\ 1&0&0&0&0 \end{bmatrix}
$$

$$
\begin{bmatrix} 0&0&0&4&1 \\ 0&0&2&0&2 \\ 0&2&1&0&0 \\ 1&0&0&0&1 \\ 0&0&0&0&1 \end{bmatrix}
\quad
\begin{bmatrix} 0&0&0&4&1 \\ 0&0&1&0&3 \\ 0&1&2&0&0 \\ 1&1&0&0&0 \\ 0&0&0&0&1 \end{bmatrix}
\quad
\begin{bmatrix} 0&0&0&3&2 \\ 0&0&3&1&0 \\ 0&2&0&0&1 \\ 0&0&0&0&2 \\ 1&0&0&0&0 \end{bmatrix}
\quad
\begin{bmatrix} 0&0&0&3&2 \\ 0&0&2&1&1 \\ 0&2&1&0&0 \\ 0&0&0&0&2 \\ 1&0&0&0&0 \end{bmatrix}
\quad
\begin{bmatrix} 0&0&0&3&2 \\ 0&0&1&1&2 \\ 0&1&2&0&0 \\ 0&1&0&0&1 \\ 1&0&0&0&0 \end{bmatrix}
$$

(Continued)

Table 11.4 (Continued)

$$\begin{bmatrix} 0 & 1 & 0 & 0 & 0 \\ 0 & 1 & 1 & 0 & 0 \\ 0 & 0 & 1 & 2 & 0 \\ 0 & 0 & 0 & 2 & 2 \\ 1 & 0 & 1 & 0 & 3 \end{bmatrix} \begin{bmatrix} 1 & 0 & 0 & 0 & 0 \\ 0 & 1 & 1 & 0 & 0 \\ 0 & 0 & 1 & 2 & 0 \\ 0 & 0 & 0 & 2 & 2 \\ 0 & 1 & 1 & 0 & 3 \end{bmatrix} \begin{bmatrix} 1 & 0 & 0 & 0 & 0 \\ 0 & 2 & 0 & 0 & 0 \\ 0 & 0 & 1 & 2 & 0 \\ 0 & 0 & 0 & 2 & 2 \\ 0 & 0 & 2 & 0 & 3 \end{bmatrix} \begin{bmatrix} 0 & 1 & 0 & 0 & 0 \\ 0 & 1 & 1 & 0 & 0 \\ 0 & 0 & 2 & 1 & 0 \\ 0 & 0 & 0 & 2 & 2 \\ 1 & 0 & 0 & 1 & 3 \end{bmatrix} \begin{bmatrix} 1 & 0 & 0 & 0 & 0 \\ 0 & 1 & 1 & 0 & 0 \\ 0 & 0 & 2 & 1 & 0 \\ 0 & 0 & 0 & 2 & 2 \\ 0 & 1 & 0 & 1 & 3 \end{bmatrix}$$

$$\begin{bmatrix} 1 & 0 & 0 & 0 & 0 \\ 0 & 2 & 0 & 0 & 0 \\ 0 & 0 & 2 & 1 & 0 \\ 0 & 0 & 0 & 2 & 2 \\ 0 & 0 & 1 & 1 & 3 \end{bmatrix} \begin{bmatrix} 1 & 0 & 0 & 0 & 0 \\ 0 & 2 & 0 & 0 & 0 \\ 0 & 0 & 3 & 1 & 0 \\ 0 & 0 & 0 & 2 & 2 \\ 0 & 0 & 0 & 2 & 3 \end{bmatrix} \begin{bmatrix} 0 & 1 & 0 & 0 & 0 \\ 0 & 1 & 1 & 0 & 0 \\ 0 & 0 & 2 & 1 & 0 \\ 0 & 0 & 0 & 3 & 1 \\ 1 & 0 & 0 & 0 & 4 \end{bmatrix} \begin{bmatrix} 1 & 0 & 0 & 0 & 0 \\ 0 & 1 & 1 & 0 & 0 \\ 0 & 0 & 2 & 1 & 0 \\ 0 & 0 & 0 & 3 & 1 \\ 0 & 1 & 0 & 0 & 4 \end{bmatrix} \begin{bmatrix} 1 & 0 & 0 & 0 & 0 \\ 0 & 2 & 0 & 0 & 0 \\ 0 & 0 & 2 & 1 & 0 \\ 0 & 0 & 0 & 3 & 1 \\ 0 & 0 & 1 & 0 & 4 \end{bmatrix}$$

$$\begin{bmatrix} 1 & 0 & 0 & 0 & 0 \\ 0 & 2 & 0 & 0 & 0 \\ 0 & 0 & 3 & 0 & 0 \\ 0 & 0 & 0 & 3 & 1 \\ 0 & 0 & 0 & 1 & 4 \end{bmatrix} \begin{bmatrix} 1 & 0 & 0 & 0 & 0 \\ 0 & 2 & 0 & 0 & 0 \\ 0 & 0 & 3 & 0 & 0 \\ 0 & 0 & 0 & 4 & 0 \\ 0 & 0 & 0 & 0 & 5 \end{bmatrix}$$

$$n = 5$$

Triangular Numbers, Matrices, and Catalan Numbers

The next example deals with *triangluar numbers* $\binom{n}{2}$ and a special class of square matrices. The first ten triangular numbers are 1, 3, 6, 10, 15, 21, 28, 36, 45, and 55.

Example 11.4 Let $g(n)$ denote the number of nonnegative integral matrices $(m_{ij})_{n \times n}$ such that:
- $m_{ij} = 0$ if $j \geq i + 2$; and
- The kth row and column sums are each $\binom{k+1}{2}$, where $1 \leq k \leq n$.

Find a formula for $g(n)$.

Solution   Table 11.5 lists the various such matrices for $1 \leq n \leq 4$.

Notice that:

$$
\begin{aligned}
g(1) &= \quad 1 = C_1 \\
g(2) &= \quad 2 = C_1 C_2 \\
g(3) &= \quad 10 = C_1 C_2 C_3. \\
g(4) &= 140 = C_1 C_2 C_3 C_4
\end{aligned}
$$

More generally, $g(n) = C_1 C_2 C_3 \ldots C_n.$[†]   ∎

Four Interesting Tidbits

For the curious-minded, we now add four interesting properties of suitable matrices. The first two were originally conjectured by Leighton and Newman, but they were all established by L. W. Shapiro:

- The number of suitable matrices with trace t is given by $\frac{(n-t-1)(3n-t-2)}{n-1}\binom{t-n-1}{n-2}$.
- The number of suitable matrices with k 1s along the main diagonal is given by
$$
\frac{2^{n-2k+1}(n-1)!}{(n-2k+1)!k!(k-1)!}
$$
- The number of suitable matrices with $a_1 = a$ is given by the *ballot number*

$$
B_{n,a} = \frac{a}{2n-a+2}\binom{2n-a+2}{n+1}
$$

[†] See R. P. Stanley, *Catalan Addendum.*

Table 11.5

$$[1] \qquad \begin{bmatrix} 0 & 1 \\ 1 & 2 \end{bmatrix} \qquad \begin{bmatrix} 1 & 0 \\ 0 & 3 \end{bmatrix}$$

$$n = 1 \qquad\qquad\qquad n = 2$$

$$\begin{bmatrix} 1 & 0 & 0 \\ 0 & 0 & 3 \\ 0 & 3 & 3 \end{bmatrix} \quad \begin{bmatrix} 0 & 1 & 0 \\ 0 & 0 & 3 \\ 1 & 2 & 3 \end{bmatrix} \quad \begin{bmatrix} 0 & 1 & 0 \\ 1 & 0 & 2 \\ 0 & 2 & 4 \end{bmatrix} \quad \begin{bmatrix} 1 & 0 & 0 \\ 0 & 1 & 2 \\ 0 & 2 & 4 \end{bmatrix} \quad \begin{bmatrix} 0 & 1 & 0 \\ 0 & 1 & 2 \\ 1 & 1 & 4 \end{bmatrix}$$

$$\begin{bmatrix} 0 & 1 & 0 \\ 1 & 1 & 1 \\ 0 & 1 & 5 \end{bmatrix} \quad \begin{bmatrix} 1 & 0 & 0 \\ 0 & 2 & 1 \\ 0 & 1 & 5 \end{bmatrix} \quad \begin{bmatrix} 0 & 1 & 0 \\ 0 & 2 & 1 \\ 1 & 0 & 5 \end{bmatrix} \quad \begin{bmatrix} 0 & 1 & 0 \\ 1 & 2 & 0 \\ 0 & 0 & 6 \end{bmatrix} \quad \begin{bmatrix} 1 & 0 & 0 \\ 0 & 3 & 0 \\ 0 & 0 & 6 \end{bmatrix}$$

$$n = 3$$

$$\begin{bmatrix} 0 & 1 & 0 & 0 \\ 1 & 2 & 0 & 0 \\ 0 & 0 & 0 & 6 \\ 0 & 0 & 6 & 4 \end{bmatrix} \begin{bmatrix} 1 & 0 & 0 & 0 \\ 0 & 3 & 0 & 0 \\ 0 & 0 & 0 & 6 \\ 0 & 0 & 6 & 4 \end{bmatrix} \begin{bmatrix} 0 & 1 & 0 & 0 \\ 1 & 1 & 1 & 0 \\ 0 & 0 & 0 & 6 \\ 0 & 1 & 5 & 4 \end{bmatrix} \begin{bmatrix} 1 & 0 & 0 & 0 \\ 0 & 2 & 1 & 0 \\ 0 & 0 & 0 & 6 \\ 0 & 1 & 5 & 4 \end{bmatrix} \begin{bmatrix} 0 & 1 & 0 & 0 \\ 0 & 2 & 1 & 0 \\ 0 & 0 & 0 & 6 \\ 1 & 0 & 5 & 4 \end{bmatrix} \begin{bmatrix} 0 & 1 & 0 & 0 \\ 1 & 0 & 2 & 0 \\ 0 & 0 & 0 & 6 \\ 0 & 2 & 4 & 4 \end{bmatrix}$$

$$\begin{bmatrix} 1 & 0 & 0 & 0 \\ 0 & 1 & 2 & 0 \\ 0 & 0 & 0 & 6 \\ 0 & 2 & 4 & 4 \end{bmatrix} \begin{bmatrix} 0 & 1 & 0 & 0 \\ 0 & 1 & 2 & 0 \\ 0 & 0 & 0 & 6 \\ 1 & 1 & 4 & 4 \end{bmatrix} \begin{bmatrix} 1 & 0 & 0 & 0 \\ 0 & 0 & 3 & 0 \\ 0 & 0 & 0 & 6 \\ 0 & 3 & 3 & 4 \end{bmatrix} \begin{bmatrix} 0 & 1 & 0 & 0 \\ 0 & 0 & 3 & 0 \\ 0 & 0 & 0 & 6 \\ 1 & 2 & 3 & 4 \end{bmatrix} \begin{bmatrix} 0 & 1 & 0 & 0 \\ 0 & 2 & 1 & 0 \\ 1 & 0 & 0 & 5 \\ 0 & 0 & 5 & 5 \end{bmatrix} \begin{bmatrix} 0 & 1 & 0 & 0 \\ 0 & 1 & 2 & 0 \\ 1 & 0 & 0 & 5 \\ 0 & 1 & 4 & 5 \end{bmatrix}$$

$$\begin{bmatrix} 0 & 1 & 0 & 0 \\ 0 & 0 & 3 & 0 \\ 1 & 0 & 0 & 5 \\ 0 & 2 & 3 & 5 \end{bmatrix} \begin{bmatrix} 0 & 1 & 0 & 0 \\ 1 & 1 & 1 & 0 \\ 0 & 1 & 0 & 5 \\ 0 & 0 & 5 & 5 \end{bmatrix} \begin{bmatrix} 1 & 0 & 0 & 0 \\ 0 & 2 & 1 & 0 \\ 0 & 1 & 0 & 5 \\ 0 & 0 & 5 & 5 \end{bmatrix} \begin{bmatrix} 0 & 1 & 0 & 0 \\ 1 & 0 & 2 & 0 \\ 0 & 1 & 0 & 5 \\ 0 & 1 & 4 & 5 \end{bmatrix} \begin{bmatrix} 1 & 0 & 0 & 0 \\ 0 & 1 & 2 & 0 \\ 0 & 1 & 0 & 5 \\ 0 & 1 & 4 & 5 \end{bmatrix} \begin{bmatrix} 0 & 1 & 0 & 0 \\ 0 & 1 & 2 & 0 \\ 0 & 1 & 0 & 5 \\ 1 & 0 & 4 & 5 \end{bmatrix}$$

$$\begin{bmatrix} 1 & 0 & 0 & 0 \\ 0 & 0 & 3 & 0 \\ 0 & 1 & 0 & 5 \\ 0 & 2 & 3 & 5 \end{bmatrix} \begin{bmatrix} 0 & 1 & 0 & 0 \\ 0 & 0 & 3 & 0 \\ 0 & 1 & 0 & 5 \\ 1 & 1 & 3 & 5 \end{bmatrix} \begin{bmatrix} 0 & 1 & 0 & 0 \\ 0 & 1 & 2 & 0 \\ 1 & 1 & 0 & 4 \\ 0 & 0 & 4 & 6 \end{bmatrix} \begin{bmatrix} 0 & 1 & 0 & 0 \\ 0 & 0 & 3 & 0 \\ 1 & 1 & 0 & 4 \\ 0 & 1 & 3 & 6 \end{bmatrix} \begin{bmatrix} 0 & 1 & 0 & 0 \\ 1 & 0 & 2 & 0 \\ 0 & 2 & 0 & 4 \\ 0 & 0 & 4 & 6 \end{bmatrix} \begin{bmatrix} 1 & 0 & 0 & 0 \\ 0 & 1 & 2 & 0 \\ 0 & 2 & 0 & 4 \\ 0 & 0 & 4 & 6 \end{bmatrix}$$

$$\begin{bmatrix} 1 & 0 & 0 & 0 \\ 0 & 0 & 3 & 0 \\ 0 & 2 & 0 & 4 \\ 0 & 1 & 3 & 6 \end{bmatrix} \begin{bmatrix} 0 & 1 & 0 & 0 \\ 0 & 0 & 3 & 0 \\ 0 & 2 & 0 & 4 \\ 1 & 0 & 3 & 6 \end{bmatrix} \begin{bmatrix} 0 & 1 & 0 & 0 \\ 0 & 0 & 3 & 0 \\ 1 & 2 & 0 & 3 \\ 0 & 0 & 3 & 7 \end{bmatrix} \begin{bmatrix} 1 & 0 & 0 & 0 \\ 0 & 0 & 3 & 0 \\ 0 & 3 & 0 & 3 \\ 0 & 0 & 3 & 7 \end{bmatrix} \begin{bmatrix} 0 & 1 & 0 & 0 \\ 1 & 2 & 0 & 0 \\ 0 & 0 & 1 & 5 \\ 0 & 0 & 5 & 5 \end{bmatrix} \begin{bmatrix} 1 & 0 & 0 & 0 \\ 0 & 3 & 0 & 0 \\ 0 & 0 & 1 & 5 \\ 0 & 0 & 5 & 5 \end{bmatrix}$$

$$\begin{bmatrix} 0 & 1 & 0 & 0 \\ 1 & 1 & 1 & 0 \\ 0 & 0 & 1 & 5 \\ 0 & 1 & 4 & 5 \end{bmatrix} \begin{bmatrix} 1 & 0 & 0 & 0 \\ 0 & 2 & 1 & 0 \\ 0 & 0 & 1 & 5 \\ 0 & 1 & 4 & 5 \end{bmatrix} \begin{bmatrix} 0 & 1 & 0 & 0 \\ 0 & 2 & 1 & 0 \\ 0 & 0 & 1 & 5 \\ 1 & 0 & 4 & 5 \end{bmatrix} \begin{bmatrix} 0 & 1 & 0 & 0 \\ 1 & 0 & 2 & 0 \\ 0 & 0 & 1 & 5 \\ 0 & 2 & 3 & 5 \end{bmatrix} \begin{bmatrix} 1 & 0 & 0 & 0 \\ 0 & 1 & 2 & 0 \\ 0 & 0 & 1 & 5 \\ 0 & 2 & 3 & 5 \end{bmatrix} \begin{bmatrix} 0 & 1 & 0 & 0 \\ 0 & 1 & 2 & 0 \\ 0 & 0 & 1 & 5 \\ 1 & 1 & 3 & 5 \end{bmatrix}$$

$$\begin{bmatrix} 1 & 0 & 0 & 0 \\ 0 & 0 & 3 & 0 \\ 0 & 0 & 1 & 5 \\ 0 & 3 & 2 & 5 \end{bmatrix} \begin{bmatrix} 0 & 1 & 0 & 0 \\ 0 & 0 & 3 & 0 \\ 0 & 0 & 1 & 5 \\ 1 & 2 & 2 & 5 \end{bmatrix} \begin{bmatrix} 0 & 1 & 0 & 0 \\ 0 & 2 & 1 & 0 \\ 1 & 0 & 1 & 4 \\ 0 & 0 & 4 & 6 \end{bmatrix} \begin{bmatrix} 0 & 1 & 0 & 0 \\ 0 & 1 & 2 & 0 \\ 1 & 0 & 1 & 4 \\ 0 & 1 & 3 & 6 \end{bmatrix} \begin{bmatrix} 0 & 1 & 0 & 0 \\ 0 & 0 & 3 & 0 \\ 1 & 0 & 1 & 4 \\ 0 & 2 & 2 & 6 \end{bmatrix} \begin{bmatrix} 0 & 1 & 0 & 0 \\ 1 & 1 & 1 & 0 \\ 0 & 1 & 1 & 4 \\ 0 & 0 & 4 & 6 \end{bmatrix}$$

$$\begin{bmatrix} 1 & 0 & 0 & 0 \\ 0 & 2 & 1 & 0 \\ 0 & 1 & 1 & 4 \\ 0 & 0 & 4 & 6 \end{bmatrix} \begin{bmatrix} 0 & 1 & 0 & 0 \\ 1 & 0 & 2 & 0 \\ 0 & 1 & 1 & 4 \\ 0 & 1 & 3 & 6 \end{bmatrix} \begin{bmatrix} 1 & 0 & 0 & 0 \\ 0 & 1 & 2 & 0 \\ 0 & 1 & 1 & 4 \\ 0 & 1 & 3 & 6 \end{bmatrix} \begin{bmatrix} 0 & 1 & 0 & 0 \\ 0 & 1 & 2 & 0 \\ 0 & 1 & 1 & 4 \\ 1 & 0 & 3 & 6 \end{bmatrix} \begin{bmatrix} 1 & 0 & 0 & 0 \\ 0 & 0 & 3 & 0 \\ 0 & 1 & 1 & 4 \\ 0 & 2 & 2 & 6 \end{bmatrix} \begin{bmatrix} 0 & 1 & 0 & 0 \\ 0 & 0 & 3 & 0 \\ 0 & 1 & 1 & 4 \\ 1 & 1 & 2 & 6 \end{bmatrix}$$

$$\begin{bmatrix} 0 & 1 & 0 & 0 \\ 0 & 1 & 2 & 0 \\ 1 & 1 & 1 & 3 \\ 0 & 0 & 3 & 7 \end{bmatrix} \begin{bmatrix} 0 & 1 & 0 & 0 \\ 0 & 0 & 3 & 0 \\ 1 & 1 & 1 & 3 \\ 0 & 1 & 2 & 7 \end{bmatrix} \begin{bmatrix} 0 & 1 & 0 & 0 \\ 1 & 0 & 2 & 0 \\ 0 & 2 & 1 & 3 \\ 0 & 0 & 3 & 7 \end{bmatrix} \begin{bmatrix} 1 & 0 & 0 & 0 \\ 0 & 1 & 2 & 0 \\ 0 & 2 & 1 & 3 \\ 0 & 0 & 3 & 7 \end{bmatrix} \begin{bmatrix} 1 & 0 & 0 & 0 \\ 0 & 0 & 3 & 0 \\ 0 & 2 & 1 & 3 \\ 0 & 1 & 2 & 7 \end{bmatrix} \begin{bmatrix} 0 & 1 & 0 & 0 \\ 0 & 0 & 3 & 0 \\ 0 & 2 & 1 & 3 \\ 1 & 0 & 2 & 7 \end{bmatrix}$$

(*Continued*)

Table 11.5 (Continued)

$$
\begin{bmatrix} 0 & 1 & 0 & 0 \\ 0 & 0 & 3 & 0 \\ 1 & 2 & 1 & 2 \\ 0 & 0 & 2 & 8 \end{bmatrix}
\begin{bmatrix} 1 & 0 & 0 & 0 \\ 0 & 0 & 3 & 0 \\ 0 & 3 & 1 & 2 \\ 0 & 0 & 2 & 8 \end{bmatrix}
\begin{bmatrix} 0 & 1 & 0 & 0 \\ 1 & 2 & 0 & 0 \\ 0 & 0 & 2 & 4 \\ 0 & 0 & 4 & 6 \end{bmatrix}
\begin{bmatrix} 1 & 0 & 0 & 0 \\ 0 & 3 & 0 & 0 \\ 0 & 0 & 2 & 4 \\ 0 & 0 & 4 & 6 \end{bmatrix}
\begin{bmatrix} 0 & 1 & 0 & 0 \\ 1 & 1 & 1 & 0 \\ 0 & 0 & 2 & 4 \\ 0 & 1 & 3 & 6 \end{bmatrix}
\begin{bmatrix} 1 & 0 & 0 & 0 \\ 0 & 2 & 1 & 0 \\ 0 & 0 & 2 & 4 \\ 0 & 1 & 3 & 6 \end{bmatrix}
$$

$$
\begin{bmatrix} 0 & 1 & 0 & 0 \\ 0 & 2 & 1 & 0 \\ 0 & 0 & 2 & 4 \\ 1 & 0 & 3 & 6 \end{bmatrix}
\begin{bmatrix} 0 & 1 & 0 & 0 \\ 1 & 0 & 2 & 0 \\ 0 & 0 & 2 & 4 \\ 0 & 2 & 2 & 6 \end{bmatrix}
\begin{bmatrix} 1 & 0 & 0 & 0 \\ 0 & 1 & 2 & 0 \\ 0 & 0 & 2 & 4 \\ 0 & 2 & 2 & 6 \end{bmatrix}
\begin{bmatrix} 0 & 1 & 0 & 0 \\ 0 & 1 & 2 & 0 \\ 0 & 0 & 2 & 4 \\ 1 & 1 & 2 & 6 \end{bmatrix}
\begin{bmatrix} 1 & 0 & 0 & 0 \\ 0 & 0 & 3 & 0 \\ 0 & 0 & 2 & 4 \\ 0 & 3 & 1 & 6 \end{bmatrix}
\begin{bmatrix} 0 & 1 & 0 & 0 \\ 0 & 0 & 3 & 0 \\ 0 & 0 & 2 & 4 \\ 1 & 2 & 1 & 6 \end{bmatrix}
$$

$$
\begin{bmatrix} 0 & 1 & 0 & 0 \\ 0 & 2 & 1 & 0 \\ 1 & 0 & 2 & 3 \\ 0 & 0 & 3 & 7 \end{bmatrix}
\begin{bmatrix} 0 & 1 & 0 & 0 \\ 0 & 1 & 2 & 0 \\ 1 & 0 & 2 & 3 \\ 0 & 1 & 2 & 7 \end{bmatrix}
\begin{bmatrix} 0 & 1 & 0 & 0 \\ 0 & 0 & 3 & 0 \\ 1 & 0 & 2 & 3 \\ 0 & 2 & 1 & 7 \end{bmatrix}
\begin{bmatrix} 0 & 1 & 0 & 0 \\ 1 & 1 & 1 & 0 \\ 0 & 1 & 2 & 3 \\ 0 & 0 & 3 & 7 \end{bmatrix}
\begin{bmatrix} 1 & 0 & 0 & 0 \\ 0 & 2 & 1 & 0 \\ 0 & 1 & 2 & 3 \\ 0 & 0 & 3 & 7 \end{bmatrix}
\begin{bmatrix} 0 & 1 & 0 & 0 \\ 1 & 0 & 2 & 0 \\ 0 & 1 & 2 & 3 \\ 0 & 1 & 2 & 7 \end{bmatrix}
$$

$$
\begin{bmatrix} 1 & 0 & 0 & 0 \\ 0 & 1 & 2 & 0 \\ 0 & 1 & 2 & 3 \\ 0 & 1 & 2 & 7 \end{bmatrix}
\begin{bmatrix} 0 & 1 & 0 & 0 \\ 0 & 1 & 2 & 0 \\ 0 & 1 & 2 & 3 \\ 1 & 0 & 2 & 7 \end{bmatrix}
\begin{bmatrix} 1 & 0 & 0 & 0 \\ 0 & 0 & 3 & 0 \\ 0 & 1 & 2 & 3 \\ 0 & 2 & 1 & 7 \end{bmatrix}
\begin{bmatrix} 0 & 1 & 0 & 0 \\ 0 & 0 & 3 & 0 \\ 0 & 1 & 2 & 3 \\ 1 & 1 & 1 & 7 \end{bmatrix}
\begin{bmatrix} 0 & 1 & 0 & 0 \\ 0 & 1 & 2 & 0 \\ 1 & 1 & 2 & 2 \\ 0 & 0 & 2 & 8 \end{bmatrix}
\begin{bmatrix} 0 & 1 & 0 & 0 \\ 0 & 0 & 3 & 0 \\ 1 & 1 & 2 & 2 \\ 0 & 1 & 1 & 8 \end{bmatrix}
$$

$$
\begin{bmatrix} 0 & 1 & 0 & 0 \\ 1 & 0 & 2 & 0 \\ 0 & 2 & 2 & 2 \\ 0 & 0 & 2 & 8 \end{bmatrix}
\begin{bmatrix} 1 & 0 & 0 & 0 \\ 0 & 1 & 2 & 0 \\ 0 & 2 & 2 & 2 \\ 0 & 0 & 2 & 8 \end{bmatrix}
\begin{bmatrix} 1 & 0 & 0 & 0 \\ 0 & 0 & 3 & 0 \\ 0 & 2 & 2 & 2 \\ 0 & 1 & 1 & 8 \end{bmatrix}
\begin{bmatrix} 0 & 1 & 0 & 0 \\ 0 & 0 & 3 & 0 \\ 0 & 2 & 2 & 2 \\ 1 & 0 & 1 & 8 \end{bmatrix}
\begin{bmatrix} 0 & 1 & 0 & 0 \\ 0 & 0 & 3 & 0 \\ 1 & 2 & 2 & 1 \\ 0 & 0 & 1 & 9 \end{bmatrix}
\begin{bmatrix} 1 & 0 & 0 & 0 \\ 0 & 0 & 3 & 0 \\ 0 & 3 & 2 & 1 \\ 0 & 0 & 1 & 9 \end{bmatrix}
$$

$$
\begin{bmatrix} 0 & 1 & 0 & 0 \\ 1 & 2 & 0 & 0 \\ 0 & 0 & 3 & 3 \\ 0 & 0 & 3 & 7 \end{bmatrix}
\begin{bmatrix} 1 & 0 & 0 & 0 \\ 0 & 3 & 0 & 0 \\ 0 & 0 & 3 & 3 \\ 0 & 0 & 3 & 7 \end{bmatrix}
\begin{bmatrix} 0 & 1 & 0 & 0 \\ 1 & 1 & 1 & 0 \\ 0 & 0 & 3 & 3 \\ 0 & 1 & 2 & 7 \end{bmatrix}
\begin{bmatrix} 1 & 0 & 0 & 0 \\ 0 & 2 & 1 & 0 \\ 0 & 0 & 3 & 3 \\ 0 & 1 & 2 & 7 \end{bmatrix}
\begin{bmatrix} 0 & 1 & 0 & 0 \\ 0 & 2 & 1 & 0 \\ 0 & 0 & 3 & 3 \\ 1 & 0 & 2 & 7 \end{bmatrix}
\begin{bmatrix} 0 & 1 & 0 & 0 \\ 1 & 0 & 2 & 0 \\ 0 & 0 & 3 & 3 \\ 0 & 2 & 1 & 7 \end{bmatrix}
$$

$$
\begin{bmatrix} 1 & 0 & 0 & 0 \\ 0 & 1 & 2 & 0 \\ 0 & 0 & 3 & 3 \\ 0 & 2 & 1 & 7 \end{bmatrix}
\begin{bmatrix} 0 & 1 & 0 & 0 \\ 0 & 1 & 2 & 0 \\ 0 & 0 & 3 & 3 \\ 1 & 1 & 1 & 7 \end{bmatrix}
\begin{bmatrix} 1 & 0 & 0 & 0 \\ 0 & 0 & 3 & 0 \\ 0 & 0 & 3 & 3 \\ 0 & 3 & 0 & 7 \end{bmatrix}
\begin{bmatrix} 0 & 1 & 0 & 0 \\ 0 & 0 & 3 & 0 \\ 0 & 0 & 3 & 3 \\ 1 & 2 & 0 & 7 \end{bmatrix}
\begin{bmatrix} 0 & 1 & 0 & 0 \\ 0 & 2 & 1 & 0 \\ 1 & 0 & 3 & 2 \\ 0 & 0 & 2 & 8 \end{bmatrix}
\begin{bmatrix} 0 & 1 & 0 & 0 \\ 0 & 1 & 2 & 0 \\ 1 & 0 & 3 & 2 \\ 0 & 1 & 1 & 8 \end{bmatrix}
$$

$$
\begin{bmatrix} 0 & 1 & 0 & 0 \\ 0 & 0 & 3 & 0 \\ 1 & 0 & 3 & 2 \\ 0 & 2 & 0 & 8 \end{bmatrix}
\begin{bmatrix} 0 & 1 & 0 & 0 \\ 1 & 1 & 1 & 0 \\ 0 & 1 & 3 & 2 \\ 0 & 0 & 2 & 8 \end{bmatrix}
\begin{bmatrix} 1 & 0 & 0 & 0 \\ 0 & 2 & 1 & 0 \\ 0 & 1 & 3 & 2 \\ 0 & 0 & 2 & 8 \end{bmatrix}
\begin{bmatrix} 0 & 1 & 0 & 0 \\ 1 & 0 & 2 & 0 \\ 0 & 1 & 3 & 2 \\ 0 & 1 & 1 & 8 \end{bmatrix}
\begin{bmatrix} 1 & 0 & 0 & 0 \\ 0 & 1 & 2 & 0 \\ 0 & 1 & 3 & 2 \\ 0 & 1 & 1 & 8 \end{bmatrix}
\begin{bmatrix} 0 & 1 & 0 & 0 \\ 0 & 1 & 2 & 0 \\ 0 & 1 & 3 & 2 \\ 1 & 0 & 1 & 8 \end{bmatrix}
$$

$$
\begin{bmatrix} 1 & 0 & 0 & 0 \\ 0 & 0 & 3 & 0 \\ 0 & 1 & 3 & 2 \\ 0 & 2 & 0 & 8 \end{bmatrix}
\begin{bmatrix} 0 & 1 & 0 & 0 \\ 0 & 0 & 3 & 0 \\ 0 & 1 & 3 & 2 \\ 1 & 1 & 0 & 8 \end{bmatrix}
\begin{bmatrix} 0 & 1 & 0 & 0 \\ 0 & 1 & 2 & 0 \\ 1 & 1 & 3 & 1 \\ 0 & 0 & 1 & 9 \end{bmatrix}
\begin{bmatrix} 0 & 1 & 0 & 0 \\ 0 & 0 & 3 & 0 \\ 1 & 1 & 3 & 1 \\ 0 & 1 & 0 & 9 \end{bmatrix}
\begin{bmatrix} 0 & 1 & 0 & 0 \\ 1 & 0 & 2 & 0 \\ 0 & 2 & 3 & 1 \\ 0 & 0 & 1 & 9 \end{bmatrix}
\begin{bmatrix} 1 & 0 & 0 & 0 \\ 0 & 1 & 2 & 0 \\ 0 & 2 & 3 & 1 \\ 0 & 0 & 1 & 9 \end{bmatrix}
$$

$$
\begin{bmatrix} 1 & 0 & 0 & 0 \\ 0 & 0 & 3 & 0 \\ 0 & 2 & 3 & 1 \\ 0 & 1 & 0 & 9 \end{bmatrix}
\begin{bmatrix} 0 & 1 & 0 & 0 \\ 0 & 0 & 3 & 0 \\ 0 & 2 & 3 & 1 \\ 1 & 0 & 0 & 9 \end{bmatrix}
\begin{bmatrix} 0 & 1 & 0 & 0 \\ 0 & 0 & 3 & 0 \\ 1 & 2 & 3 & 0 \\ 0 & 0 & 0 & 10 \end{bmatrix}
\begin{bmatrix} 1 & 0 & 0 & 0 \\ 0 & 0 & 3 & 0 \\ 0 & 3 & 3 & 0 \\ 0 & 0 & 0 & 10 \end{bmatrix}
\begin{bmatrix} 0 & 1 & 0 & 0 \\ 1 & 2 & 0 & 0 \\ 0 & 0 & 4 & 2 \\ 0 & 0 & 2 & 8 \end{bmatrix}
\begin{bmatrix} 1 & 0 & 0 & 0 \\ 0 & 3 & 0 & 0 \\ 0 & 0 & 4 & 2 \\ 0 & 0 & 2 & 8 \end{bmatrix}
$$

$$
\begin{bmatrix} 0 & 1 & 0 & 0 \\ 1 & 1 & 1 & 0 \\ 0 & 0 & 4 & 2 \\ 0 & 1 & 1 & 8 \end{bmatrix}
\begin{bmatrix} 1 & 0 & 0 & 0 \\ 0 & 2 & 1 & 0 \\ 0 & 0 & 4 & 2 \\ 0 & 1 & 1 & 8 \end{bmatrix}
\begin{bmatrix} 0 & 1 & 0 & 0 \\ 0 & 2 & 1 & 0 \\ 0 & 0 & 4 & 2 \\ 1 & 0 & 1 & 8 \end{bmatrix}
\begin{bmatrix} 0 & 1 & 0 & 0 \\ 1 & 0 & 2 & 0 \\ 0 & 0 & 4 & 2 \\ 0 & 2 & 0 & 8 \end{bmatrix}
\begin{bmatrix} 1 & 0 & 0 & 0 \\ 0 & 1 & 2 & 0 \\ 0 & 0 & 4 & 2 \\ 0 & 2 & 0 & 8 \end{bmatrix}
\begin{bmatrix} 0 & 1 & 0 & 0 \\ 0 & 1 & 2 & 0 \\ 0 & 0 & 4 & 2 \\ 1 & 1 & 0 & 8 \end{bmatrix}
$$

$$
\begin{bmatrix} 0 & 1 & 0 & 0 \\ 0 & 2 & 1 & 0 \\ 1 & 0 & 4 & 1 \\ 0 & 0 & 1 & 9 \end{bmatrix}
\begin{bmatrix} 0 & 1 & 0 & 0 \\ 0 & 1 & 2 & 0 \\ 1 & 0 & 4 & 1 \\ 0 & 1 & 0 & 9 \end{bmatrix}
\begin{bmatrix} 0 & 1 & 0 & 0 \\ 1 & 1 & 1 & 0 \\ 0 & 1 & 4 & 1 \\ 0 & 0 & 1 & 9 \end{bmatrix}
\begin{bmatrix} 1 & 0 & 0 & 0 \\ 0 & 2 & 1 & 0 \\ 0 & 1 & 4 & 1 \\ 0 & 0 & 1 & 9 \end{bmatrix}
\begin{bmatrix} 0 & 1 & 0 & 0 \\ 1 & 0 & 2 & 0 \\ 0 & 1 & 4 & 1 \\ 0 & 1 & 0 & 9 \end{bmatrix}
\begin{bmatrix} 1 & 0 & 0 & 0 \\ 0 & 1 & 2 & 0 \\ 0 & 1 & 4 & 1 \\ 0 & 1 & 0 & 9 \end{bmatrix}
$$

(Continued)

Table 11.5 (Continued)

$$
\begin{bmatrix} 0 & 1 & 0 & 0 \\ 0 & 1 & 2 & 0 \\ 0 & 1 & 4 & 1 \\ 1 & 0 & 0 & 9 \end{bmatrix}
\begin{bmatrix} 0 & 1 & 0 & 0 \\ 0 & 1 & 2 & 0 \\ 1 & 1 & 4 & 0 \\ 0 & 0 & 0 & 10 \end{bmatrix}
\begin{bmatrix} 0 & 1 & 0 & 0 \\ 1 & 0 & 2 & 0 \\ 0 & 2 & 4 & 0 \\ 0 & 0 & 0 & 10 \end{bmatrix}
\begin{bmatrix} 1 & 0 & 0 & 0 \\ 0 & 1 & 2 & 0 \\ 0 & 2 & 4 & 0 \\ 0 & 0 & 0 & 10 \end{bmatrix}
\begin{bmatrix} 0 & 1 & 0 & 0 \\ 1 & 2 & 0 & 0 \\ 0 & 0 & 5 & 1 \\ 0 & 0 & 1 & 9 \end{bmatrix}
\begin{bmatrix} 1 & 0 & 0 & 0 \\ 0 & 3 & 0 & 0 \\ 0 & 0 & 5 & 1 \\ 0 & 0 & 1 & 9 \end{bmatrix}
$$

$$
\begin{bmatrix} 0 & 1 & 0 & 0 \\ 1 & 1 & 1 & 0 \\ 0 & 0 & 5 & 1 \\ 0 & 1 & 0 & 9 \end{bmatrix}
\begin{bmatrix} 1 & 0 & 0 & 0 \\ 0 & 2 & 1 & 0 \\ 0 & 0 & 5 & 1 \\ 0 & 1 & 0 & 9 \end{bmatrix}
\begin{bmatrix} 0 & 1 & 0 & 0 \\ 0 & 2 & 1 & 0 \\ 0 & 0 & 5 & 1 \\ 1 & 0 & 0 & 9 \end{bmatrix}
\begin{bmatrix} 0 & 1 & 0 & 0 \\ 0 & 2 & 1 & 0 \\ 1 & 0 & 5 & 0 \\ 0 & 0 & 0 & 10 \end{bmatrix}
\begin{bmatrix} 0 & 1 & 0 & 0 \\ 1 & 1 & 1 & 0 \\ 0 & 1 & 5 & 0 \\ 0 & 0 & 0 & 10 \end{bmatrix}
\begin{bmatrix} 1 & 0 & 0 & 0 \\ 0 & 2 & 1 & 0 \\ 0 & 1 & 5 & 0 \\ 0 & 0 & 0 & 10 \end{bmatrix}
$$

$$
\begin{bmatrix} 0 & 1 & 0 & 0 \\ 1 & 2 & 0 & 0 \\ 0 & 0 & 6 & 0 \\ 0 & 0 & 0 & 10 \end{bmatrix}
\begin{bmatrix} 1 & 0 & 0 & 0 \\ 0 & 3 & 0 & 0 \\ 0 & 0 & 6 & 0 \\ 0 & 0 & 0 & 10 \end{bmatrix}
$$

In particular,

$$B_{n,1} = \frac{1}{2n+1}\binom{2n+1}{n+1}$$

$$= C_n$$

- The number N of palindromic suitable matrices is given by

$$N = \begin{cases} 0 & \text{if } n \text{ is even} \\ \dfrac{2}{n+1}\dbinom{n-1}{(n-1)/2} & \text{otherwise} \end{cases}$$

$$= \begin{cases} 0 & \text{if } n \text{ is even} \\ C_{\frac{n-1}{2}} & \text{otherwise} \end{cases}$$

Catalan Goes to the World Series

Here, we present a delightful application of Catalan numbers to the World Series, studied in 1993 by L. W. Shapiro and by W. Hamilton of the University of the District of Columbia. Although it is a best-of-seven series, we assume that the series ends when one of the two teams wins n games and that there are no draws. Suppose team A wins each game with constant probability p and loses each game with probability $q = 1 - p$. We would like to compute the expected number E_n of games to be played.

Table 11.6 shows the values of E_n for $1 \le n \le 5$.

Table 11.6 Expected Values

n	E_n
1	1
2	$2(1 + pq)$
3	$3(1 + pq + 2p^2q^2)$
4	$4(1 + pq + 2p^2q^2 + 5p^3q^3)$
5	$5(1 + pq + 2p^2q^2 + 5p^3q^3 + 14p^4q^4)$

It exhibits an interesting pattern: The coefficient of $(pq)^k$ in the expression E_n/n gives the Catalan number C_k, where $0 \le k \le n$. For example,

$$\frac{E_5}{5} = \sum_{k=0}^{4} C_k(pq)^k$$

More generally, Shapiro and Hamilton established the following result.

Theorem 11.1 Suppose two teams, A and B, play a series of games, and the series ends when one team wins n games. Let p be the probability that team A wins a game, and let $q = 1 - p$. Let E_n denote the expected number of games to be played. Then

$$\frac{E_n}{n} = \sum_{k=0}^{n-1} C_k (pq)^k$$

Proof   The proof consists of three parts. First, we express $\frac{E_n}{n}$ as a sum and then use the sum to derive a recurrence relation for $\frac{E_{n+1}}{n+1}$. Finally, we employ iteration to develop the desired formula for $\frac{E_n}{n}$. The second part uses Pascal's identity, and the facts

$$\sum_{i=1}^{n+1} b_i - \sum_{i=1}^{n} b_i = b_{n+1}$$

and $p + q = 1$. The third part uses the fact that

$$2\binom{2n}{n} - \binom{2n+1}{n} = 2 \cdot \frac{(2n)!}{n!n!} - \frac{(2n+1)!}{n!(n+1)!}$$

$$= \frac{(2n)!}{(n+1)!n!}[2(n+1) - (2n+1)]$$

$$= \frac{(2n)!}{(n+1)!n!}$$

$$= C_n$$

(1) To express E_n as a sum:

$$E_n = n(p^n + q^n) + (n+1)\binom{n}{1}(p^n q + pq^n) + (n+2)\binom{n+1}{2}(p^n q^2 + p^2 q^n)$$

$$+ \cdots + (2n-1)\binom{2n-2}{n-1}(p^n q^{n-1} + p^{n-1} q^n)$$

$$= \sum_{k=0}^{n-1} (n+k)\binom{n-1+k}{k}(p^n q^k + p^k q^n)$$

$$= \sum_{k=0}^{n-1} (n+k)\frac{(n-1+k)!}{k!(n-1)!}(p^n q^k + p^k q^n)$$

$$= \sum_{k=0}^{n-1} n \cdot \frac{(n+k)!}{k!n!} \left(p^n q^k + p^k q^n \right)$$

$$= n \sum_{k=0}^{n-1} \binom{n+k}{k} \left(p^n q^k + p^k q^n \right)$$

(2) To derive a recursive formula for $\frac{E_{n+1}}{n+1}$:

$$\frac{E_{n+1}}{n+1} = \sum_{k=0}^{n} \binom{n+k+1}{k} \left(p^{n+1} q^k + p^k q^{n+1} \right)$$

$$= \sum_{k=0}^{n} \binom{n+k+1}{k} \left[p^n q^k (1-q) + p^k q^n (1-p) \right]$$

$$= \sum_{k=0}^{n} \binom{n+k+1}{k} \left(p^n q^k + p^k q^n \right)$$

$$- \sum_{k=0}^{n} \binom{n+k+1}{k} \left(p^n q^{k+1} + p^{k+1} q^n \right)$$

$$= \sum_{k=0}^{n} \binom{n+k}{k} \left(p^n q^k + p^k q^n \right) + \sum_{k=1}^{n} \binom{n+k}{k-1} \left(p^n q^k + p^k q^n \right)$$

$$- \sum_{j=0}^{n+1} \binom{n+j}{j-1} \left(p^n q^j + p^j q^n \right)$$

$$= \sum_{k=0}^{n} \binom{n+k}{k} \left(p^n q^k + p^k q^n \right) - \binom{2n+1}{n} \left(p^n q^{n+1} + p^{n+1} q^n \right)$$

$$= \sum_{k=0}^{n-1} \binom{n+k}{k} \left(p^n q^k + p^k q^n \right) + 2 \binom{2n}{n} p^n q^n$$

$$- \binom{2n+1}{n} \left[\left(p^n q^n (p+q) \right) \right]$$

$$= \sum_{k=0}^{n-1} \binom{n+k}{k} \left(p^n q^k + p^k q^n \right) + 2 \binom{2n}{n} p^n q^n - \binom{2n+1}{n} p^n q^n$$

$$= \frac{E_n}{n} + \left[2 \binom{2n}{n} - \binom{2n+1}{n} \right] p^n q^n$$

$$= \frac{E_n}{n} + C_n p^n q^n$$

(3) To derive an explicit formula for $\frac{E_n}{n}$; we have:

$$\frac{E_n}{n} = \frac{E_{n-1}}{n-1} + C_{n-1}p^{n-1}q^{n-1}$$

$$\sum_{k=2}^{n} \frac{E_k}{k} = \sum_{k=2}^{n-1} \frac{E_{k-1}}{k-1} + \sum_{k=2}^{n} C_{k-1}p^{k-1}q^{k-1}$$

$$= \frac{E_1}{1} + \sum_{k=2}^{n-1} \frac{E_k}{k} + \sum_{k=1}^{n-1} C_k p^k q^k$$

That is,

$$\frac{E_n}{n} = E_1 + \sum_{k=1}^{n-1} C_k p^k q^k$$

$$= 1 + \sum_{k=1}^{n-1} C_k (pq)^k$$

$$= \sum_{k=0}^{n-1} C_k (pq)^k$$

as desired.   ∎

Catalan numbers occur in the theory of lattices also. Those and additional occurrences in combinatorics, lattice paths, trees, and polyominoes can be found in Stanley's *Enumerative Combinatorics,* vol. 2 and his *Catalan Addendum.*

All Is Not That Well

In each example pursued in this and the preceding chapters, everything worked out beautifully: Patterns held up and the answers were always Catalan numbers. But we know from experience that patterns can fail, so a conclusion arrived at by inductive reasoning must be suspect, unless and until it can be confirmed with a mathematical proof.

The next three examples involve Catalan numbers but illustrate the danger in totally relying on inductive reasoning. The first example was studied in 1958 by F. L. Miksa, L. Moser, and M. Wyman of the University of Alberta, Canada.

Example 11.5 Suppose there are n differently colored tennis balls and at most n identical plastic containers for them. We would like to distribute the balls into the containers in such a way that:

- No container can hold more than three balls, and
- The order of the balls in any container is immaterial.

Find the number of distributions possible.

Solution   Let d_n denote the number of possible distributions and let $A, B, C, D, \ldots$ denote the various balls. (Notice that when there are four or more balls, we need to use at least two containers.) Figure 11.6 shows the various possibilities for $0 \le n \le 4$.

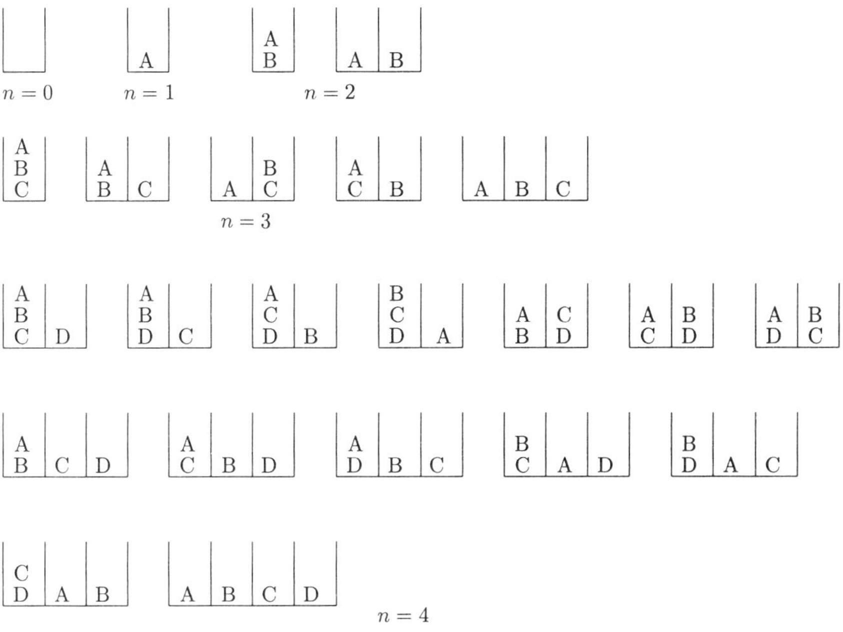

Figure 11.6

Thus we have $d_0 = 1 = C_0, d_1 = 1 = C_1, d_2 = 2 = C_2, d_3 = 5 = C_3$, and $d_4 = 14 = C_4$. Consequently, we are tempted to conjecture that $D_n = C_n$ for every integer $n \ge 0$. Unfortunately, however, the pattern fails with $n = 5$. Miksa et al. showed that $d_5 = 46 \ne 42 = C_5$, and $d_6 = 166 \ne 132 = C_6$. ∎

For the curious-minded, the correct sequence $\{d_n\}$ generated in this example is

$$1, 1, 2, 5, 14, 46, 166, 652, 2780, 12644, 61136, 312676, \ldots$$

This is sequence numbered 579 in Sloane's *Handbook of Integer Sequences*.

To see this, let $G(n, r)$ denote the number of ways n distinguishable objects can be placed into at most n indistinguishable boxes, if each box can hold at most r

objects. Then Miksa et al established the following recursive definition:

$$G(n, 1) = 1$$

$$G(n, r) = G(n - 1, r) + \binom{n-1}{1} G(n - 2, r) + \binom{n-1}{2} G(n - 2, r)$$

$$+ \cdots + G(n - r, r)$$

Using this recursive definition and differentiation, it is easy to develop the following exponential generating function for $G(n, r)$: $e^{x + \frac{x^2}{2!} + \cdots + \frac{x^r}{r!}}$; that is, $G(n, r)$ is the coefficient of $\frac{x^n}{n!}$ in the power series expansion of $e^{x + \frac{x^2}{2!} + \cdots + \frac{x^r}{r!}}$.

For example, $G(5, 3) = $ coefficient of $\frac{x^5}{5!}$ in the expansion of $e^{x + \frac{x^2}{2!} + \frac{x^3}{3!}}$, where

$$e^{x + \frac{x^2}{2} + \frac{x^3}{6}} = 1 + \ldots + \frac{1}{2!}\left(x + \frac{x^2}{2} + \frac{x^3}{6}\right)^2 + \frac{1}{3!}\left(x + \frac{x^2}{2} + \frac{x^3}{6}\right)^3$$

$$+ \frac{1}{4!}\left(x + \frac{x^2}{2} + \frac{x^3}{6}\right)^4 + \frac{1}{5!}\left(x + \frac{x^2}{2} + \frac{x^3}{6}\right)^5 + \ldots$$

$$= \ldots + \frac{x^2}{2}\left(1 + \frac{x}{2} + \frac{x^2}{6}\right)^2 + \frac{x^3}{6}\left(1 + \frac{x}{2} + \frac{x^2}{6}\right)^3$$

$$+ \frac{x^4}{24}\left(1 + \frac{x}{2} + \frac{x^2}{6}\right)^4 + \frac{x^5}{120} + \ldots$$

$$= \ldots + \frac{x^2}{2}\left(\ldots + \frac{x^3}{6} + \ldots\right) + \frac{x^3}{6}\left(\ldots + \frac{5}{4}x^2 + \cdots\right)$$

$$+ \frac{x^4}{24}(\ldots + 2x + \ldots) + \frac{x^5}{120} + \ldots$$

$$= \ldots + \left(\frac{1}{12} + \frac{5}{24} + \frac{2}{24} + \frac{1}{120}\right)x^5 + \ldots$$

$$= \ldots + \frac{46}{5!}x^5 + \ldots$$

Therefore,

$$G(5, 3) = \text{coefficient of } \frac{x^5}{5!} \text{ in the expansion}$$

$$= 46$$

as expected.

The next example was studied in 1968 by J. E. Koehler of Seattle University, Seattle, Washington.

Example 11.6 Find the number of ways of making n folds using a strip of $n + 1$ stamps, without distinguishing between the top and the bottom, and front and back, or left and right, where $n \geq 0$.

Solution   Let f_n denote the number of folds. Figure 11.7 shows the various possibilities for $0 \leq n \leq 4$. It follows from the figure that $f_0 = 1 = C_0, f_1 = 1 = C_1, f_2 = 2 = C_2, f_3 = 5 = C_3$, and $f_4 = 14 = C_4$. Once again, it is tempting to believe that $f_n = C_n$ for every n.

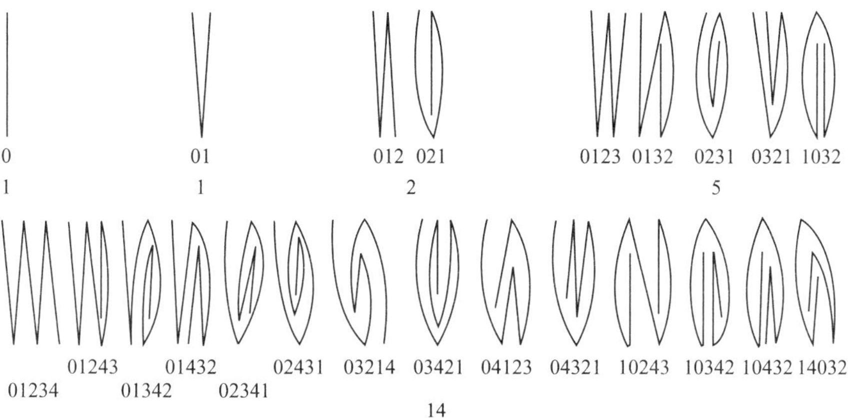

Figure 11.7

Again, the pattern breaks down with $n = 5$. Koehler showed that $f_5 = 39 \neq 42 = C_5$, and $f_6 = 120 \neq 132 = C_6$.   ∎

The correct sequence $\{f_n\}$ generated in this example is

$$1, 1, 2, 5, 14, 39, 120, 358, 1176, 3527, 11622, 36627, 121622, \ldots$$

This is sequence numbered 576 in Sloane's handbook.

The next and last example in this short list of pattern failures deals with group theory in abstract algebra.

Example 11.7 Find the number of nonisomorphic groups of order 2^n, where $n \geq 0$.

Solution   Let g_n denote the number of nonisomorphic groups of order 2^n. It appears from Table 11.7[†] that $g_n = C_n$ for every $n \geq 0$. Once again, the pattern fails when $n = 5$. In fact, $g_5 = 51 \neq 42 = C_5$, and $g_6 = 267 \neq 132 = C_6$.

[†] See H. Hall and J. K. Senior, *The Groups of Order 2^n* ($n \leq 6$), Macmillan, New York, 1964; and the articles by E. Rodemich and E. A. O'Brien.

Table 11.7 Number of
Nonisomorphic Simple Groups

n	0	1	2	3	4	5
g_n	1	1	2	5	14	?

Interestingly, the sequence $\{g_n\}$ is

$$1, 1, 2, 5, 14, 51, 267, 2328, 56092, \ldots$$

This is sequence numbered 581 in Sloane's handbook.

Conclusion? Although pattern recognition is very important in mathematics, we need to be careful of blindly following patterns; it can lead us to dangerous conclusions.

12

Pascal's Triangle and Catalan Numbers

In the preceding chapters, we encountered a number of occurrences of Catalan numbers C_n and several explicit formulas for C_n. All formulas, except Segner's recursive formula, involve binomial coefficients. Consequently, the many treasures of Pascal's triangle include the ability to extract Catalan numbers from it.

To this end, consider Pascal's triangle in Figure 12.1, where the CBCs $\binom{2n}{n}$ are circled for easy identification.

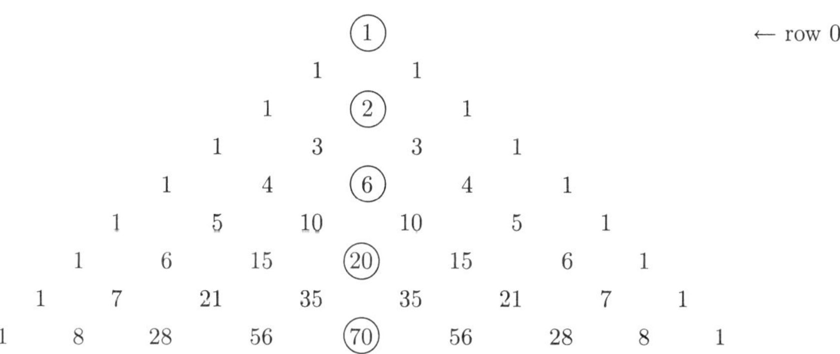

Figure 12.1 Pascal's Triangle

There are several ways Catalan numbers can be read from this array.

- The most obvious way to compute C_n is by using the explicit formula $C_n = \frac{1}{n+1}\binom{2n}{n}$, that is, by dividing the CBC by $n + 1$. Thus, by dividing the circled numbers by one more than one-half of their row numbers, we can compute the various Catalan numbers.

For example, to compute C_4, we go to row 8; read $\binom{8}{4} = 70$; and then divide 70 by $4 + 1 = 5 : \frac{70}{5} = 14$.

- Since

$$
\begin{aligned}
C_n &= \frac{(2n)!}{(n+1)!n!} \\
&= \frac{1}{n} \cdot \frac{(2n)!}{(n-1)!(n+1)!} \\
&= \frac{1}{n}\binom{2n}{n-1}
\end{aligned}
$$

C_n can also be computed by dividing the term immediately to the left (or right) of the CBC by one-half of the row number. See the boxed numbers in Figure 12.2.

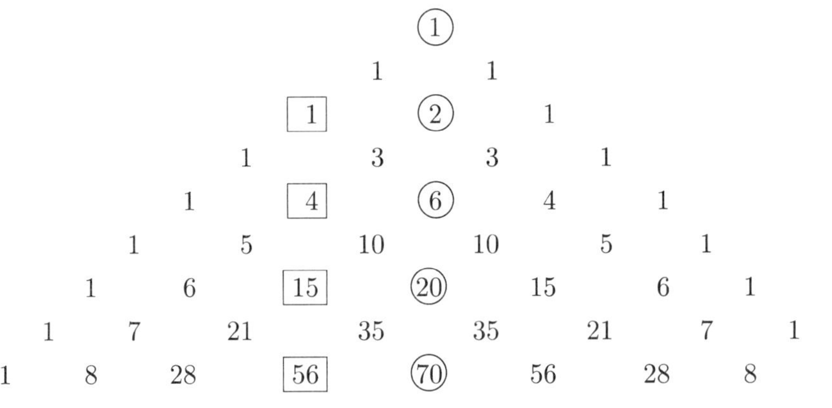

Figure 12.2

For example, $C_4 = \frac{1}{4}\binom{8}{3} = \frac{1}{4} \cdot 56 = 14$.

- In Chapter 9, we found that

$$
C_n = \binom{2n}{n} - \binom{2n}{n-1}
$$

So C_n can be computed by subtracting from the CBC its adjacent neighbor on the left (or right); see the circled and boxed numbers in Figure 12.3.

For example, $C_4 = \binom{8}{4} - \binom{8}{3} = 70 - 56 = 14$.

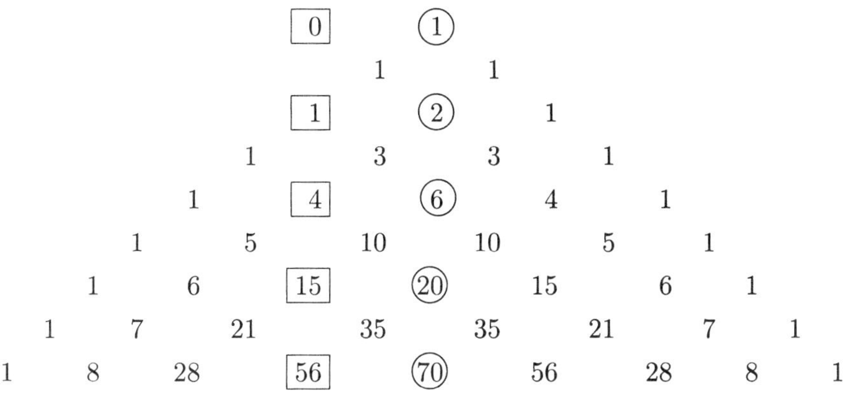

Figure 12.3

- Notice that

$$\binom{2n}{n} - \binom{2n}{n-2} = \frac{(2n)!}{n!n!} - \frac{(2n)!}{(n-2)!(n+2)!}$$

$$= \frac{(2n)!}{n!(n+1)!}\left[(n+1) - \frac{n(n-1)}{n+2}\right]$$

$$= \frac{(2n)!}{n!(n+1)!} \cdot \frac{2(2n+1)}{n+2}$$

$$= \frac{(2n)!(2n+1)(2n+2)}{n!(n+1)!(n+2)}$$

$$= \frac{(2n+2)!}{(n+1)!(n+2)!}$$

$$= C_{n+1}$$

Consequently, every Catalan number, except C_0, can be obtained from Pascal's triangle by subtracting $\binom{2n}{n-2}$ from the CBC $\binom{2n}{n}$; see the circled and boxed numbers in Figure 12.4.

For instance, suppose we would like to compute C_5 (notice that here $n = 4$). Go to row 8, read $\binom{8}{4}$ and $\binom{8}{2}$, and subtract the latter from the former:

$$C_5 = \binom{8}{4} - \binom{8}{2}$$

$$= 70 - 28$$

$$= 42$$

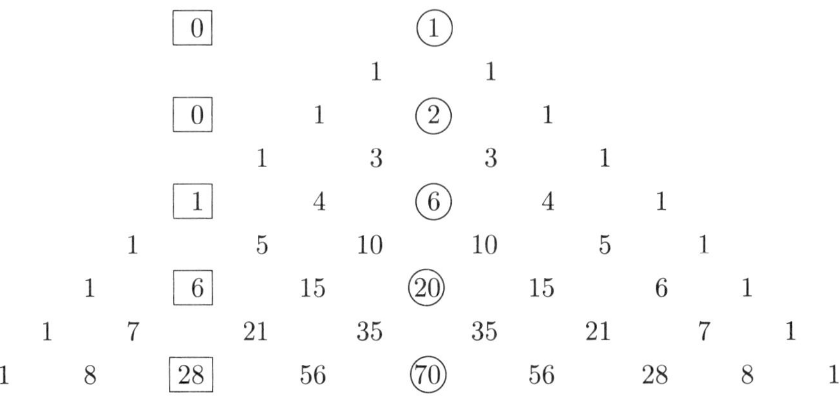

Figure 12.4

- Odd-numbered rows in Pascal's triangle can be employed to compute Catalan numbers. In Chapter 6, we found that

$$C_n = \frac{1}{2n+1} \binom{2n+1}{n}$$

Since each odd-numbered row contains an even number of terms, there are two middle terms in such a row. Dividing each middle term by its row number yields a Catalan number. See the boxed numbers in Figure 12.5.

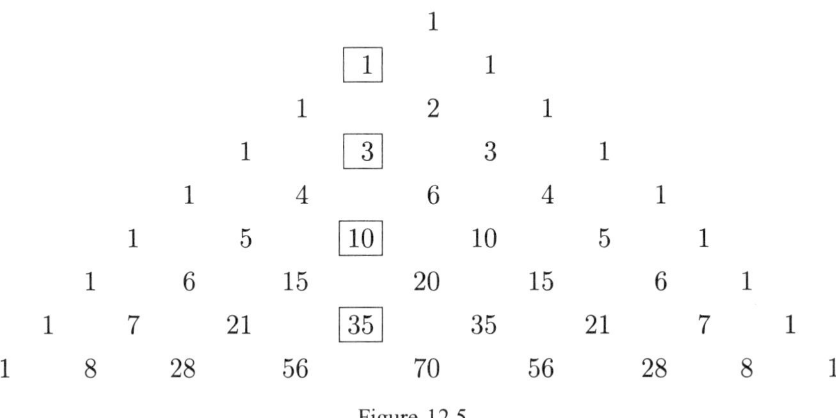

Figure 12.5

For example, consider row 9:

$$1 \quad 9 \quad 36 \quad 84 \quad \boxed{126} \quad 126 \quad 84 \quad 36 \quad 9 \quad 1$$

↑
a middle term

One of the two middle terms is 126; dividing it by the row number 9 yields C_4:

$$C_4 = \frac{126}{9} = 14$$

- C_n can also be computed by subtracting $\binom{2n-1}{n-2}$ from $\binom{2n-1}{n-1}$, where $n \geq 1$:

$$\binom{2n-1}{n-1} - \binom{2n-1}{n-2} = \frac{(2n-1)!}{(n-1)!n!} - \frac{(2n-1)!}{(n-2)!(n+1)!}$$

$$= \frac{(2n-1)!}{(n+1)!n!}\left[n(n+1) - n(n-1)\right]$$

$$= \frac{(2n-1)!}{(n+1)!n!}(2n)$$

$$= \frac{(2n)!}{(n+1)!n!}$$

$$= C_n$$

Thus, C_n can be computed from Pascal's triangle by subtracting from the first middle term in row $2n-1$ the element immediately to its left. See the boxed numbers in Figure 12.6.

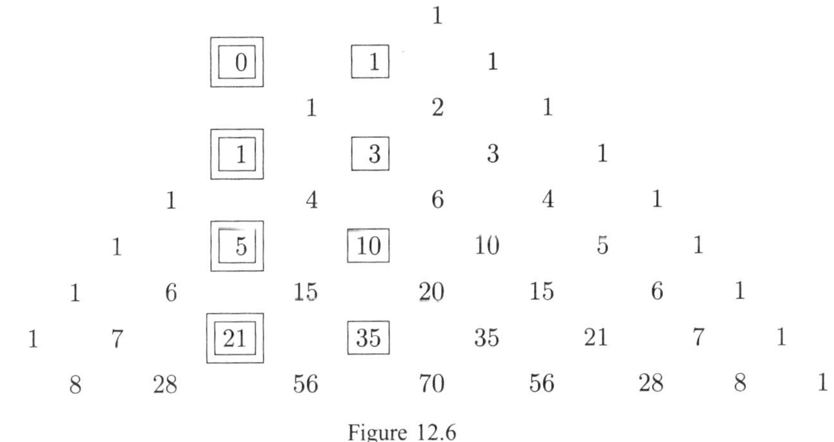

Figure 12.6

For example,

$$C_4 = \binom{7}{3} - \binom{7}{2}$$

$$= 35 - 21$$

$$= 14$$

- Here is yet another way of computing C_n:

$$2\binom{2n}{n} - \binom{2n+1}{n} = \frac{2(2n)!}{n!n!} - \frac{(2n+1)!}{n!(n+1)}$$

$$= \frac{(2n)!}{(n+1)!n!}[2(n+1) - (2n+1)]$$

$$= \frac{(2n)!}{(n+1)!n!}$$

$$= C_n$$

Accordingly, each Catalan number C_n can be obtained by subtracting a middle term in row $2n+1$ from twice the CBC in row $2n$. See the circled and boxed numbers in Figure 12.7.

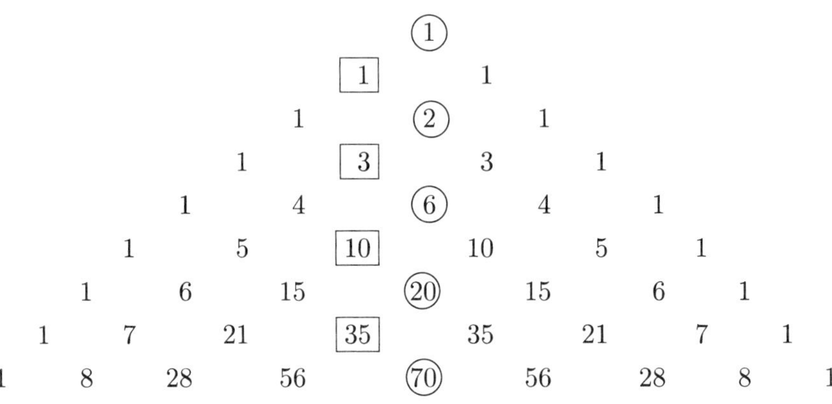

Figure 12.7

For example,

$$C_3 = 2\binom{6}{3} - \binom{7}{3}$$

$$= 2 \cdot 20 - 35$$

$$= 5$$

- Here is a method due to Guy for computing C_n:

$$\binom{2n+1}{n+1} - 2\binom{2n}{n+1} = \frac{(2n+1)!}{(n+1)!n!} - 2 \cdot \frac{(2n)!}{(n+1)!9n-1)!}$$

$$= (2n+1)C_n - (2n)C_n$$

$$= C_n$$

Accordingly, C_n can be computed by subtracting from the second middle term $\binom{2n+1}{n+1}$ in row $2n + 1$ twice the element $\binom{2n}{n+1}$ immediately to the right of the CBC in row $2n$. See the boxed numbers in Figure 12.8.

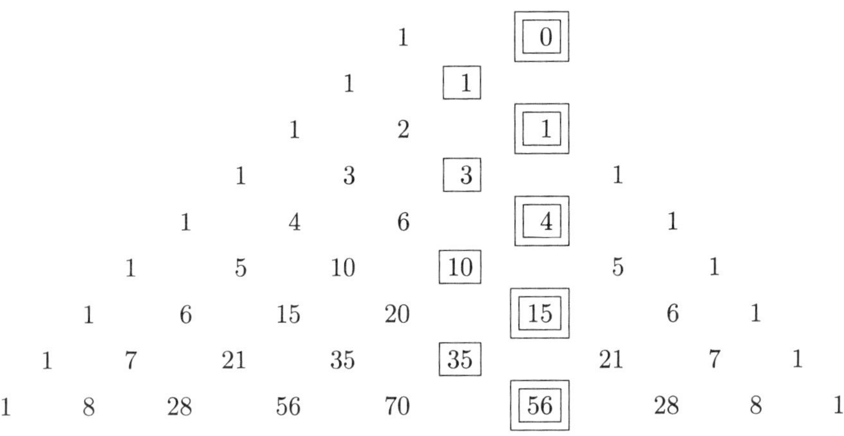

Figure 12.8

For example,

$$C_3 = \binom{7}{4} - 2\binom{6}{4}$$
$$= 35 - 2 \cdot 15$$
$$= 5$$

By virtue of symmetry, C_n can also be computed as

$$C_n = \binom{2n + 1}{n} - 2\binom{2n}{n - 1}$$

Touchard's Recursive Formula

In 1928, Jacques Touchard of France developed yet another recursive formula for C_n that uses the binomial coefficients in row n of Pascal's triangle:

$$C_{n+1} = \sum_{r=0}^{\lfloor n/2 \rfloor} \binom{n}{2r} 2^{n-2r} C_r$$

This can be established using the generating function $C(x)$ of C_n, as follows.

From Chapter 5, we have:

$$C(x) = \sum_{n \geq 0} C_n x^n$$

$$= \frac{1 - \sqrt{1 - 4x}}{2x}$$

Then

$$C\left[\frac{x^2}{(1 - 2x)^2}\right] = \frac{1 - 2x}{x} \cdot \frac{1 - 2x - \sqrt{1 - 4x}}{2x}$$

$$\frac{x}{1 - 2x} C\left[\frac{x^2}{(1 - 2x)^2}\right] = \frac{1 - 2x - \sqrt{1 - 4x}}{2x}$$

$$= C(x) - 1$$

$$= \sum_{n \geq 1} C_n x^n$$

$$= \sum_{n \geq 0} C_{n+1} x^{n+1} \tag{12.1}$$

But

$$\frac{x}{1 - 2x} C\left[\frac{x^2}{(1 - 2x)^2}\right] = \sum_{r \geq 0} C_r \frac{x^{2r+1}}{(1 - 2x)^{2r+1}}$$

$$= \sum_{r \geq 0} C_r \left[\sum_{r \geq 0} \binom{2r + 1 + k - 1}{k} (2x)^k\right] x^{2r+1}$$

$$= \sum_{r \geq 0} C_r \left[\sum_{k \geq 0} \binom{2r + k}{k} 2^k x^{2r+k+1}\right]$$

$$\text{Coefficient of } x^{n+1} \text{ on the RHS} = \sum_{0 \leq 2k \leq n} \binom{n}{n - 2r} 2^{n-2r} C_r$$

$$= \sum_{r=0}^{\lfloor n/2 \rfloor} \binom{n}{2r} 2^{n-2r} C_r \tag{12.2}$$

The identity follows by equating the coefficients of x^{n+1} from equations (12.1) and (12.2).

Accordingly, C_{n+1} can be computed recursively as the weighted sum of the binomial coefficients $\binom{n}{2r}$ in even positions in row n and using $2^{n-2r} C_r$ as weights.

For example,

$$C_5 = \sum_{r=0}^{2} \binom{4}{2r} 2^{4-2r} C_r$$

$$= \binom{4}{0} 2^4 C_0 + \binom{4}{2} 2^2 C_1 + \binom{4}{4} 2^0 C_2$$

$$= \boxed{1} \cdot 16 \cdot 1 + \boxed{6} \cdot 4 \cdot 1 + \boxed{1} \cdot 1 \cdot 2$$

$$= 42$$

```
                            1
                       1         1
                   1        2        1
               1       3        3       1
          [1]      4       [6]      4      [1]          ←row 4
        1      5      10        10      5      1
      1     6      15      20       15      6      1
    1    7     21      35       35     21     7     1
  1    8    28     56      70      56     28     8    1
```

Figure 12.9

See Figure 12.9, where the binomial coefficients in even positions in row 4 are boxed in:

Binomial coefficients	:	1	6	1
Weights	:	$2^4 C_0$	$2^2 C_1$	$2^0 C_2$
Weighted sum	:	$16 \cdot 1$ $+$	$4 \cdot 6$ $+$	$2 \cdot 1$
C_5	:	42		

Catalan Polynomials

Interestingly, in 1973, using Segner's recursive formula, John Riordan of Rockefeller University, New York City, introduced a family of polynomials $c_n(x)$, which we call *Catalan polynomials*. They are defined recursively as follows:

$$c_0(x) = 1, \quad c_1(x) = x$$

$$c_n(x) = \sum_{i=0}^{n-1} c_{n-1-i}(x), \quad n \geq 2$$

For example,

$$c_2(x) = c_0(x)c_1(x) + c_1(x)c_0(x)$$
$$= 1 \cdot x + x \cdot 1$$
$$= 2x$$

and

$$c_3(x) = c_0(x)c_2(x) + c_1(x)c_1(x) + c_2(x)c_0(x)$$
$$= 1 \cdot 2x + x \cdot 2 + 2x \cdot 1$$
$$= x^2 + 4x$$

Notice that $c_n(1) = C_n$; this follows from the recursive definition of $c_n(x)$.

Using generating functions and partial derivatives, Riordan developed an explicit formula for $c_n(x)$:

$$c_n(x) = \sum_{r=0}^{\lfloor (n-1)/2 \rfloor} \binom{n-1}{2r} 2^{n-2r-1} C_r x^r$$

Consequently,

$$c_n(1) = \sum_{r=0}^{\lfloor (n-1)/2 \rfloor} \binom{n-1}{2r} 2^{n-2r-1} C_r$$

That is,

$$C_n = \sum_{r=0}^{\lfloor (n-1)/2 \rfloor} \binom{n-1}{2r} 2^{n-2r-1} C_r$$

This is indeed Touchard's recursive formula for C_n.

Another Combinatorial Recursive Formula

There is another combinatorial recursive formula for C_n, as the following theorem shows. It can be established using strong induction.

Theorem 12.1 Let $n \geq 1$. Then

$$C_n = \sum_{r=1}^{\lfloor (n+1)/2 \rfloor} (-1)^{r-1} \binom{n-r+1}{r} C_{n-r}$$

∎

Thus, Pascal's triangle can once again be used to compute C_n in a totally different way: Multiply the binomial coefficients along the northeast diagonal beginning at $\binom{n}{1}$ with $C_{n-1}, C_{n-2}, \ldots$; alternately change signs; then compute the resulting sum. See Figure 12.10.

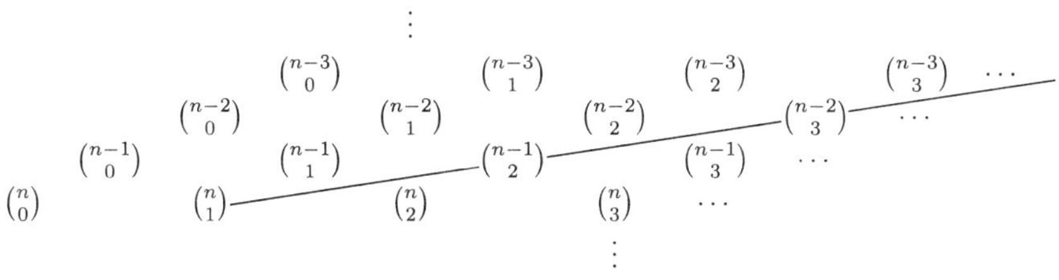

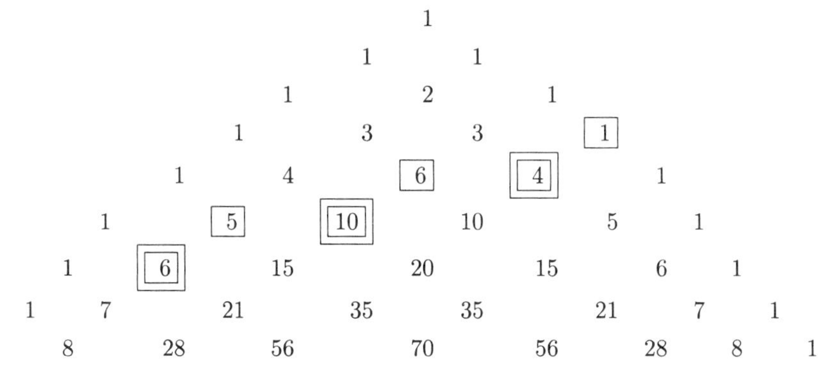

Figure 12.10

For example,

$$C_5 = \binom{5}{1} C_4 - \binom{4}{2} C_3 + \binom{3}{3} C_2$$
$$= \boxed{5} \cdot 14 - \boxed{6} \cdot 5 + \boxed{1} \cdot 2$$
$$= 42$$

and

$$C_6 = \binom{6}{1} C_5 - \binom{5}{2} C_4 + \binom{4}{3} C_3$$
$$= \boxed{\boxed{6}} \cdot 42 - \boxed{\boxed{10}} \cdot 14 + \boxed{\boxed{4}} \cdot 5$$
$$= 132$$

See the single-boxed and double-boxed diagonals in Figure 12.11.

Figure 12.11

The formula in Theorem 12.1 reminds us of Lucas's formula[†] for computing the nth Fibonacci number F_n using the binomial coefficients along a northeast diagonal:

$$F_n = \sum_{r=0}^{\lfloor (n+1)/2 \rfloor} \binom{n-r+1}{r}$$

The next theorem provides another combinatorial identity linking Catalan numbers with binomial coefficients that are vertically located in Pascal's triangle. It was developed in 1987 by D. Jonah when he was attending a summer program at Hope College, Holland, Michigan.

Before presenting the theorem, we make two useful observations:

- Recall from Example 9.1 that the number of paths from the origin to the lattice point (n, n) that do not cross the northeast diagonal $y = x$ is the Catalan number C_n. So the number of paths from (1,0) to the lattice point $(n + 1, n)$ that lie below the line $y = x$ is also C_n. Each such path can be obtained by horizontally shifting an original path from the origin to (n, n) by one unit eastward. This process is clearly reversible. See Figure 12.12.

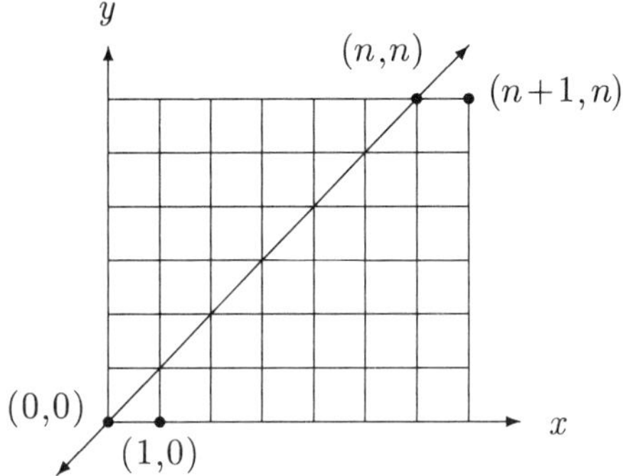

Figure 12.12 Lattice-Walking

- The number of paths from the lattice point (a, b) to the lattice point (c, d) on the cartesian plane equals the number of paths from (0,0) to $(c - a, d - b)$, by virtue of vertical and horizontal translations, namely,

$$\frac{((c - a) + (d - b)!}{(c - a)!(d - b)!} = \binom{c + d - a - b}{c - a}$$

See Figure 12.13.

† See T. Koshy, *Fibonacci and Lucas Numbers with Applications.*

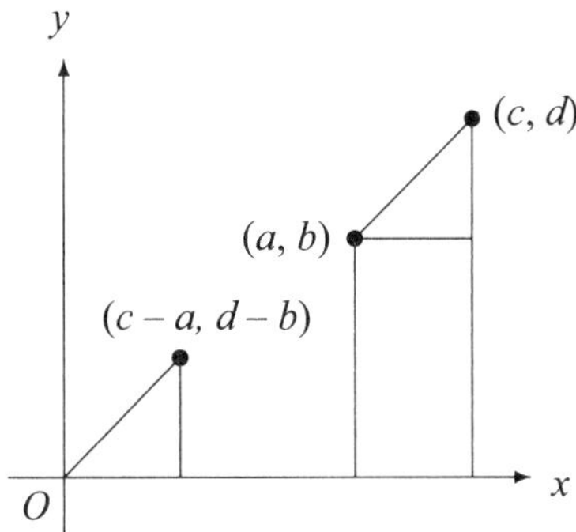

Figure 12.13 Lattice-Walking

We are now ready for Jonah's identity. The proof is based on the one given in 1990 by P. Hilton and J. Pedersen. It employs both the addition and multiplication principles.

Theorem 12.2 (*Jonah's Theorem*)   Let $n \geq 2r$. Then

$$\binom{n+1}{r} = \binom{n}{r} C_0 + \binom{n-2}{r-1} C_1 + \binom{n-4}{r-2} C_2 + \ldots + \binom{n-2r}{0} C_r$$

Proof We will determine the number of paths from $(0,0)$ to the lattice point $(r, n - r + 1)$ on the cartesian plane in two different ways. There are

$$\binom{r + (n - r + 1)}{r} = \binom{n+1}{r}$$

such paths.

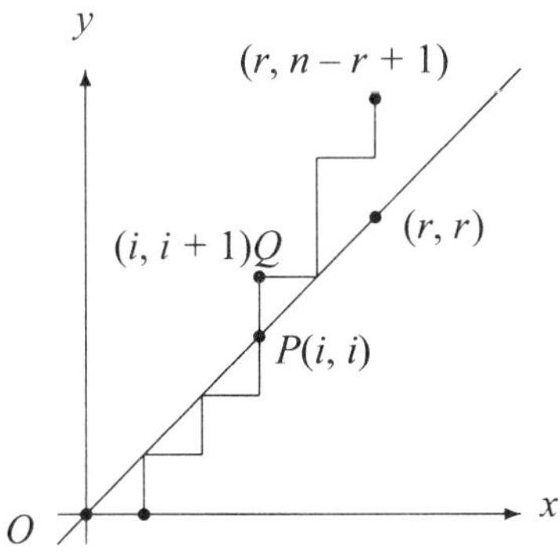

Figure 12.14 Lattice-Walking

Since $n \geq 2r$, every such path from the origin must cross the line $y = x$ at least once. Let $P(i, i)$ be the first point, where the path crosses the line $y = x$ (see Figure 12.14). There are C_i paths from the origin to P that do *not* cross the line $y = x$. There are exactly $\binom{n-2i}{r-i}$ paths from $Q(i, i+1)$ to $(r, n-i+1)$. By the multiplication principle, there are $\binom{n-2i}{r-i} C_i$ paths from the origin to $(r, n-r+1)$ that pass through $P(i, i)$. Since this is true for each i, by the addition principle, the total number of paths from the origin to $(r, n-r+1)$ is given by $\sum_{i=0}^{r} \binom{n-2i}{r-i} C_i$. Thus

$$\binom{n+1}{r} = \sum_{i=0}^{r} \binom{n-2i}{r-i} C_i$$

as desired.　∎

For example, let $n = 7$ and $r = 3$. Then:

$$\sum_{i=0}^{3} \binom{7-2i}{3-i} C_i = \binom{7}{3} C_0 + \binom{5}{2} C_1 + \binom{3}{1} C_2 + \binom{1}{0} C_3$$

$$= \boxed{35} \cdot 1 + \boxed{10} \cdot 1 + \boxed{3} \cdot 2 + \boxed{1} \cdot 5$$

$$= \boxed{\boxed{56}}$$

$$= \binom{8}{3}$$

See the boxed and double-boxed numbers in Figure 12.15.

```
                                1
                        [1]            1
                    1            2            1
                1        [3]            3            1
            1        4            6            4            1
        1        5       [10]          10           5            1
    1        6        15           20           15           6        1
  1      7       21        [35]          35           21           7        1
1      8      28      [[56]]          70           56           28          8      1
```

Figure 12.15

Similarly,

$$\sum_{i=0}^{4} \binom{9-2i}{4-i} C_i = \binom{9}{4} C_0 + \binom{7}{3} C_1 + \binom{5}{2} C_2 + \binom{3}{1} C_3 + \binom{1}{0} C_4$$

$$= 126 \cdot 1 + 35 \cdot 1 + 10 \cdot 2 + 3 \cdot 5 + 1 \cdot 14$$

$$= 210$$

$$= \binom{10}{4}$$

In 1990, Hilton and Pedersen generalized Jonah's identity, as the next theorem shows.

Theorem 12.3 (*Hilton and Pedersen*, 1990) Let n be an arbitrary rational number and r any integer ≥ 0. Then

$$\binom{n+1}{r} = \binom{n}{r} C_0 + \binom{n-2}{r-1} C_1 + \binom{n-4}{r-2} C_2 + \cdots + \binom{n-2r}{0} C_r \quad \blacksquare$$

For example, let $n = 5/2$ and $r = 3$. Then

$$\binom{5/2}{3} = \frac{(5/2)(3/2)(1/2)}{3!} = \frac{5}{16}; \quad \binom{1/2}{2} = \frac{(1/2)(-1/2)}{2!} = -\frac{1}{8}$$

$$\binom{-3}{1} = -\frac{3}{2}; \qquad\qquad \binom{-7}{0} = 1$$

$$\binom{7/2}{3} = \frac{(7/2)(5/2)(3/2)}{3!} = \frac{35}{16}$$

$$\sum_{i=0}^{3} \binom{5/2-2i}{3-i} C_i = \binom{5/2}{3} C_0 + \binom{1/2}{2} C_1 + \binom{-3/2}{1} C_2 + \binom{-7/2}{0} C_3$$

$$= \tfrac{5}{16} \cdot 1 - \tfrac{1}{8} \cdot 1 - \tfrac{3}{2} \cdot 2 + 1 \cdot 5$$

$$= \tfrac{35}{16}$$

$$= \binom{7/2}{3}$$

13

Divisibility Properties

This chapter presents a few divisibility properties of Catalan numbers. First, we identify Catalan numbers that are odd and then those that are prime.

Parity of Catalan Numbers

Table 13.1 gives the first eighteen Catalan numbers. Odd numbers among them are marked with asterisks.

Table 13.1 First Eighteen Catalan Numbers

n	0	1	2	3	4	5	6	7	8
C_n	1^*	1^*	$2^{\dagger}$	$5^{*\dagger}$	14	42	132	429^*	1430

n	9	10	11	12	13	14	15	16	17
C_n	4862	16796	58786	208012	742900	2674440	9694845^*	35357670	129644790

It follows from the table that when $n \leq 17$, C_n is odd for $n = 0, 1, 3, 7,$ and 15, all of which are of the form $2^m - 1$. When $m > 0$, such numbers are known as *Mersenne numbers*, named after French mathematician and Franciscan monk Marin Mersenne.

The following theorem identifies odd Catalan numbers.

Theorem 13.1 (*Koshy and Salmassi*, 2004) For $n > 0$, C_n is odd if and only if n is a Mersenne number.

Proof It follows from Segner's recurrence relation that

$$
C_n = \begin{cases} 2\left(C_0 C_{n-1} + C_1 C_{n-2} + \cdots + C_{\frac{n}{2}-1} C_{\frac{n}{2}}\right) & \text{if } n \text{ is even} \\[2ex] 2\left(C_0 C_{n-1} + C_1 C_{n-2} + \cdots + C_{\frac{n-3}{2}} C_{\frac{n+1}{2}}\right) + C_{\frac{n-1}{2}}^2 & \text{otherwise} \end{cases}
$$

> **Marin Mersenne** (1588–1648), "best known as the priest-scientist who facilitated the cross-fertilization of the most eminent minds of his time," was born in Soultière, France. He was baptized on the same day and christened as Marin since it was the feast of the Nativity of Mary. After attending the College de Mans and the Jesuit College at La Flêche, he went to Paris to study theology and became a Minim friar in 1611. Science began to dominate his religious thought, and in 1624 he accepted the Copernican theory that the sun, not the Earth, was the center of the solar system. Mersenne corresponded with many scientists and philosophers including René Descartes (1596–1650); his residence became a meeting place for such eminent thinkers as Pierre de Fermat (1601?–1665), Girard Desargues (1593–1662?), and Fr. Pierre Gassendi. Mersenne even came to the defense of Descartes and Galileo when their works were attacked by the Church. He also made important contributions to music and acoustics.

Consequently, for $n > 0$, C_n is odd if and only if both n and $C_{\frac{n-1}{2}}$ are odd. The same argument implies that C_n is odd if and only if $\frac{n-1}{2}$ and $C_{\frac{n-3}{4}}$ are both odd or $\frac{n-1}{2} = 0$. Continuing this finite descent, it follows that C_n is odd if and only if $C_{\frac{n-(2^m-1)}{2^m}}$ is odd, where $m \geq 1$. But the least value of k for which C_k is odd is $k = 0$. Thus the sequence of these *if and only if* statements terminates when $\frac{n-(2^m-1)}{2^m} = 0$; that is, when $n = 2^m - 1$, a Mersenne number. $\blacksquare$

Primality of Catalan Numbers

Returning to Table 13.1, we make another observation: Exactly two of these Catalan numbers are prime; they are identified with a dagger. The following theorem confirms this observation.

Theorem 13.2 (*Koshy and Salmasi*, 2004) The only prime Catalan numbers are C_2 and C_3.

Proof   It follows from the explicit formula (5.3) for C_n that $(n+2)C_{n+1} = (4n+2)C_n$. Assume that C_n is prime for some n. It follows that if $n > 3$, then $\frac{n+2}{C_n} < 1$; so $C_n > n + 2$. Consequently, $C_n \mid C_{n+1}$, so $C_{n+1} = kC_n$ for some positive integer k. Then $4n + 2 = k(n + 2)$, whence $1 \leq k \leq 3$ and thus $n \leq 4$. It follows that C_2 and C_3 are the only Catalan numbers that are prime. $\blacksquare$

Divisibility Properties of Catalan Numbers

We close this chapter with a few basic divisibility properties of Catalan numbers.

Theorem 13.3 (*Koshy and Salmassi*, 2004) $C_n | n + 2$ if and only if $n \leq 3$, where $n \geq 0$.

Proof   If $n \leq 3$, then clearly, $C_n | n + 2$.

On the other hand, let $n > 3$. Then, by the explicit formula,

$$C_n = \frac{(n + 2)(n + 3) \cdots (2n)}{n!}$$

$$\begin{aligned}
\frac{n + 2}{C_n} &= \frac{n!}{(n + 3)(n + 4) \cdots (2n - 1)(2n)} \\
&= \frac{1 \cdot 2 \cdot 3 \cdot 4 \cdots (n - 1)n}{(n + 3)(n + 4) \cdots (2n - 1)(2n)} \\
&= \frac{3 \cdot 4 \cdot 5 \cdots (n - 1)}{(n + 3)(n + 4) \cdots [n + (n - 1)]}
\end{aligned}$$

The RHS is clearly not an integer, so $\frac{n+2}{C_n}$ is not an integer; that is, $C_n \nmid n + 2$. Thus, if $n > 3$, then $C_n \nmid n + 2$.   ∎

Theorem 13.2 now follows from this result, as the following corollary shows.

Corollary 13.1 The only prime Catalan numbers are C_2 and C_3.

Proof   We have:

$$C_{n+1} = \frac{2(2n + 1)}{n \mid 2} C_n, \quad n \geq 0$$

$$(n + 2)C_{n+1} = 2(2n + 1)C_n$$

Assume that C_n is a prime. Then either $C_n | n + 2$ or $C_n | C_{n+1}$.

Case 1   Suppose $C_n | n + 2$. Then, by Theorem 13.3, $n \leq 3$. Since $0 \leq n \leq 3$ and C_n is a prime, there are only two such values of n, namely, $n = 2$ and $n = 3$. Correspondingly, $C_2 = 2$ and $C_3 = 5$ are primes.

Case 2   Suppose $C_n | C_{n+1}$. Then $\frac{C_{n+1}}{C_n}$ is a positive ineger k. So

$$\frac{2(2n + 1)}{n + 2} = k$$

That is,

$$4n + 2 = nk + 2k$$

$$n(4 - k) = 2k - 2$$

This implies that $1 \le k < 4$. When $k = 1, n = 0; C_0 = 1$ is not a prime. When $k = 2, n = 1; C_1 = 1$ is not a prime. When $k = 3, n = 4; C_4 = 14$ is still not a prime.

Thus, if C_n is a prime, then $n = 2$ or $n = 3$. In other words, $C_2 = 2$ and $C_3 = 5$ are the only two Catalan numbers that are prime. ∎

We now enumerate a few divisiblity properties discovered in 1973 by R. Alter and K. K. Kubota of the University of Kentucky, Lexington, where p is a prime and $k \ge 1$:

- $p \nmid C_{p^k-1}$.
- Suppose $3 \mid C_{n-2}$ and $3 \nmid C_{n-1}$.
  - $n \equiv 0 \pmod 9$
  - $3 \nmid \prod_{i=0}^{5} C_{n+i-1}$
  - $3 \mid C_{n+5}$
- Let $p > 3$. Suppose $p \mid C_{n-2}$ and $p \nmid C_{n-1}$. Then
  - $n \equiv 0 \pmod p$
  - $p \nmid \prod_{i=0}^{(p+1)/2} C_{n+i-1}$
  - $p \mid C_{n+(p+1)/2}$
- Catalan numbers that are not divisible by an odd prime p occur in blocks of length 6 if $p = 3$, and of length $\frac{p+3}{2}$ if $p > 3$.
- Suppose $p \nmid C_{n-1}$ and $p \mid C_n$. Then $n \equiv (p+1)/2 \pmod p$, where p is odd.
- Suppose $3 \nmid C_{n-1}$. Then $3 \mid C_n$ if and only if $n \equiv 5 \pmod 9$, where $n \ge 5$.

The proofs are omitted in the interest of brevity.

14

A Catalan Triangle

In 1976, L. W. Shapiro of Howard University, Washington, D.C., introduced a triangular array of numbers $B(n, r)$, which he called a *Catalan triangle*. The numbers $B(n, r)$ are defined recursively as follows, where $n \geq 1$:

$$B(n, r) = \begin{cases} 1 & \text{if } r = 1 = n \\ B(n-1, r-1) + 2B(n-1, r) + B(n-1, r+1) & \text{if } 1 \leq r \leq n \\ 0 & \text{otherwise} \end{cases}$$

$$(14.1)$$

For example, $B(2, 1) = 2; B(2, 2) = 1;$ and $B(3, 2) = 4$.

Since $B(1, 1)$ is an integer, it follows by the recurrence relation that $B(n, r)$ is always an integer.

The recursive definition of $B(n, r)$ can be used to construct the triangular array in Table 14.1.

It appears from the triangular array that column 1 is made up of Catalan numbers; that is, $B(n, 1) = C_n$. This was established by Shapiro.

Notice that

$$B(n, 1) = B(n-1, 0) + 2B(n-1, 1) + B(n-1, 2)$$

$$= 0 + 2B(n-1, 1) + B(n-1, 2)$$

$$C_n = 2C_{n-1} + B(n-1, 2)$$

Thus $B(n, 2) = C_{n+1} - 2C_n$, where $n \geq 1$.

Table 14.1 Catalan Triangle

r \\ n	1	2	3	4	5	6
1	1					
2	2	1				
3	5	4	1			
4	14	14	6	1		
5	42	48	27	8	1	
6	132	165	110	44	10	1

Catalan numbers   Cayley numbers

For example,

$$B(5,2) = C_6 - 2C_5$$
$$= 132 - 2 \cdot 42$$
$$= 48$$

It follows from the recurrence relation that

$$B(n,n) = B(n-1, n-1) + 2B(n-1, n) + B(n-1, n+1)$$
$$= B(n-1, n-1) + 0 + 0$$
$$= B(n-1, n-1)$$

Since $B(1,1) = 1$, it follows by induction that $B(n,n) = 1$ for every $n \geq 1$. In other words, the main diagonal in Table 14.1 consists of 1s.

Again, by the recurrence relation,

$$B(n, n-1) = B(n-1, n-2) + 2B(n-1, n-1) + B(n-1, n)$$
$$= B(n-1, n-2) + 2 \cdot 1 + 0$$
$$= B(n-1, n-2) + 2$$

Because $B(2,1) = 2$, it follows by induction that $B(n, n-1) = 2(n-1)$, where $n \geq 2$. That is, the diagonal immediately below the main diagonal consists of even numbers; each is twice the preceding row number.

An Explicit Formula for $B(n,2)$

The fact that $B(n,2) = C_{n+1} - 2C_n$ can be employed to develop a delightful formula for $B(n,2)$ in terms of a binomial coefficient, where $n \geq 2$.

Recalling that $C_n = \frac{1}{n+1}\binom{2n}{n}$, we have:

$$
\begin{aligned}
B(n, 2) &= C_{n+1} - 2C_n \\
&= \frac{(2n+2)!}{(n+2)!(n+1)!} - 2 \cdot \frac{(2n)!}{(n+1)!n!} \\
&= \frac{(2n)!}{(n+1)!n!}\left[\frac{(2n+1)(2n+2)}{(n+1)(n+2)} - 2\right] \\
&= \frac{2(2n)!}{(n+1)!n!}\left(\frac{2n+1-n-2}{n+2}\right) \\
&= \frac{2(2n)!}{(n+1)!n!} \cdot \frac{n-1}{n+2} \\
&= \frac{2(2n)!}{n(n+2)!(n-2)!} \\
&= \frac{2}{n}\binom{2n}{n-2}
\end{aligned}
$$

For example,

$$
\begin{aligned}
B(5, 2) &= \frac{2}{5}\binom{10}{3} \\
&= \frac{2}{5}\cdot 120 \\
&= 48
\end{aligned}
$$

See Table 14.1.

$B(n, 3)$ in Terms of C_n

As in the case of $B(n, 2)$, $B(n, 3)$ also can be expressed in terms of Catalan numbers.

Since $B(n, 2) = B(n-1, 1) + 2B(n-1, 2) + B(n-1, 3)$, it follows that:

$$
\begin{aligned}
B(n-1, 3) &= B(n, 2) - B(n-1, 1) - 2B(n-1, 2) \\
&= (C_{n+1} - 2C_n) - C_{n-1} - 2(C_n - 2C_{n-1}) \\
&= C_{n+1} - 4C_n + 3C_{n-1}
\end{aligned}
$$

That is,

$$
B(n, 3) = C_{n+2} - 4C_{n+1} + 3C_n
$$

where $n \geq 3$.

The same technique can be used to derive similar formulas for $B(n, r)$, where $n \geq r$:

$$B(n, 1) = C_n$$
$$B(n, 2) = C_{n+1} - 2C_n$$
$$B(n, 3) = C_{n+2} - 4C_{n+1} + 3C_n$$
$$B(n, 4) = C_{n+3} - 6C_{n+2} + 10C_{n+1} - 4C_n$$
$$B(n, 5) = C_{n+4} - 8C_{n+2} + 21C_{n+2} - 20C_{n+1} + 5C_n$$

Just as we expressed $B(n, 2)$ in terms of binomial coefficients, these formulas can be employed to develop similar formulas for $B(n, r)$, where $n \geq r$:

$$B(n, 1) = \frac{1}{n} \binom{2n}{n-1} \tag{14.2}$$

$$B(n, 2) = \frac{2}{n} \binom{2n}{n-2} \tag{14.3}$$

$$B(n, 3) = \frac{3}{n} \binom{2n}{n-3} \tag{14.4}$$

More generally, we have the following result.

Theorem 14.1 Let $n \geq r$. Then

$$B(n, r) = \frac{r}{n} \binom{2n}{n-r}$$

Proof Since $\frac{1}{1}\binom{1}{1-1} = 1, B(1, 1) = 0$. If $r > n$, then $\binom{2n}{n-r} = 0$; so $B(n, r) = 0$. Thus, the initial and boundary conditions are satisfied by $\frac{r}{n}\binom{2n}{n-r}$.

It remains to show that $\frac{r}{n}\binom{2n}{n-r}$ satisfies the recurrence relation (14.1). To this end, we have:

$$\frac{r-1}{n-1}\binom{2n-2}{(n-1)-(r-1)} + \frac{2r}{n-1}\binom{2n-2}{(n-1)-r}$$
$$+ \frac{r+1}{n-1}\binom{2n-2}{(n-1)-(r+1)} = \frac{r-1}{n-1}\binom{2n-2}{n-r} + \frac{2r}{n-1}\binom{2n-2}{n-r-1}$$

$$+ \frac{r+1}{n-1} \binom{2n-2}{n-r-2}$$

$$= \frac{r-1}{n-1} \cdot \frac{(2n-2)!}{(n-r)!(n+r-2)!}$$

$$+ \frac{2r}{n-1} \cdot \frac{(2n-2)!}{(n-r-1)!(n+r-1)!}$$

$$+ \frac{r-1}{n-1} \cdot \frac{(2n-2)!}{(n-r-2)!(n+r)!}$$

$$= \frac{4n^2 r - 6nr + 2r}{(n-1)(2n)(2n-1)} \binom{2n}{n-r}$$

$$= \frac{2r(n-1)(2n-1)}{(2n)(n-1)(2n-1)} \binom{2n}{n-r}$$

$$= \frac{r}{n} \binom{2n}{n-r}$$

Thus $\frac{r}{n}\binom{2n}{n-r}$ satisfies the recursive definition (14.1); so $\frac{r}{n}\binom{2n}{n-r} = B(n,r)$, as desired. $\blacksquare$

This theorem yields the following results, as found earlier.

Corollary 14.1

$$B(n,1) = C_n$$

$$B(n, n-1) = 2(n-1)$$

Proof

$$B(n,1) = \frac{1}{n}\binom{2n}{n-1}$$

$$= C_n$$

and

$$B(n, n-1) = \frac{n-1}{n}\binom{2n}{1}$$

$$= \frac{(2n)(n-1)}{n}$$

$$= 2(n-1)$$

$\blacksquare$

Corollary 14.2 Let $n \geq r \geq 1$. Then

$$\frac{r}{n} \Bigg| \binom{2n}{n-r}$$

Proof This follows because $B(n,r)$ is an integer. ∎

This property was proposed as a problem in 1934 by G. Birkhoff of Harvard University.

A Geometric Interpretation

Before pursuing a few additional properties of $B(n,r)$, we present a fascinating geometric interpretation of $B(n,r)$. But first, a few simple definitions.

Consider the lattice points in the first quadrant on the cartesian plane. From each lattice point $v_i = (a_i, b_i)$, we can walk one unit to the next lattice point to the right or above: $v_{i+1} = (a_i + 1, b_i)$ or $v_{i+1} = (a_i, b_i + 1)$. The *length* of the path $v_0 - v_1 - v_2 \ldots v_{n-1} - v_n$ is n.

Consider two paths $v_0 - v_1 - v_2 \ldots v_{n-1} - v_n$ and $w_0 - w_1 - w_2 \ldots w_{n-1} - w_n$ with the same length n, where $v_n = (a_n, b_n)$ and $w_n = (c_n, d_n)$. Then $|a_n - c_n|$ is the *distance* between the two paths; thus the distance is the absolute difference of the x-coordiantes of the terminal points of the two paths. The two paths *intersect* if $v_i = w_i$ for some i, where $0 < i \leq n$.

For example, consider the pairs of paths in Figure 14.1. The distance between the two paths in Figure 14.1(a) is 1; that between the paths in (b) is 2; and that between the paths in (c) is 3.

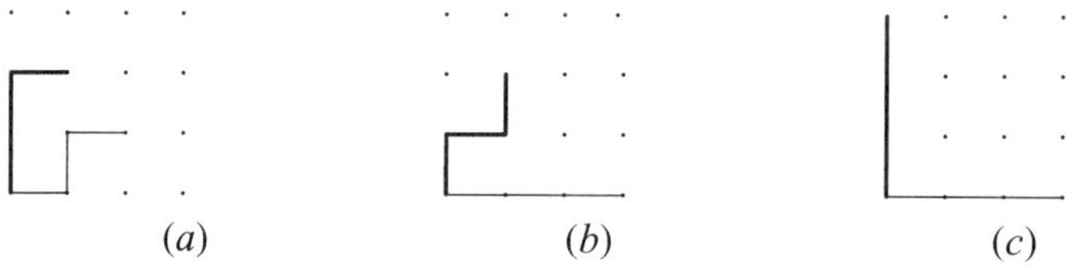

(a) (b) (c)

Figure 14.1

Let $A(n,r)$ denote the number of pairs of nonintersecting paths (originating at the origin), where $1 \leq r \leq n$. Figure 14.2 shows the possible paths for several pairs of values of n and r.

An interesting observation: $A(n,r) = B(n,r)$ for $1 \leq r \leq n \leq 3$ and $A(4,3) = 6 = B(4,3)$.

More generally, we claim that $A(n,r) = B(n,r)$ for all values of n and r. To see this, consider two paths of length $n-1$ with distance r. This pair can be extended

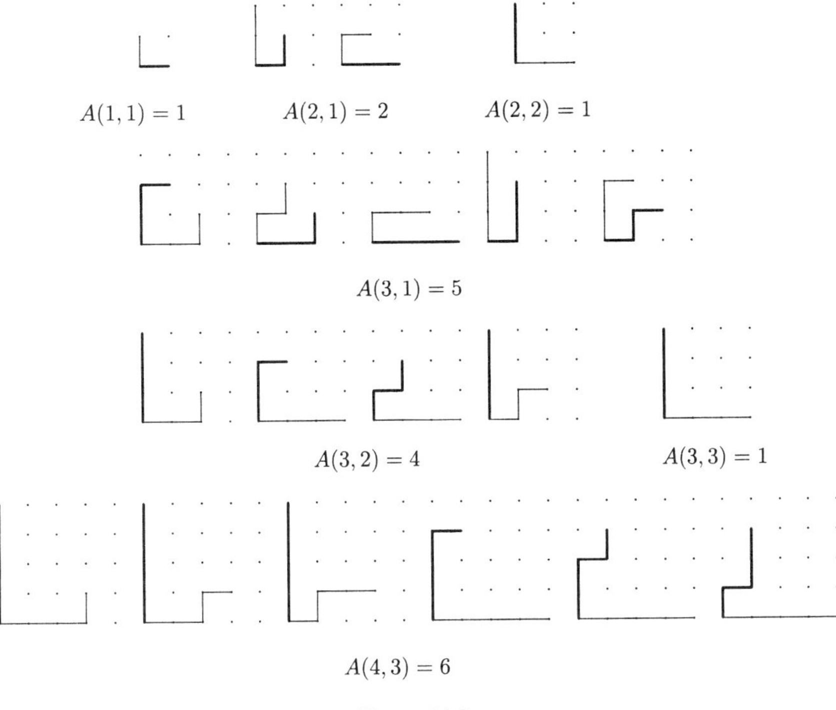

$$A(1,1) = 1 \qquad A(2,1) = 2 \qquad A(2,2) = 1$$

$$A(3,1) = 5$$

$$A(3,2) = 4 \qquad\qquad\qquad A(3,3) = 1$$

$$A(4,3) = 6$$

Figure 14.2

to four pairs of length n: one pair with distance $r + 1$, two pairs with distance r, and one pair with distance $r - 1$. Consequently,

$$A(n,r) = A(n-1,r-1) + 2A(n-1,r) + A(n-1,r+1)$$

Further, $A(n,0) = 0$ if $n \geq 1$, and $A(n,r) = 0$ if $n < r$. Thus $A(n,r)$ satisfies exactly the same recursive definition as $B(n,r)$; so $A(n,r) = B(n,r)$.

We now establish a few properties of Catalan triangle.

Theorem 14.2

$$\sum_{r=1}^{n} B(n,r) = \frac{1}{2}\binom{2n}{n}$$

Proof   Let

$$S_n = \sum_{r=1}^{n} B(n,r)$$

Since each pair of nonintersecting paths of length $n - 1$ extends to four pairs of length n, $S_n = 4S_{n-1} - C_{n-1}$, where the Catalan number C_{n-1} accounts for the paths of length $n - 1$ that intersect at step n.

To establish the summation formula, it suffices to show that $\frac{1}{2}\binom{2n}{n}$ satisfies the recurrence relation:

$$4S_{n-1} - C_{n-1} = 4\left[\frac{1}{2}\binom{2n-2}{n-1}\right] - \frac{1}{n}\binom{2n-2}{n-1}$$

$$= 2\binom{2n-2}{n-1} - \frac{1}{n}\binom{2n-2}{n-1}$$

$$= \frac{2n-1}{n}\binom{2n-2}{n-1}$$

$$= \frac{(2n-1)!}{n(n-1)!(n-1)!}$$

$$= \frac{1}{2}\binom{2n}{n}$$

$$= S_n$$

Thus

$$\sum_{r=1}^{n} B(n,r) = \frac{1}{2}\binom{2n}{n} \qquad \blacksquare$$

It follows by this theorem that the sum of the entries in row n of Catalan triangle equals $\frac{1}{2}\binom{2n}{n}$.

For example,

$$\sum_{r=1}^{5} B(5,r) = 42 + 48 + 27 + 8 + 1$$

$$= 126$$

$$= \frac{1}{2}\binom{10}{5}$$

Theorem 14.3

$$B(n,r) = \sum_{j=1}^{n-r+1} C_j B(n-j, r-1)$$

Proof   Consider two nonintersecting paths of length n with distance r. At some step, say, the $(n-j)$th step, they must have distance $r-1$ for the last time. For the remaining j steps, the two paths never come closer than at the $(n-j+1)$th step. Since the two paths can be continued in C_j ways for every

j, by the addition and multiplication principles, the total number of pairs of nonintersecting paths of length n with distance r is given by

$$\sum_{j=1}^{n-r+1} C_j B(n-j, r-1)$$

That is,

$$B(n,r) = \sum_{j=1}^{n-r+1} C_j B(n-j, r-1) \qquad \blacksquare$$

For example, let $n = 5$ and $r = 3$. Then:

$$\sum_{j=1}^{3} C_j B(5-j, 2) = C_1 \cdot B(4,2) + C_2 \cdot B(3,2) + C_3 \cdot B(2,2)$$

$$= 1 \cdot 14 + 2 \cdot 4 + 5 \cdot 1$$

$$= \boxed{27}$$

$$= B(5,3)$$

See Table 14.2.

Table 14.2

r \ n	1	2	3	4	5	6
1	1					
2	2	1				
3	5	4	1			
4	14	14	6	1		
5	42	48	27	8	1	
6	132	165	110	44	10	1

↑
Catalan numbers

Theorem 14.3 can be interpreted as follows. Each element $B(n,r)$ can be obtained by computing the weighted sum of the top $n - r + 1$ elements in column $r - 1$ and above row n, where the weights are the Catalan numbers

$C_1, C_2, \ldots C_{n-r+1}$:

$$B(n,r) = C_1 \cdot B(n-1, r-1) + C_2 \cdot B(n-2, r-1) + \cdots + C_{n-r+1} \cdot B(r-1, r-1)$$

Using this formula, we have

$$
\begin{aligned}
B(2,2) &= C_1 \cdot B(1,1) \\
&= C_1 C_1 \\
B(3,2) &= C_1 \cdot B(2,1) + C_2 \cdot B(1,1) \\
&= C_1 C_2 + C_2 C_1 \\
&= \sum_{i+j=3} C_i C_j
\end{aligned}
$$

and

$$
\begin{aligned}
B(3,3) &= C_1 \cdot B(2,2) \\
&= C_1 C_1 C_1 \\
&= \sum_{i+j+k=3} C_i C_j C_k
\end{aligned}
$$

Similarly, it can be shown that

$$
\begin{aligned}
B(4,2) &= \sum_{i+j=4} C_i C_j \\
B(4,3) &= \sum_{i+j+k=4} C_i C_j C_k \\
B(4,4) &= \sum_{i+j+k+l=4} C_i C_j C_k C_l
\end{aligned}
$$

where $i, j, k \geq 1$.

Table 14.3

n \ r	1	2	3	4
1	C_1			
2	C_2	$\sum_{i+j=2} C_i C_j$		
3	C_3	$\sum_{i+j=3} C_i C_j$	$\sum_{i+j+k=3} C_i C_j C_k$	
4	C_4	$\sum_{i+j=4} C_i C_j$	$\sum_{i+j+k=4} C_i C_j C_k$	$\sum_{i+j+k+l=4} C_i C_j C_k C_l$

Using this technique, the Catalan triangle can be rewritten completely in terms of Catalan numbers, as in Table 14.3, where $i, j, k, l \geq 1$.

Accordingly, we have the following result.

Theorem 14.4

$$B(n, r) = \sum_{i_1 + i_2 + \cdots + i_k = n} C_{i_1} C_{i_2} \cdots C_{i_k}$$

where each index $i_j \geq 1$.   ■

The next theorem shows that this process is reversible; that is, the elements $B(n, r)$ can be used to compute C_j.

Theorem 14.5

$$\sum_{r=1}^{\min(m,n)} B(m, r)B(n, r) = C_{m+n-1}$$

Proof   Consider all C_{m+n-1} pairs of paths which intersect for the first time after $m + n$ steps. At step m, any such pair must have some distance r and the remaining n steps can be considered a pair of nonintersecting paths of length n and also of the same distance k. Since $1 \leq r \leq \min(m, n)$, the result follows by the addition and multiplication principles.   ■

For example, let $m = 6$ and $n = 4$. Then:

$$\sum_{r=1}^{\min(6,4)} B(6, r)B(4, r) = \sum_{r=1}^{4} B(6, r)B(4, r)$$

$$= B(6, 1)B(4, 1) + B(6, 2)B(4, 2) + B(6, 3)B(4, 3)$$

$$+ B(6, 4)B(4, 4)$$

$$= 132 \cdot 14 + 165 \cdot 14 + 110 \cdot 6 + 44 \cdot 1$$

$$= 4,862$$

$$= C_9$$

$$= C_{6+4-1}$$

Theorem 14.5 can be interpreted as follows: *The dot product of every two adjacent rows in the Catalan triangle is a Catalan number.* For example, $(5 \ 4 \ 1 \ 0) \cdot (14 \ 14 \ 6 \ 1) = 132 = C_6$.

Theorem 14.5 has an interesting byproduct.

Corollary 14.3

$$\sum_{r=1}^{n} B^2(n,r) = C_{2n-1}$$

$$\sum_{r=1}^{n} B(n,r)B(n+1,r) = C_{2n}$$ ∎

For example,

$$\begin{aligned}
\sum_{r=1}^{6} B^2(n,r) &= B^2(6,1) + B^2(6,2) + B^2(6,3) + B^2(6,3) + B^2(6,4) \\
&\quad + B^2(6,5) + B^2(6,6) \\
&= 132^2 + 165^2 + 110^2 + 44^2 + 10^2 + 1^2 \\
&= 58,786 \\
&= C_{11}
\end{aligned}$$

and

$$\begin{aligned}
\sum_{r=1}^{4} B(4,r)B(5,r) &= B(4,1)B(5,1) + B(4,2)B(5,2) + B(4,3)B(5,3) \\
&\quad + B(4,4)B(5,4) \\
&= 14 \cdot 42 + 14 \cdot 48 + 6 \cdot 27 + 1 \cdot 8 \\
&= 1,430 \\
&= C_8
\end{aligned}$$

Theorem 14.4 coupled with Corollary 14.3 yields the following result.

Corollary 14.4

$$\sum_{r=1}^{n} \left(\sum_{i_1+i_2+\cdots+i_r} C_{i_1} C_{i_2} \cdots C_{i_r} \right)^2 = C_{2n-1}$$ ∎

We now turn to a new family of positive integers called *Cayley numbers*.

Cayley Numbers

Returning to the Catalan triangle in Table 14.1, notice that the second column consists of the integer sequence

$$1 \quad 4 \quad 14 \quad 48 \quad 165 \quad 572 \quad 2002 \ldots$$

They are called *Cayley numbers* and are named after Arthur Cayley, who discovered them in 1891 while investigating the dissection of convex polygons into small polygons by drawing nonintersecting diagonals.

Interestingly, Cayley numbers can be generated from the power series for Catalan numbers:

$$(1 + 2x + 5x^2 + 14x^3 + 42x^4 + \cdots)^2 = 1 + 4x + 14x^2 + 48x^3 + 165x^4 + \cdots .$$

This follows by Theorem 14.3. Thus

$$\frac{1 - 2x - \sqrt{1 - 4x}}{2x^2} = 1 + 4x + 14x^2 + 48x^3 + 165x^4 + \cdots$$

More generally, $\left(\sum_{i=1}^{\infty} C_i x^{i-1} \right)^n$ is the generating function of the nth column of the Catalan triangle, by virtue of Theorem 14.3.

We close this chapter with an interesting application to random walks.

Random Walks

Suppose there is a cop and a robber at the origin on the cartesian plane, and they walk from lattice point to lattice point in the first quadrant. At each lattice point, each flips a fair coin to determine if he should proceed east or north. Assuming that they move simultaneously, we would like to compute the probability that they meet after leaving the origin.

Since the cop and the robber have 2^n equally likely paths of length n available, there are 4^n pairs of paths of length n. By Theorem 14.2, $2\left[\frac{1}{2}\binom{2n}{n}\right] = \binom{2n}{n}$ of them do not meet; here the extra factor 2 accounts for the fact that we have labeled the two paths in each pair.

The probability that the cop and the robber will not meet is given by

$$\lim_{n \to \infty} \frac{\binom{2n}{n}}{4^n} = \lim_{n \to \infty} \frac{(2n)!}{4^n n! n!}$$

$$= \lim_{n \to \infty} \frac{\sqrt{4n\pi}(2ne^{-1})^{2n}}{\left[\sqrt{2n\pi}(ne^{-1})^n\right]^2 \cdot 4^n}$$

$$= \lim_{n \to \infty} \frac{1}{\sqrt{n\pi}}$$

$$= 0$$

where we have used the Stirling's formula $m! \approx m^m e^{-m} \sqrt{2n\pi}$.

Thus, the probability that the robber will meet the cop again is $1 - 0 = 1$.

15

A Family of Binary Words

Recall that in Example 6.12, we investigated binary words consisting of n 1s and n 0s such that the number of 1s in each subword when scanned from left to right is greater than or equal to the number of 0s in it; the number of such binary words is the Catalan number C_n. In other words, the number of strings consisting of n 1s and $n - 1$s such that every parial sum is ≥ 0 is C_n; see Example 6.10.

In lieu of using the same number of 0s, suppose we use k 0s, where $0 \leq k \leq n$. Let $A(n, k)$ denote the number of such arrangements. In this chapter, we investigate the various properties of $A(n, k)$, including an explicit formula, and deduce C_n as a special case of $A(n, k,)$. In the process, we will encounter the Pascal-like triangle we constructed in Table 8.1. This general case was studied in 1996 by D. F. Bailey of Trinity University, San Antonio, Texas.

We begin with the basic properties of $A(n, k)$ in the following lemma.

Lemma 15.1

1. $A(n, 0) = 1, \quad n \geq 0$
2. $A(n, 1) = n, \quad n \geq 1$
3. $A(n + 1, k) = A(n + 1, k - 1) + A(n, k), \quad 1 \leq k \leq n$
4. $A(n + 1, n + 1) = A(n + 1, n), \quad n \geq 1$

Proof

1. When $k = 0$, we can form only one word, namely, the word consisting of n 1s. So, $A(n, 0) = 1$.

2. Suppose $k = 1$. Then every word consists of $n + 1$ bits with exactly one 0. So the 0 can be anywhere in positions 2 through n; that is, the position of the 0 has n choices; each choice yields a new word. Consequently, $A(n, 1) = n$.

3. Consider a binary word consisting of n 1s and k 0s, a total of $n + k$ bits. Let us denote it by $a_1 a_2 \cdots a_{n+k} a_{n+k+1}$.

Case 1   Suppose $a_{n+k+1} = 1$.
Such a word looks like this:

$$\underbrace{- \, - \, - \, \cdots \, - \, - \, \overset{\displaystyle 1}{-}}_{n + k \text{ bits}}$$

By definition, there are $A(n, k)$ such binary words with the desired property.

Case 2   Suppose $a_{n+k+1} = 0$.
Such a word looks like this:

$$\underbrace{- \, - \, - \, \cdots \, - \, - \, \overset{\displaystyle 0}{-}}_{n + 1 \text{ 1s and } k - 1 \text{ 0s}}$$

The remaining $n + k$ positions are filled with $n + 1$ 1s and $k - 1$ 0s. By definition, there are $A(n + 1, k - 1)$ binary words with the given property. By the addition principle, the total number of binary words with the given property equals $A(n, k) + A(n + 1, k - 1)$. That is,

$$A(n + 1, k) = A(n + 1, k - 1) + A(n, k)$$

4. To establish this result, consider a binary word consisting of $n + 1$ 1s and $n + 1$ 0s with the desired property. Notice that the last element in the word must be 0:

$$\underbrace{- \, - \, - \, \cdots \, - \, - \, \overset{\displaystyle 0}{-}}_{n + 1 \text{ 1s and } n \text{ 0s}}$$

Consequently, the substring consisting of $n + 1$ 1s and n 0s must have the desired property; by definition, there are $A(n + 1, n)$ such binary words. Thus $A(n + 1, n + 1) = A(n + 1, n)$, where $n \geq 1$.   ∎

Interestingly, the properties in Lemma 15.1 can be used to construct a Pascal-like triangle, as in Table 15.1. Property 1 implies that column 0 is made up of 1s, and property 2 indicates that column 1 consists of positive integers, when $n \geq 1$. Property 3 is a Pascal-like identity; it shows that the element in row n and column k equals the sum of the previous entry in the same row and the element just above it. For instance, $75 + 95 = 165$. Property 4 indicates that the last two elements in each row are equal.

Notice that this is exactly the same triangular array we encountered in Table 8.1, where we studied the number of possible paths of a rook from the upper left-hand

Table 15.1

n \ j	0	1	2	3	4	5	6	7
0	1							
1	1	1						
2	1	2	2					
3	1	3	5	5				
4	1	4	9	14	14			
5	1	5	14	28	42	42		
6	1	6	20	48	90	132	132	
7	1	7	27	75	165	297	429	429
8	1	8	35	110	275	572	1001	1430	1430

Catalan numbers

corner to the lower right-hand corner without crossing the northeast diagonal. The same triangular array was developed in 1962 by P. Lafer and C. T. Long of Washington State University in their investigation of Euler's triangulation problem in Example 5.1.

We need three more lemmas before we can develop an explicit formula for $A(n, k)$.

Lemma 15.2 Let $1 \leq k \leq n$. Then

$$A(n, k) = \sum_{i=k}^{n} A(i, k - 1)$$

Proof (*by induction*)   Since $A(1, 1) = A(1, 0)$, by Lemma 15.1, the result is true when $n = 1$.

Now, assume that the formula is true for an arbitrary integer $n \geq 2$. Consider $A(n + 1, k)$.

Case 1   Let $k \leq n$.
Then, by Lemma 15.1 and the inductive hypothesis,

$$A(n + 1, k) = A(n + 1, k - 1) + A(n, k)$$

$$= A(n + 1, k - 1) + \sum_{i=k}^{n} A(i, k - 1)$$

$$= \sum_{i=k}^{n+1} A(i, k - 1)$$

Case 2   Let $k = n + 1$.

Then, by Lemma 15.1,

$$A(n + 1, n + 1) = A(n + 1, n)$$

$$= \sum_{i=k}^{n+1} A(i, k - 1)$$

In both cases, the formula works for $n + 1$.

Thus, by induction, the formula holds for every $n \geq 1$.   ∎

For example,

$$A(7, 3) = \sum_{i=3}^{7} A(i, 2)$$

$$= A(3, 2) + A(4, 2) + A(5, 2) + A(6, 2) + A(7, 2)$$

$$= 5 + 9 + 14 + 20 + 27$$

$$= 75$$

See Table 15.1.

The next theorem gives an explicit formula for $A(n, 2)$.

Theorem 15.1   Let $n \geq 2$. Then

$$A(n, 2) = \frac{(n - 1)(n + 2)}{2}$$

Proof   By Lemma 15.2,

$$A(n, 2) = \sum_{i=2}^{n} A(i, 1)$$

$$= \sum_{i=2}^{n} i$$

$$= \frac{n(n + 1)}{2} - 1$$

$$= \frac{(n - 1)(n + 2)}{2}$$ ∎

For example,

$$A(7,2) = \frac{6 \cdot 9}{2}$$

$$= 27$$

See Table 15.1.

The next theorem gives us an explicit formula for $A(n,3)$.

Theorem 15.2 Let $n \geq 3$. Then

$$A(n,3) = \frac{(n-2)(n+2)(n+3)}{6}$$

Proof   By Theorem 15.1,

$$A(n,3) = \sum_{i=3}^{n} A(i,2)$$

$$= \sum_{i=3}^{n} \left[\frac{i(i+1)}{2} - 1 \right]$$

$$= \frac{1}{2} \sum_{i=3}^{n} i^2 + \frac{1}{2} \sum_{i=3}^{n} i - (n-2)$$

$$= \frac{1}{2} \sum_{i=1}^{n} i^2 + \frac{1}{2} \sum_{i=1}^{n} i - (n-2) - 4$$

$$= \frac{n(n+1)(2n+1)}{12} + \frac{n(n+1)}{4} - (n+2)$$

$$= \frac{n(n+1)(n+2)}{6} - (n+2)$$

$$= \frac{(n+2)(n^2+n-6)}{6}$$

$$= \frac{(n-2)(n+2)(n+3)}{6}$$ ∎

For example,

$$A(5,3) = \frac{3 \cdot 7 \cdot 8}{6}$$

$$= 28$$

See Table 15.1.

Using Theorem 15.2 and the same technique as in its proof, we can derive an explicit formula for $A(n, 4)$.

Theorem 15.3

$$A(n, 4) = \frac{(n-3)(n+2)(n+3)(n+4)}{24}$$

Proof   Using Theorem 15.2, we have:

$$A(n, 4) = \sum_{i=4}^{n} A(i, 3)$$

$$6A(n, 4) = \sum_{i=4}^{n} (i^3 + 3i^2 - 4i - 12)$$

$$= \sum_{i=1}^{n} (i^3 + 3i^2 - 4i - 12) - \sum_{i=1}^{3} (i^3 + 3i^2 - 4i - 12)$$

$$= \left[\frac{n^2(n+1)^2}{4} + 3 \cdot \frac{n(n+1)(2n+1)}{6} - 4 \cdot \frac{n(n+1)}{2} - 12n \right]$$
$$\quad - (36 + 42 - 24 - 36)$$

$$= \frac{n(n+1)}{4} [n(n+1) + 2(2n+1) - 8] - 12n - 18$$

$$= \frac{n(n+1)}{4} (n^2 + 5n - 6) - 12n - 18$$

$$24A(n, 4) = n^4 + 6n^3 - n^2 - 54n - 72$$

$$= (n-3)(n+2)(n+3)(n+4)$$

$$A(n, 4) = \frac{(n-3)(n+2)(n+3)(n+4)}{24} \qquad\blacksquare$$

For example,

$$A(6, 4) = \frac{3 \cdot 8 \cdot 9 \cdot 10}{24}$$

$$= 90$$

See Table 15.1.

This theorem yields two interesting byproducts. Because $A(n, 4)$ is always an integer, it follows that

$$n^4 + 6n^3 - n^2 - 54n - 72 \equiv 0 \ (\text{mod } 24)$$

That is,

$$n^2(n^2 + 6n - 1) \equiv 6n \ (\text{mod } 24)$$

So

$$n^3(n + 6) \equiv n(n + 6) \ (\text{mod } 24)$$

In particular, if n and 24 are relatively prime, then so are $n + 6$ and 24; the converse is also true. So $n^2 \equiv 1 \ (\text{mod } 24)$. In other words, those least positive residues modulo 24 that are relatively prime to 24 are self-invertible. Using Euler's φ-function* there are exactly $\varphi(24) = \varphi(2^3 \cdot 3) = 24 \left(1 - \frac{1}{2}\right)\left(1 - \frac{1}{3}\right) = 8$ such residues; they are given by $\pm 1, \pm 5, \pm 7,$ and ± 11, that is, 1, 5, 7, 11, 13, 17, 19, and 23.

It follows from Theorems 15.2 and 15.3 that an explicit formula for $A(n, k)$ can be developed, provided we know the summation formulas for $\sum_{i=1}^{n} i^j$, where $1 \leq j \leq k - 1$.

For example, the following summation formulas can be employed to derive explicit formulas for $A(n, k)$, where $5 \leq k \leq 8$:

$$\sum_{i=1}^{n} i^4 = \frac{n(n + 1)(2n + 1)(3n^2 + 3n - 1)}{30}$$

$$\sum_{i=1}^{n} i^5 = \frac{n^2(n + 1)^2(2n^2 + 2n - 1)}{12}$$

$$\sum_{i=1}^{n} i^6 = \frac{n(n + 1)(2n + 1)(3n^4 + 6n^3 - 3n + 1)}{42}$$

$$\sum_{i=1}^{n} i^7 = \frac{n^2(n + 1)^2(3n^4 + 6n^3 - n^2 - 4n + 2)}{24}$$

We need one more result before we move to the explicit formula for $A(n, k)$.

Lemma 15.4

$$\sum_{i=k+1}^{n} (i+1-k)(i+2)(i+3) \cdots (i+k) = \frac{1}{k+1}(n-k)(n+2)(n+3) \cdots (n+1+k)$$

* See T. Koshy, *Elementary Number Theory with Applications*, 2nd ed., Academic Press, Burlington, Massachusetts, 2007.

Proof (*by induction*)   When $n = k + 1$,

$$\text{LHS} = (k + 2 - 3)(k + 3)(k + 4) \cdots (2k + 1)$$

$$= 2(k + 3)(k + 4) \cdots (2k + 1)$$

$$\text{RHS} = \frac{1}{k + 1}(k + 1 - k)(k + 3)(k + 4) \cdots (2k + 1)(2k + 2)$$

$$= 2(k + 3)(k + 4) \cdots (2k + 1)$$

$$= \text{LHS}$$

So the result is true when $n = k + 1$.

Suppose the result is true for an arbitrary ineger $n \geq k + 1$. Then

$$\sum_{i=k+1}^{n+1} (i + 1 - k)(i + 2)(i + 3) \cdots (i + k)$$

$$= \sum_{i=k+1}^{n} (i + 1 - k)(i + 2)(i + 3) \cdots (i + k)$$

$$+ (n + 2 - k)(n + 3)(n + 4) \cdots (n + 1 + k)$$

$$= \frac{1}{k + 1}(n - k)(n + 2)(n + 3) \cdots (n + 1 + k)$$

$$+ (n + 2 - k)(n + 3)(n + 4) \cdots (n + 1 + k)$$

$$= \frac{(n + 3)(n + 4) \cdots (n + 1 + k)}{k + 1} [(n - k)(n + 2) + (n + 2 - k)(k + 1)]$$

$$= \frac{(n + 3)(n + 4) \cdots (n + 1 + k)}{k + 1} \cdot (n + 1 - k)(n + 2 + k)$$

$$= \frac{1}{k + 1}(n + 1 - k)(n + 3)(n + 4) \cdots (n + 1 + k)(n + 2 + k)$$

Thus, by induction, the formula works for every integer $n \geq k + 1$.   ∎

The formulas in Theorems 15.1, 15.2, and 15.3 reveal an interesting and predictable pattern:

$$A(n, 2) = \frac{(n - 1)(n + 2)}{2!}$$

$$A(n, 3) = \frac{(n - 2)(n + 2)(n + 3)}{3!}$$

$$A(n, 4) = \frac{(n - 3)(n + 2)(n + 2)(n + 3)(n + 4)}{4!}$$

More generally, we conjecture that

$$A(n, k) = \frac{(n + 1 - k)(n + 2)(n + 3) \cdots (n + k)}{k!}$$

The next theorem establishes this main result by induction.

Theorem 15.4 Let $n \geq k \geq 1$. Then

$$A(n, k) = \frac{(n + 1 - k)(n + 2)(n + 3) \cdots (n + k)}{k!}$$

Proof (*by induction*)   When $k = 1$,

$$\text{RHS} = n = A(n, 1) = \text{LHS}$$

So the result is true when $k = 1$.

Now, assume it is true for an arbitrary integer $k \geq 1$. Then, by Lemma 15.2,

$$A(n, k + 1) = \sum_{i=k+1}^{n} A(i, k)$$

$$= \sum_{i=k+1}^{n} \frac{(i + 1 - k)(i + 2)(i + 3) \cdots (i + k)}{k!}$$

$$= \frac{1}{k!} \cdot \frac{1}{k + 1}(n - k)(n + 2)(n + 3) \cdots (n + 1 + k)$$

$$= \frac{1}{(k + 1)!} [(n + 1) - (k + 1)] (n + 2)(n + 3) \cdots [n + (k + 1)]$$

Thus, by induction, the formula works for every integer k, where $1 \leq k \leq n$. $\blacksquare$

For example,

$$A(7,4) = \frac{(7+1-4)(7+2)(7+3)(7+4)}{4!}$$

$$= \frac{4 \cdot 9 \cdot 10 \cdot 11}{4!}$$

$$= 165$$

See Table 15.1.

Suppose we let $k = n$ in Theorem 15.4. Then

$$A(n,n) = \frac{(n+2)(n+3)\cdots(2n)}{n!}$$

$$= \frac{1}{n+1}\binom{2n}{n}$$

$$= C_n$$

Accordingly, we have the following result, as we observed in Example 6.12.

Corollary 15.1

$$A(n,n) = C_n$$

■

16

Tribinomial Coefficients

This chapter investigates binomial coefficients for triangular numbers t_m, where $m \geq 1$ and their relationships to Catalan numbers. We call them tribinomial coefficients. They were studied by W. Hansell around 1974 and are denoted by $\left[\begin{smallmatrix} n \\ r \end{smallmatrix}\right]$.

Tribinomial Coefficients

Recall that the binomial coefficient $\binom{n}{r}$ is given by

$$\binom{n}{r} = \frac{n!}{r!(n-r)!}$$

$$= \frac{n(n-1)\cdots(n-r+1)}{r(r-1)\cdots 2 \cdot 1}$$

Using the same idea, binomial coefficients for triangular numbers, that is, *tribinomial coefficients* $\left[\begin{smallmatrix} n \\ r \end{smallmatrix}\right]$ are defined as follows:

$$\begin{bmatrix} n \\ r \end{bmatrix} = \frac{t_n^*}{t_r^* t_{n-r}^*} \tag{16.1}$$

$$= \frac{t_n t_{n-1} \cdots t_{n-r+1}}{t_r t_{r-1} \cdots t_2 t_1} \tag{16.2}$$

where $\left[\begin{smallmatrix} n \\ 0 \end{smallmatrix}\right] = \left[\begin{smallmatrix} n \\ n \end{smallmatrix}\right] = 1$, and $t_m^* = t_m t_{m-1} \cdots t_2 t_1$. The definition of t_m^* corresponds to that of $m! = m(m-1) \cdots 2 \cdot 1$.

In 1974, Verner E. Hoggatt (1921–1980) of what was then San Jose State College, California, claimed that the numbers $\left[\begin{smallmatrix} n \\ r \end{smallmatrix}\right]$ are integers. We will confirm this later.

For instance,

$$\begin{bmatrix} 5 \\ 3 \end{bmatrix} = \frac{t_5 t_4 t_3}{t_3 t_2 t_1}$$

$$= \frac{15 \cdot 10 \cdot 6}{6 \cdot 3 \cdot 1}$$

$$= 50$$

An Explicit Formula

Since $t_m = \frac{m(m+1)}{2}$, we have:

$$\begin{bmatrix} n \\ r \end{bmatrix} = \frac{\dfrac{(n+1)n}{2} \cdot \dfrac{n(n-1)}{2} \cdots \dfrac{(n-r+2)(n-r+1)}{2}}{\dfrac{(r+1)r}{2} \cdot \dfrac{r(r-1)}{2} \cdots \dfrac{3 \cdot 2}{2} \cdot \dfrac{2 \cdot 1}{2}}$$

$$= \frac{(n+1)n^2(n-1)^2 \cdots (n-r+2)^2(n-r+1)}{(r+1)r^2(r-1)^2 \cdots 3^3 \cdot 2^2 \cdot 1^2}$$

$$= \frac{(n+1)n(n-1)\cdots(n-r+1)}{(r+1)!} \cdot \frac{n(n-1)\cdots(n-r+2)}{r!}$$

$$= \frac{(n+1)!}{(n-r)!(r+1)!} \cdot \frac{n!}{r!(n-r)!} \cdot \frac{1}{n-r+1}$$

$$= \frac{1}{n-r+1} \binom{n+1}{r+1} \binom{n}{r} \tag{16.3}$$

For example,

$$\begin{bmatrix} 5 \\ 3 \end{bmatrix} = \frac{1}{3} \binom{6}{4} \binom{5}{3}$$

$$= \frac{15 \cdot 10}{3}$$

$$= 50$$

as expected.

$$
\begin{array}{ccccccccccccc}
& & & & & & \boxed{1} & & & & & & \leftarrow \text{ row } 0\\
& & & & & 1 & & 1 & & & & &\\
& & & & 1 & & \boxed{3} & & 1 & & & &\\
& & & 1 & & 6 & & 6 & & 1 & & &\\
& & 1 & & 10 & & \boxed{20} & & 10 & & 1 & &\\
& 1 & & 15 & & 50 & & 50 & & 15 & & 1 &\\
1 & & 21 & & 105 & & \boxed{175} & & 105 & & 21 & & 1
\end{array}
$$

$\nearrow$

t_n

Figure 16.1 Tribinomial Triangle

The various tribinomial coefficents $\left[\begin{smallmatrix} n \\ r \end{smallmatrix}\right]$, can be arranged to yield the tribinomial array in Figure 16.1.

This array manifests two obvious properties:

- Northeast diagonal 1 consists of triangular numbers. This is so, since

$$
\begin{aligned}
\begin{bmatrix} n \\ 1 \end{bmatrix} &= \frac{1}{n}\binom{n+1}{2}\binom{n}{1}\\
&= \frac{(n+1)n}{2}\\
&= t_n
\end{aligned}
$$

- The entries in each row are symmetrically placed about the vertical line through the middle; that is,

$$
\begin{bmatrix} n \\ r \end{bmatrix} = \begin{bmatrix} n \\ n-r \end{bmatrix}
$$

This is so, since

$$
\begin{aligned}
\begin{bmatrix} n \\ n-r \end{bmatrix} &= \frac{1}{r+1}\binom{n+1}{n-r+1}\binom{n}{n-r}\\
&= \frac{1}{r+1}\frac{(n+1)!}{(n-r+1)!\,r!}\binom{n}{r}\\
&= \frac{1}{n-r+1}\frac{(n+1)!}{(r+1)!(n-r)!}\binom{n}{r}
\end{aligned}
$$

$$= \frac{1}{n-r+1} \binom{n+1}{r+1} \binom{n}{r}$$

$$= \begin{bmatrix} n \\ r \end{bmatrix}$$

We now investigate recurrence relations satisfied by the tribinomial coefficient.

Recurrence Relations

Notice that

$$\begin{bmatrix} n \\ r \end{bmatrix} t_r = \frac{t_n^*}{t_r^* t_{n-r}^*} \cdot t_r$$

$$= \frac{t_n^*}{t_{r-1}^* t_{n-r}^*}$$

$$= \frac{t_{n-1}^*}{t_{r-1}^* t_{n-r}^*} \cdot t_n$$

$$= \begin{bmatrix} n-1 \\ r-1 \end{bmatrix} t_n$$

Thus $\begin{bmatrix} n \\ r \end{bmatrix}$ satisfies the recurrence relation

$$\begin{bmatrix} n \\ r \end{bmatrix} = \begin{bmatrix} n-1 \\ r-1 \end{bmatrix} \frac{t_n}{t_r} \tag{16.4}$$

[This reminds us of the combinatorial identity $\binom{n}{r} = \frac{n}{r}\binom{n-1}{r-1}$.]

It follows from identity (16.4) that

$$\begin{bmatrix} 2n \\ n \end{bmatrix} = \frac{t_{2n}}{t_n} \begin{bmatrix} 2n-1 \\ n-1 \end{bmatrix}$$

$$= \frac{2(2n+1)}{n+1} \begin{bmatrix} 2n-1 \\ n-1 \end{bmatrix}$$

$$(n+1) \begin{bmatrix} 2n \\ n \end{bmatrix} = 2(2n+1) \begin{bmatrix} 2n-1 \\ n-1 \end{bmatrix} \tag{16.5}$$

But $(n+1, 2n+1) = 1$, where (a, b) denotes the greatest common divisor of the positive integers a and b. So $n+1 | 2 \begin{bmatrix} 2n-1 \\ n-1 \end{bmatrix}$ and $2n+1 | \begin{bmatrix} 2n \\ n \end{bmatrix}$.

For example, let $n = 3$. Then $4 | 2 \begin{bmatrix} 5 \\ 3 \end{bmatrix}$ and $7 | \begin{bmatrix} 6 \\ 3 \end{bmatrix}$; see Figure 16.1.

Similarly, it can be shown that

$$\begin{bmatrix} n \\ r \end{bmatrix} = \begin{bmatrix} n-1 \\ r \end{bmatrix} \frac{t_n}{t_{n-r}} \tag{16.6}$$

For example,

$$\begin{bmatrix} 6 \\ 4 \end{bmatrix} = \begin{bmatrix} 5 \\ 3 \end{bmatrix} \frac{t_6}{t_4} = 50 \cdot \frac{21}{10} = 105$$

$$= \begin{bmatrix} 5 \\ 4 \end{bmatrix} \frac{t_6}{t_2} = 15 \cdot \frac{21}{3} = 105$$

Recurrence relation (16.4) can be used to establish the following divisibility property for tribinomials:

$$\frac{t_m}{(t_m, t_n)} \, \bigg| \, \begin{bmatrix} m \\ n \end{bmatrix} \tag{16.7}$$

To establish this, we let:

$$d = (t_m, t_n)$$

$$= At_m + Bt_n$$

for some integers A and B by the euclidean algorithm. Then:

$$\begin{bmatrix} m \\ n \end{bmatrix} d = A \begin{bmatrix} m \\ n \end{bmatrix} t_m + B \begin{bmatrix} m \\ n \end{bmatrix} t_n$$

$$= A \begin{bmatrix} m \\ n \end{bmatrix} t_m + B \begin{bmatrix} m-1 \\ n-1 \end{bmatrix} t_m$$

$$= \left(A \begin{bmatrix} m \\ n \end{bmatrix} + B \begin{bmatrix} m \\ n \end{bmatrix} \right) t_m$$

So

$$t_m \, \bigg| \, \begin{bmatrix} m \\ n \end{bmatrix} d$$

That is,

$$\frac{t_m}{(t_m, t_n)} \, \bigg| \, \begin{bmatrix} m \\ n \end{bmatrix}$$

as desired.

This property corresponds to Hermite's divisibility property (1.8):

$$\frac{m}{(m,n)} \;\Bigg|\; \binom{m}{n}$$

In particular, let $m = 2n$. Then

$$\frac{t_{2n}}{(t_{2n}, t_n)} \;\Bigg|\; \left[\begin{array}{c} 2n \\ n \end{array} \right]$$

But

$$(t_{2n}, t_n) = \left(\frac{2n(2n+1)}{2}, \frac{n(n+1)}{2} \right)$$

$$= \begin{cases} n & \text{if } n \text{ is odd} \\ n/2 & \text{otherwise} \end{cases}$$

Thus $\frac{t_{2n}}{n} \big| \left[\begin{array}{c} 2n \\ n \end{array} \right]$ if n is odd; and $\frac{2t_{2n}}{n} \big| \left[\begin{array}{c} 2n \\ n \end{array} \right]$ if n is even. That is, $2n + 1 \mid \left[\begin{array}{c} 2n \\ n \end{array} \right]$ if n is odd, and $2(2n + 1) \mid \left[\begin{array}{c} 2n \\ n \end{array} \right]$ otherwise. [Notice from equation (1.9) that the former is true even if n is even.]

For example, let $n = 6$. We have $t_{12} = \frac{12 \cdot 13}{2} = 78$, and

$$\left[\begin{array}{c} 12 \\ 6 \end{array} \right] = \frac{t_{12} t_{11} t_{10} t_9 t_8 t_7}{t_6 t_5 t_4 t_3 t_2 t_1} = 226{,}512$$

Clearly, $2(2n+1) = 26$, and $26 \mid 226512$. On the other hand, $\left[\begin{array}{c} 12 \\ 6 \end{array} \right] = 175, 7 \mid 175$, but $2 \cdot 7 \nmid 175$.

Unfortunately, Hermite's second divisibility property (1.9) that $\frac{m-n+1}{(m+1,n)} \mid \binom{m}{n}$ does *not* have an analogous result for tribinomials. For example, when $m = 5$ and $n = 2$,

$$\frac{t_{m-n+1}}{(t_{m+1}, t_n)} = \frac{t_4}{(t_6, t_2)} = \frac{10}{(21, 3)} = \frac{10}{3}$$

is not even an integer.

Hoggatt-Hansell Identity for Tribinomials

The Hoggatt-Hansell identity

$$\binom{n-1}{r-1}\binom{n}{r+1}\binom{n+1}{r} = \binom{n-1}{r}\binom{n+1}{r+1}\binom{n}{r-1} \tag{16.8}$$

for ordinary binomial coefficients holds for tribinomials also, as Hoggatt and Hansell showed in 1971:

$$\begin{bmatrix} n-1 \\ r-1 \end{bmatrix}\begin{bmatrix} n \\ r+1 \end{bmatrix}\begin{bmatrix} n+1 \\ r \end{bmatrix} = \begin{bmatrix} n-1 \\ r \end{bmatrix}\begin{bmatrix} n+1 \\ r+1 \end{bmatrix}\begin{bmatrix} n \\ r-1 \end{bmatrix} \qquad (16.9)$$

This can be established as follows:

$$\begin{aligned}
\text{LHS} &= \frac{t^*_{n-1}}{t^*_{r-1}t^*_{n-r}} \cdot \frac{t^*_{n}}{t^*_{r+1}t^*_{n-r-1}} \cdot \frac{t^*_{n+1}}{t^*_{r}t^*_{n-r+1}} \\[2mm]
&= \frac{t^*_{n-1}}{t^*_{r}t^*_{n-r-1}} \cdot \frac{t^*_{n+1}}{t^*_{r+1}t^*_{n-r}} \cdot \frac{t^*_{n}}{t^*_{r-1}t^*_{n-r+1}} \\[2mm]
&= \begin{bmatrix} n-1 \\ r \end{bmatrix}\begin{bmatrix} n+1 \\ r+1 \end{bmatrix}\begin{bmatrix} n \\ r-1 \end{bmatrix} \\[2mm]
&= \text{RHS}
\end{aligned}$$

See Figure 16.2. The product of the six tribinomial coefficents is a square.

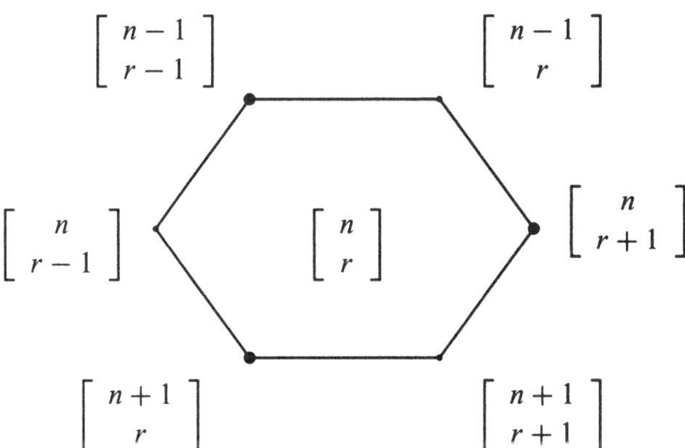

Figure 16.2

For example,

$$\begin{bmatrix} 5 \\ 3 \end{bmatrix}\begin{bmatrix} 6 \\ 5 \end{bmatrix}\begin{bmatrix} 7 \\ 4 \end{bmatrix} = 50 \cdot 21 \cdot 490$$

$$= 15 \cdot 196 \cdot 175$$

$$= \begin{bmatrix} 5 \\ 4 \end{bmatrix}\begin{bmatrix} 7 \\ 5 \end{bmatrix}\begin{bmatrix} 6 \\ 3 \end{bmatrix}$$

See Figure 16.3.

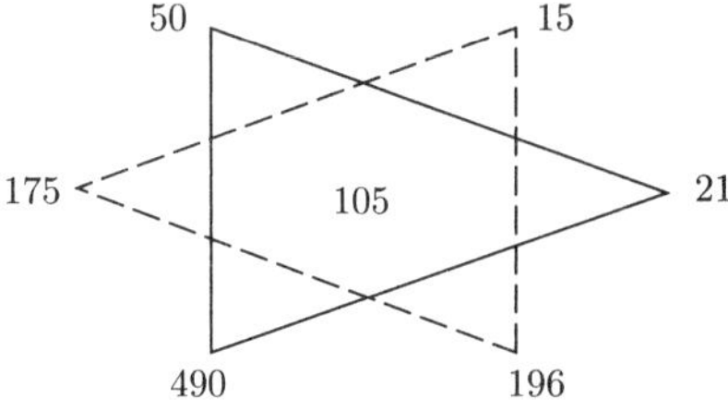

Figure 16.3

In fact, just as H. W. Gould of the University of West Virginia generalized (16.8) to the identity

$$\binom{n-a}{r-a}\binom{n}{r+a}\binom{n+a}{r} = \binom{n-a}{r}\binom{n+a}{r+a}\binom{n}{r-a}$$

formula (16.9) can be generalized as

$$\begin{bmatrix} n-a \\ r-a \end{bmatrix}\begin{bmatrix} n \\ r+a \end{bmatrix}\begin{bmatrix} n+a \\ r \end{bmatrix} = \begin{bmatrix} n-a \\ r \end{bmatrix}\begin{bmatrix} n+a \\ r+a \end{bmatrix}\begin{bmatrix} n \\ r-a \end{bmatrix} \tag{16.10}$$

This also can be established algebraically:

$$\begin{bmatrix} n-a \\ r-a \end{bmatrix}\begin{bmatrix} n \\ r+a \end{bmatrix}\begin{bmatrix} n+a \\ r \end{bmatrix} = \frac{t^*_{n-a}}{t^*_{r-a}t^*_{n-r}} \cdot \frac{t^*_n}{t^*_{r+a}t^*_{n-r-a}} \cdot \frac{t^*_{n+a}}{t^*_r t^*_{n+a-r}}$$

$$= \frac{t^*_{n-a}}{t^*_r t^*_{n-r-a}} \cdot \frac{t^*_{n+a}}{t^*_{r+a}t^*_{n-r}} \cdot \frac{t^*_{n+a}}{t^*_{r-a}t^*_{n-r+a}}$$

$$= \begin{bmatrix} n-a \\ r \end{bmatrix}\begin{bmatrix} n+a \\ r+a \end{bmatrix}\begin{bmatrix} n \\ r-a \end{bmatrix}$$

as desired.

For example, let $n = 6, a = 2$, and $r = 3$. Then

$$\begin{bmatrix} 4 \\ 1 \end{bmatrix}\begin{bmatrix} 6 \\ 5 \end{bmatrix}\begin{bmatrix} 8 \\ 3 \end{bmatrix} = 10 \cdot 21 \cdot 1176$$

$$= 10 \cdot 1176 \cdot 21$$

$$= \begin{bmatrix} 4 \\ 3 \end{bmatrix}\begin{bmatrix} 8 \\ 5 \end{bmatrix}\begin{bmatrix} 6 \\ 1 \end{bmatrix}$$

See Figure 16.4.

$$
\begin{array}{c|ccccc}
 & 1 & 2 & 3 & 4 & 5 \\
\hline
4 & \bullet & & \star & & \\
5 & & & & & \\
6 & \star & & & & \bullet \\
7 & & & & & \\
8 & & & \bullet & \star &
\end{array}
$$

Figure 16.4

Central Tribinomial Coefficients

It follows from the explicit formula (16.3) for $\left[\begin{array}{c} n \\ r \end{array}\right]$ that the *central tribinomial coefficient* (CTC) is given by

$$
\begin{aligned}
\left[\begin{array}{c} 2n \\ n \end{array}\right] &= \frac{1}{2n-n+1}\binom{2n+1}{n+1}\binom{2n}{n} \\
&= (2n+1)\cdot\frac{1}{n+1}\binom{2n}{n}\cdot\frac{1}{n+1}\binom{2n}{n} \\
&= (2n+1)C_n^2
\end{aligned}
\tag{16.11}
$$

where C_n denotes the nth Catalan number.

For example,

$$
\begin{aligned}
\left[\begin{array}{c} 6 \\ 3 \end{array}\right] &= 5C_3^2 \\
&= 5\cdot 5^2 \\
&= 175
\end{aligned}
$$

See Figure 16.1.

Next we investigate the parity of the CTCs.

Parity of the CTCs

It follows from property (16.11) that $\left[\begin{array}{c} 2n \\ n \end{array}\right]$ is odd if and only if C_n is odd. But C_n is odd if and only if either $n = 0$ or n is a Mersenne number $2^m - 1$ (see Chapter 13). Consequently, $\left[\begin{array}{c} 2n \\ n \end{array}\right]$ is odd if and only if either $n = 0$ or n is a Mersenne number.

For example, let $n = 7 = 2^3 - 1$, a Mersenne number. Then

$$\begin{bmatrix} 2n \\ n \end{bmatrix} = \begin{bmatrix} 14 \\ 7 \end{bmatrix}$$
$$= 15C_7^2$$
$$= 15 \cdot 429$$
$$= 6,435$$

is an odd integer.

On the other hand, let $n = 12$, *not* a Mersenne number. Then

$$\begin{bmatrix} 2n \\ n \end{bmatrix} = \begin{bmatrix} 24 \\ 12 \end{bmatrix}$$
$$= 25C_{12}^2$$
$$= 25 \cdot 208012$$
$$= 55,200,300$$

is an even integer.

Square CTCs

Since $\begin{bmatrix} 2n \\ n \end{bmatrix}$ contains a square factor, it is tempting to ask if there are central tribinomial coefficients that are squares. To investigate this, it is well known[†] that if N is a square, then $N \equiv 0, 1, 4,$ or $7 \pmod 9$. It follows from (16.11) that $\begin{bmatrix} 2n \\ n \end{bmatrix}$ is a square if and only if $2n + 1$ is a square. So if $n \not\equiv 0, 3, 4,$ or $6 \pmod 9$, then $\begin{bmatrix} 2n \\ n \end{bmatrix}$ is not a square.

For example, becasue $5 \not\equiv 0, 3, 4,$ or $6 \pmod 9$, $\begin{bmatrix} 10 \\ 5 \end{bmatrix}$ is not a square. In fact,

$$\begin{bmatrix} 10 \\ 5 \end{bmatrix} = 11C_5^2$$
$$= 11 \cdot 42^2$$
$$= 19,404$$

[†] See T. Koshy, *Elementary Number Theory with Applications,* 2nd ed., Academic Press, Burlington, Massachusetts.

is not a square, whereas

$$\left[\begin{array}{c} 8 \\ 4 \end{array}\right] = 9C_4^2$$
$$= 9 \cdot 14^2$$
$$= 42^2$$

a square.

Row Sums

The tribinomial array in Figure 16.1 contains yet anotherintriguing hidden treasure. To see this, we compute the various row sums; see Figure 16.5; each sum is a Catalan number. In fact, the nth row sum is C_{n+1}, where $n \geq 0$; that is,

$$\sum_{r=0}^{n} \left[\begin{array}{c} n \\ r \end{array}\right] = C_{n+1}$$

n \ r	0	1	2	3	4	5	6	row sums
0	1							1
1	1	1						2
2	1	3	1					5
3	1	6	6	1				14
4	1	10	20	10	1			42
5	1	15	50	50	15	1		132
6	1	21	105	175	105	21	1	429

$$\uparrow \qquad\qquad\qquad\qquad \nwarrow \qquad\qquad\qquad \uparrow$$
$$t_n \qquad\qquad\qquad\qquad\quad t_n \qquad\qquad\qquad C_{n+1}$$

Figure 16.5

For example,

$$\sum_{r=0}^{4} \left[\begin{array}{c} 4 \\ r \end{array}\right] = \left[\begin{array}{c} 4 \\ 0 \end{array}\right] + \left[\begin{array}{c} 4 \\ 1 \end{array}\right] + \left[\begin{array}{c} 4 \\ 2 \end{array}\right] + \left[\begin{array}{c} 4 \\ 3 \end{array}\right] + \left[\begin{array}{c} 4 \\ 4 \end{array}\right]$$
$$= 1 + 10 + 20 + 10 + 1$$
$$= 42$$

Salmassi's Proof

We now confirm that the nth row sum is C_{n+1}; the proof presented here was given by my colleague M. Salmassi in 2005. It employs the following combinatorial identities:*

$$\binom{a}{b} = \frac{a}{b}\binom{a-1}{b-1} \tag{16.12}$$

$$\sum_k \binom{r}{a+k}\binom{s}{b-k} = \binom{r+s}{a+b} \tag{16.13}$$

By identity (16.12), we have

$$\frac{1}{n+1}\binom{n+1}{n-r+1} = \frac{1}{n-r+1}\binom{n}{n-r}$$

$$= \frac{1}{n-r+1}\binom{n}{r}$$

So by formula (16.3),

$$\sum_{r=0}^{n}\left[\begin{array}{c} n \\ r \end{array}\right] = \sum_{r=0}^{n}\frac{1}{n-r+1}\binom{n}{r}\binom{n+1}{r+1}$$

$$= \sum_{r=0}^{n}\frac{1}{n+1}\binom{n+1}{n-r+1}\binom{n+1}{r+1}$$

$$= \frac{1}{n+1}\sum_{r=0}^{n}\binom{n+1}{r+1}\binom{n+1}{n-r+1}$$

$$= \frac{1}{n+1}\sum_{r=0}^{n}\binom{n+1}{1+r}\binom{n+1}{(n+1)-r}$$

$$= \frac{1}{n+1}\binom{(n+1)+(n+1)}{1+(n+1)}$$

$$= \frac{1}{n+1}\binom{2n+2}{n+2}$$

$$= \frac{1}{n+2}\binom{2n+2}{n+1}$$

$$= C_{n+1}$$

as desired.

---

* See R. L. Graham et al., *Concrete Mathematics*, Addison-Wesley, Reading, Massachusetts, 1990, p. 174.

Finally, Hoggatt showed that there is yet another way of evaluating $\left[\begin{smallmatrix} n \\ r \end{smallmatrix}\right]$. To see this, notice that

$$\begin{vmatrix} \dbinom{n}{r} & \dbinom{n}{r+1} \\ \dbinom{n+1}{r} & \dbinom{n+1}{r+1} \end{vmatrix} = \binom{n}{r}\binom{n+1}{r+1} - \binom{n}{r+1}\binom{n+1}{r}$$

$$= \binom{n}{r}\binom{n+1}{r+1} - \frac{n-r}{n-r+1}\binom{n}{r}\binom{n+1}{r+1}$$

$$= \binom{n}{r}\binom{n+1}{r+1}\left[1 - \frac{n-r}{n-r+1}\right]$$

$$= \frac{1}{n-r+1}\binom{n}{r}\binom{n+1}{r+1}$$

$$= \begin{bmatrix} n \\ r \end{bmatrix} \tag{16.14}$$

Since binomial coefficients are integers, this result confirms that the numbers $\left[\begin{smallmatrix} n \\ r \end{smallmatrix}\right]$ are indeed integers.

For example,

$$\begin{vmatrix} \dbinom{4}{2} & \dbinom{4}{3} \\ \dbinom{5}{2} & \dbinom{5}{3} \end{vmatrix} = \begin{vmatrix} 6 & 4 \\ 10 & 10 \end{vmatrix}$$

$$= 60 - 40$$

$$= 20$$

$$= \begin{bmatrix} 4 \\ 2 \end{bmatrix}$$

Formula (16.14) implies that the tribinomial array $A = (a_{ij})$ can be constructed directly from Pascal's triangle $P = (p_{ij})$:

$$a_{ij} = \begin{vmatrix} p_{ij} & p_{i,j+1} \\ p_{i+1,j} & p_{i+1,j+1} \end{vmatrix}$$

$$= p_{ij}p_{i+1,j+1} - p_{i+1,j}p_{i,j+1}$$

That is, the value of a_{ij} is given by the 2×2 determinant with the corresponding element p_{ij} in Pascal's triangle as its leading element.

For example, from Pascal's triangle in Figure 16.6, $p_{42}p_{53} - p_{52}p_{43} = \boxed{6} \cdot 10 - 10 \cdot 4 = \boxed{\boxed{20}} = a_{42}$; see the tribinomial array in Figure 16.6.

$$
\begin{array}{cccccc}
1 & & & & & \\
1 & 1 & & & & \\
1 & 2 & 1 & & & \\
1 & 3 & 3 & 1 & & \\
1 & 4 & \boxed{6} & 4 & 1 & \\
1 & 5 & 10 & 10 & 5 & 1
\end{array}
\qquad\qquad
\begin{array}{cccccc}
1 & & & & & \\
1 & 1 & & & & \\
1 & 3 & 1 & & & \\
1 & 6 & 6 & 1 & & \\
1 & 10 & \boxed{\boxed{20}} & 10 & 1 & \\
1 & 15 & 50 & 50 & 15 & 1
\end{array}
$$

Pascal's Triangle   Tribinomial Array

Figure 16.6

Interestingly, we can extend definition (16.2) to include negative integers $-n$, where $n > 0$, just as we can in the case of ordinary binomial coefficients. To this end, first notice that $t_{-n} = \dfrac{-n(-n+1)}{2} = \dfrac{n(n-1)}{2} = t_{n-1}$. Then

$$
\begin{aligned}
\begin{bmatrix} -n \\ r \end{bmatrix} &= \frac{t_{-n}t_{-n-1}\cdots t_{-n-r+1}}{t_r \cdots t_1} \\[2mm]
&= \frac{t_{-n}t_{-(n+1)}\cdots t_{-(n+r-1)}}{t_r \cdots t_1} \\[2mm]
&= \frac{t_{n+r-2}\cdots t_n t_{n-1}}{t_r \cdots t_1} \\[2mm]
&= \frac{t^*_{n+r-2}}{t^*_r t^*_{n-2}} \\[2mm]
&= \begin{bmatrix} n+r-2 \\ r \end{bmatrix}
\end{aligned}
\qquad (16.15)
$$

again an integer.

For example,

$$
\begin{bmatrix} -5 \\ 3 \end{bmatrix} = \begin{bmatrix} 6 \\ 3 \end{bmatrix} = 175
$$

Figure 16.7 shows the tribinomial array extended upward.
Using formulas (16.15) and (16.3), it follows that

$$
\begin{bmatrix} -n \\ r \end{bmatrix} = \begin{bmatrix} n+r-2 \\ r \end{bmatrix}
$$

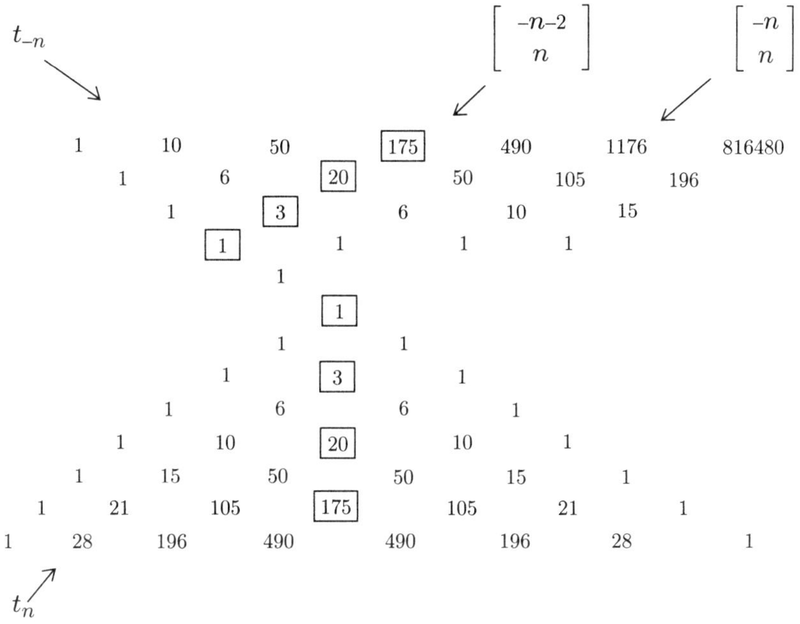

$$t_{-n}$$

$$\begin{bmatrix} -n-2 \\ n \end{bmatrix} \qquad \begin{bmatrix} -n \\ n \end{bmatrix}$$

$$
\begin{array}{ccccccccc}
1 & & 10 & & 50 & & \boxed{175} & & 490 & & 1176 & & 816480 \\
 & 1 & & 6 & & \boxed{20} & & 50 & & 105 & & 196 & \\
 & & 1 & & \boxed{3} & & 6 & & 10 & & 15 & & \\
 & & & \boxed{1} & & 1 & & 1 & & 1 & & & \\
 & & & & 1 & & & & & & & & \\
 & & & & \boxed{1} & & & & & & & & \\
 & & & & 1 & & 1 & & & & & & \\
 & & & 1 & & \boxed{3} & & 1 & & & & & \\
 & & 1 & & 6 & & 6 & & 1 & & & & \\
 & 1 & & 10 & & \boxed{20} & & 10 & & 1 & & & \\
1 & & 15 & & 50 & & 50 & & 15 & & 1 & & \\
1 & 21 & & 105 & & \boxed{175} & & 105 & & 21 & & 1 & \\
1 & 28 & & 196 & & 490 & & 490 & & 196 & & 28 & & 1 \\
\end{array}
$$

$$t_n$$

Figure 16.7 Extended Tribinomial Triangle

$$= \frac{1}{(n+r-2)-r+1} \binom{n+r-1}{r+1} \binom{n+r-2}{r}$$

$$= \frac{1}{n-1} \binom{n+r-1}{r+1} \binom{n+r-2}{r} \tag{16.16}$$

Notice that:

$$
\begin{vmatrix}
\binom{-n}{r} & \binom{-n}{r+1} \\
\binom{-n+1}{r} & \binom{-n+1}{r+1}
\end{vmatrix}
$$

$$
=
\begin{vmatrix}
(-1)^r \binom{n+r-1}{r} & (-1)^{r+1} \binom{n+r}{r+1} \\
(-1)^r \binom{n+r-2}{r} & (-1)^{r+1} \binom{n+r-1}{r+1}
\end{vmatrix}
$$

$$
= -
\begin{vmatrix}
\binom{n+r-1}{r} & \binom{n+r}{r+1} \\
\binom{n+r-2}{r} & \binom{n+r-1}{r+1}
\end{vmatrix}
$$

$$= -\left[\binom{n+r-1}{r}\binom{n+r-1}{r+1} - \binom{n+r-2}{r}\binom{n+r-1}{r+1}\right]$$

$$= -\left[\binom{n+r-1}{r+1}\binom{n+r-2}{r}\frac{n+r-1}{n-1} - \binom{n+r-2}{r}\binom{n+r-1}{r+1}\frac{n+r}{n-1}\right]$$

$$= -\binom{n+r-1}{r+1}\binom{n+r-2}{r}\left[\frac{n+r-1}{n-1} - \frac{n+r}{n-1}\right]$$

$$= \frac{1}{n-1}\binom{n+r-1}{r+1}\binom{n+r-2}{r}$$

$$= \left[\begin{array}{c} -n \\ r \end{array}\right]$$

So each $\left[\begin{smallmatrix} -n \\ r \end{smallmatrix}\right]$, an integer, can be computed from the extended Pascal's triangle. For example,

$$\left[\begin{array}{c} -5 \\ 4 \end{array}\right] = \left| \begin{array}{cc} \binom{-5}{4} & \binom{-5}{5} \\ \binom{-4}{4} & \binom{-4}{5} \end{array} \right|$$

$$= \left| \begin{array}{cc} \binom{8}{4} & -\binom{9}{5} \\ \binom{7}{4} & -\binom{8}{5} \end{array} \right|$$

$$= -\left| \begin{array}{cc} 70 & 126 \\ 35 & 56 \end{array} \right|$$

$$= 490$$

In particular, we have

$$\bullet \quad \left[\begin{array}{c} -n \\ 1 \end{array}\right] = \frac{1}{n-1}\binom{n}{2}\binom{n-1}{1} = \binom{n}{2} = t_{n-1}$$

$$\bullet \quad \left[\begin{array}{c} -n \\ 0 \end{array}\right] = \left[\begin{array}{c} n-2 \\ 0 \end{array}\right] = 1$$

$$\bullet \quad \left[\begin{array}{c} -n \\ 0 \end{array}\right] = \left[\begin{array}{c} 2n-2 \\ n \end{array}\right]$$

$$\qquad = \frac{1}{n-1}\binom{2n-1}{n+1}\binom{2n-2}{n}$$

$$\begin{aligned}
&= \frac{1}{n-1} \cdot \frac{(2n)!}{(n+1)!n!} \cdot \frac{(2n-2)!}{n!(n-1)!} \cdot \frac{(n-1)(n-2)(n-1)}{2n} \\
&= \frac{(n-1)(n-2)}{2n} C_n C_{n-1} \\
&= \frac{t_{n-1} C_n C_{n-1}}{n}
\end{aligned}$$

where $n \geq 2$.

For example,

$$\begin{bmatrix} -5 \\ 5 \end{bmatrix} = \frac{t_4 C_5 C_4}{5} = \frac{10 \cdot 42 \cdot 14}{5} = 1176$$

It is known that C_n is odd if and only if either $n = 0$ or n is a Mersenne number $2^m - 1$, where $m \geq 1$. Let n be a Mersenne number ≥ 3, so $m \geq 2$. Then $t_n = (2^m - 1)2^{m-1}$ is even. Because n is a Mersenne number, C_n is odd and hence C_{n-1} is even. Therefore,

$$\frac{t_{n-1} C_n C_{n-1}}{n} = \begin{bmatrix} -n \\ n \end{bmatrix}$$

is even. Thus, if n is a Mersenne number ≥ 3, then $\begin{bmatrix} -n \\ n \end{bmatrix}$ is an even integer.

Since $\begin{bmatrix} 2m \\ m \end{bmatrix} = (2m+1)C_m^2$, we also have

$$\begin{aligned}
\begin{bmatrix} -(n+2) \\ n \end{bmatrix} &= \begin{bmatrix} 2n \\ n \end{bmatrix} \\
&= (2n+1)C_n^2
\end{aligned}$$

where $n \geq 0$. See the boxed numbers in Figure 16.7.

Recurrence Relations for the Upper Half

Recurrence relations (16.4) and (16.6) can be extended to the upper half of the tribinomial triangle also:

$$\begin{aligned}
\begin{bmatrix} -n \\ r \end{bmatrix} &= \begin{bmatrix} -(n+1) \\ r-1 \end{bmatrix} \frac{t_{-n}}{t_r} \\
&= \begin{bmatrix} n+r-2 \\ r-1 \end{bmatrix} \frac{t_{n-1}}{t_r}
\end{aligned} \qquad (16.17)$$

and

$$\begin{bmatrix} -n \\ r \end{bmatrix} = \begin{bmatrix} -(n+1) \\ r \end{bmatrix} \frac{t_{-n}}{t_{-n-r}}$$

$$= \begin{bmatrix} n+r-1 \\ r \end{bmatrix} \frac{t_{n-1}}{t_{n+r-1}} \qquad (16.18)$$

where $n > 0$ in both cases.

For example,

$$\begin{bmatrix} -5 \\ 3 \end{bmatrix} = \begin{bmatrix} 6 \\ 2 \end{bmatrix} \frac{t_4}{t_3}$$

$$= 105 \cdot \frac{10}{6}$$

$$= 175$$

Generalized Tribinomials

Definitions (16.1) and (16.2) can be extended to construct generalized tribinomials $\begin{bmatrix} n \\ r \end{bmatrix}_k$, where k is a fixed positive integer and $\begin{bmatrix} n \\ r \end{bmatrix}_1 = \begin{bmatrix} n \\ r \end{bmatrix}$. To this end, let $U_n = t^*_{nk} = t_{nk} t_{(n-1)k} \cdots t_k$. Then

$$\begin{bmatrix} n \\ r \end{bmatrix}_k = \frac{U^*_n}{U^*_r U^*_{n-r}}$$

$$= \frac{U_n \cdots U_{n-r+1}}{U_r \cdots U_1}$$

$$= \begin{bmatrix} n-1 \\ r-1 \end{bmatrix}_k \frac{t_{nk}}{t_{rk}} \qquad (16.19)$$

$$= \begin{bmatrix} n-1 \\ r \end{bmatrix}_k \frac{t_{nk}}{t_{(n-r)k}} \qquad (16.20)$$

where $\begin{bmatrix} n \\ 0 \end{bmatrix}_k = 1 = \begin{bmatrix} n \\ n \end{bmatrix}_k$.

Unfortunately, $\begin{bmatrix} n \\ r \end{bmatrix}_k$ need not be an integer when $k > 1$. For instance, $\begin{bmatrix} 2 \\ 1 \end{bmatrix}_2 = \frac{10}{3}$ is not an integer. See Figure 16.8.

$$
\begin{array}{ccccccccc}
 & & & & 1 & & & & \\
 & & & 1 & & 1 & & & \\
 & & 1 & & \frac{10}{3} & & 1 & & \\
 & 1 & & 7 & & 7 & & 1 & \\
1 & & 12 & & \frac{126}{5} & & \frac{126}{5} & & 1
\end{array}
$$

Figure 16.8 Tribinomial Array with $k = 2$

17

Generalized Catalan Numbers

Catalan numbers are a special case of a larger class of numbers $C(n, k)$, defined by

$$C(n, k) = \frac{1}{kn + 1} \binom{(k + 1)n}{n} \tag{17.1}$$

where $k \geq 0$. Clearly, $C(n, 1) = C_n$.

As in the case of ordinary Catalan numbers, every *generalized Catalan number* $C(n, k)$ is an integer. To see this, notice that:

$$
\binom{kn + n}{n} - \binom{kn + n}{n - 1} = \frac{(kn + n)!}{n!(kn)!} - \frac{(kn + n)!}{(n - 1)!(kn + 1)!}
$$

$$
= \frac{(kn + n)!}{(kn + 1)!n!}[(kn + 1) - n]
$$

$$
= \frac{(kn + n)!}{(kn + 1)!n!}[(k - 1)n + 1]
$$

$$
= \frac{1}{kn + 1} \binom{kn + n}{n}[(k - 1)n + 1]
$$

Because the LHS is an integer, it follows that the RHS is also an integer. But $kn + 1$ and $(k - 1)n + 1$ are relatively prime. So, $kn + 1 \mid \binom{kn + n}{n}$; that is, $C(n, k)$ is an integer for every $n, k \geq 0$.

Notice that $C(n, k)$ can also be written as

$$C(n, k) = \frac{1}{n} \binom{(k + 1)n}{n - 1}$$

When $k = 0, C(n, 0) = \binom{n}{n}$. The corresponding sequence consists of 1s: 1, 1, 1, 1, ...

Table 17.1 shows the various sequences corresponding to $k = 0, 1, 2, 3$, and 4.

Table 17.1

$k = 0$:	$1, 1, 1, 1, 1, \ldots, \binom{n}{n}, \ldots$
$k = 1$:	$1, 1, 2, 5, 14, \ldots, \dfrac{1}{n+1}\binom{2n}{n}, \ldots$
$k = 2$:	$1, 1, 3, 12, 55, \ldots, \dfrac{1}{2n+1}\binom{3n}{n}, \ldots$
$k = 3$:	$1, 1, 4, 22, 140, \ldots, \dfrac{1}{3n+1}\binom{4n}{n}, \ldots$
$k = 4$:	$1, 1, 5, 35, 285, \ldots, \dfrac{1}{4n+1}\binom{5n}{n}, \ldots$

Because $C(n, k) = \frac{1}{kn+1}\binom{(k+1)n}{n}$, the various values can be computed by following the diagonal originating at the apex 1 in Pascal's triangle and passing through the element $\binom{(k+1)n}{n}$ in row $(k + 1)n$. See Table 17.2.

Table 17.2 Pascal's Triangle

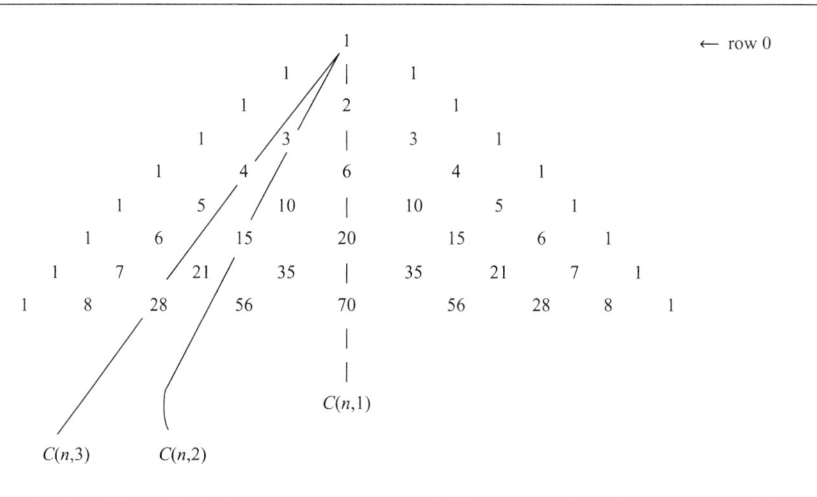

Example 6.10 Revisited

The following example investigates a generalization of Example 6.10.

Example 17.1 Find the number of $(mn+1)$-tuples $a_0 a_1 \ldots a_{mn}$ of 1s and $(1-m)$s such that every partial sum is positive and $a_0 + a_1 + \cdots + a_{mn} = 1$, where $m \geq 2$.

Solution Suppose there are k occurrences of 1s and hence $(mn + 1 - k)$ occurrences of $(1 - m)$s in each tuple. Then:

$$k \cdot 1 + (mn + 1 - k)(1 - m) = 1$$
$$k + mn + 1 - k - m(mn + 1 - k) = 1$$
$$m(n - mn - 1 + k) = 0$$
$$n - mn - 1 + k = 0$$
$$k = mn - n + 1$$

Thus each tuple consists of $mn - n + 1$ 1s and $mn + 1 - k = (mn+1) - (mn-n+1) = n$ $(1-m)$s. So the total number of such tuples that can be formed is

$$\frac{(mn + 1)!}{(mn - n + 1)!n!} = \binom{mn + 1}{n}$$

Let N denote the number of $(mn + 1)$-tuples with the given property. There are $mn + 1$ cyclic shifts of an $(mn + 1)$-tuple. By Raney's lemma, exactly one of them has positive partial sums. So the total number of $(mn + 1)$-tuples consisting of $mn - n + 1$ 1s and n $(1 - m)$s equals $(mn + 1)N$. Thus

$$(mn + 1)N = \binom{mn + 1}{n}$$
$$N = \frac{1}{mn + 1}\binom{mn + 1}{n}$$
$$= \frac{1}{(m - 1)n + 1}\binom{mn}{n}$$

∎

Fuss Numbers

Numbers

$$F(m, n) = \frac{1}{mn + 1}\binom{mn + 1}{n}$$

were studied by N. I. Fuss in 1791, forty-seven years before Catalan investigated the parenthesization problem. We call them *Fuss numbers*.[†]

Notice that

$$F(2, n) = \frac{1}{2n + 1}\binom{2n + 1}{n}$$

$$= C_n$$

Although $F(2, n) = C_n$ is an integer, it is not obvious that every Fuss number is an integer. We now establish that this is indeed the case. To this end, notice that:

$$F(m, n) = (mn + 1)F(m, n) - mnF(m, n)$$

$$= \binom{mn + 1}{n} - \frac{mn}{mn + 1}\binom{mn + 1}{n}$$

$$= \binom{mn + 1}{n} - \frac{mn}{mn + 1} \cdot \frac{(mn + 1)!}{n!(mn - n + 1)!}$$

$$= \binom{mn + 1}{n} - m \cdot \frac{(mn)!}{(n - 1)!(mn - n + 1)!}$$

$$= \binom{mn + 1}{n} - m\binom{mn}{n - 1}$$

Since the RHS is an integer, it follows that every Fuss number is indeed an integer, as desired.

[†] R. L. Graham et al. called them *Fuss-Catalan numbers*.

Appendix A

This appendix presents some basics of bijections and combinatorics, recursion, generating functions,* and congruences,† and the principle of mathematical induction. They are frequently used throughout the discussion of Catalan numbers.

A.1 Bijections

A concept used throughout mathematics is the function. Bijections are a special class of functions.

A function $f : A \to B$ is *injective* if distinct input values in A produce distinct output values in B; that is, $a_1 \neq a_2$ implies that $f(a_1) \neq f(a_2)$, where $a_1, a_2, \in A$. Equivalently, f is injective if $f(a_1) = f(a_2)$ implies that $a_1 = a_2$. For example, the function $g : \mathbf{N} \to \mathbf{N}$, defined by $f(n) = 2n$, is injective, where $\mathbf{N} = \{1, 2, 3, \ldots\}$; but the function $h : \mathbf{N} \to \mathbf{N}$, defined by $h(n) = 2$, is not injective.

A function $f : A \to B$ is *surjective* if every element b in B occurs as the output for some element a in A; that is, given any $b \in B$, there is a suitable element $a \in A$ such that $f(a) = b$. For example, the above function g is not surjective, but the function $h : \mathbf{N} \to \{3, 4, 5, \ldots\}$, defined by $h(n) = n + 2$, is surjective.

A function $f : A \to B$ is *bijective* if it is both injective and surjective. A bijection is also called a *one-to-one correspondence*. For example, there is a bijection between the set of fingers on our left hand and that on our right hand.

---

* For a detailed discussion, see T. Koshy, *Discrete Mathematics with Applications*, Elsevier/Academic Press, Boston, Massachusetts, 2004.

† For a detailed discussion, see T. Koshy, *Elementary Number Theory with Applications*, 2nd ed., Academic Press, Burlington, Massachusetts, 2007.

A function $f : A \to B$ is a bijection if and only if $|A| = |B|$, where $|S|$ denotes the cardinality of the set S. Consequently, bijections are very important in mathematics, just as they are in the real world.

A.2   Fundamental Counting Principles

The whole of combinatorics hinges on three fundamental counting priciples: the inclusion-exclusion principle, the addition principle, and the multiplication principle.

Inclusion-Exclusion Principle Suppose a task A can be done in m different ways, task B in n different ways, and both can be done in k different ways. Then task A or B can be done in $m + n - k$ different ways. This can be written symbolically as follows: $|A \cup B| = |A| + |B| - |A \cap B|$, where $|T|$ denotes the number of different ways task T can be done.

For example, there are $4 + 26 - 2 = 28$ different ways a king or a black card can be selected from a standard deck of playing cards.

Suppose two tasks cannot occur simultaneously. Such tasks are *mutually exclusive*. For example, selecting a black queen and selecting a red king are mutually exclusive tasks.

Addition Principle Let A and B be mutually exclusive tasks. Suppose task A can be done in m different ways and task B in n different ways. Then task A or B can be done in $m + n$ different ways. Symbolically, $|A \cup B| = |A| + |B|$, where $A \cap B = \theta$.

For example, a red queen or a black king can be drawn in $2 + 2 = 4$ different ways from a standard deck of playing cards.

Both the addition principle and the inclusion-exclusion principle can be extended to any finite number of tasks.

Multiplication Principle Suppose a task A is made up of two subtasks, subtask A_1 followed by subtask A_2. Suppose subtask A_1 can be done in m_1 different ways and subtask A_2 in m_2 different ways for each way subtask A_1 can be done. Then task A can be done in $m_1 m_2$ different ways. Symbolically, $|A \times B| = |A| \cdot |B|$.

For example, we can form $5 \cdot 26 = 130$ two-letter words that begin with a vowel—$a, e, i, o,$ or u.

The multiplication priciple also can be extended to any finite number of subtasks in an obvious way.

A.3 Recursion

Recursion is one of the most elegant problem-solving techniques. It is so powerful a tool that most programming languages support it.

Consider, for example, the well-known *handshake problem*: "There are n guests at a party. Each person shakes hands with everybody else exactly once. How many handshakes are made?"

If we decide to solve a problem such as this, the solution may not be obvious. However, it is possible that the problem could be defined in terms of a simpler version of itself. Such a definition is an *inductive definition*. Consequently, the given problem can be solved provided the simpler version can be solved. This idea can be used to define the number of handshakes $h(n)$ made by the n guests as follows:

$$h(n) = h(n-1) + (n-1), \quad n \geq 2$$

where $h(1) = 0$.

More generally, we have the following definition.

Recursive Definition of a Function

Let $a \in W$ and $X = \{a, a+1, a+2, \ldots\}$, where $W = \{0, 1, 2, 3, \ldots\}$. An *inductive definition* of a function f with domain X consists of three parts:

- **Basis step**: A few initial values $f(a), f(a+1), \ldots, f(a+k-1)$ are specified. Equations that specify such initial values are initial conditions.
- **Recursive step**: A formula to compute $f(n)$ from the k preceding functional values $f(n-1), f(n-2), \ldots, f(n-k)$ is made. Such a formula is a recurrence relation (or recursive formula).
- **Terminal step**: Only values thus obtained are valid functional values. (For convenience, we drop this clause from the recursive definition.)

In a *recursive definition* of f, $f(n)$ may be defined using the values $f(k)$, where $k \neq n$, so not all recursively defined functions can be defined inductively.

Thus, the recursive definition of f consists of a finite number of initial conditions and a recurrence relation.

The simplest example of a recursive definition is that of the *factorial function* f. It is defined by $f(n) = n \cdot f(n-1)$, where $f(0) = 1$.

Recursion can be employed to find the minimum and maximum of three or more real numbers. For instance, $\min\{w, x, y, z\} = \min\{w, \{\min\{x, \min\{y, z\}\}\}\}$; $\max\{w, x, y, z\}$ can be evaluated similarly. For example,

$$\min\{23, 5, -6, 47, 31\} = \min\{23, \min\{5, \min\{-6, \min\{47, 31\}\}\}\} = -6$$

and

$$\max\{23, 5, -6, 47, 31\} = \max\{23, \max\{5, \max\{-6, \max\{47, 31\}\}\}\} = 47$$

A.4   Generating Functions

Generating functions provide a powerful tool for solving recursively defined and combinatorial problems. They were invented in 1718 by French mathematician Abraham De Moivre (1667–1754), when he used them to solve[†] the *Fibonacci recurrence relation $F_n = F_{n-1} + F_{n-2}$.*

To begin with, notice that $\frac{x^6-1}{x-1} = 1 + x + x^2 + x^3 + x^4 + x^5$. Accordingly, the function $f(x) = \frac{x^6-1}{x-1}$ is called the generating function of the sequence of coefficients 1, 1, 1, 1, 1, 1 in the polynomial on the right-hand side (RHS).

More generally, we make the following definition.

Generating Function

Let $a_0, a_1, \ldots, a_n, \ldots$ be a sequence $\{a_n\}$ of real numbers. Then the function

$$f(x) = a_0 + a_1 x + a_2 x^2 + \cdots + a_n x^n + \cdots$$

is the generating function for the $\{a_n\}$. The RHS is a *formal power series* and x is just a dummy variable.

For example,

$$f(x) = 1 + 2x + 3x^2 + 4x^3 + \cdots + (n+1)x^n + \cdots$$

is the generating function for the sequence of positive integers, and

$$g(x) = 1 + 3x + 6x^2 + 10x^3 + \cdots + \frac{n(n+1)}{2}x^n + \cdots$$

is the generating function of the sequence of *triangular numbers* $1, 3, 6, 10, \cdots$.

Generating functions for the finite sequence $a_0, a_1, \ldots, a_n$ can be defined by letting $a_i = 0$ for $i > n$. Thus, $f(x) = a_0 + a_1 x + a_2 x^2 + \cdots + a_n x^n$ is the generating function for the finite sequence $a_0, a_1, \ldots, a_n$.

Just as two ordinary functions can be equal, so can two generating functions. For instance, the generating functions $f(x) = \sum_{n=0}^{\infty} a_n x^n$ and $g(x) = \sum_{n=0}^{\infty} a_n x^n$ are *equal* if $a_n = b_n$ for every $n \geq 0$.

[†] See author's, *Fibonacci and Lucas Numbers with Applications*, Wiley, New York, 2001.

As in the case of real-valued functions, generating functions can be added and multiplied. For example, let $f(x) = \sum\limits_{n=0}^{\infty} a_n x^n$ and $g(x) = \sum\limits_{n=0}^{\infty} a_n x^n$. Then:

$$f(x) + g(x) = \sum_{n=0}^{\infty} (a_n + b_n) x^n$$

$$f(x)g(x) = \sum_{n=0}^{\infty} \left(\sum_{i=0}^{n} a_i b_{n-i} \right) x^n$$

A.5   Congruences

The congruence relation is one of the most remarkable relations in number theory. It was introduced and developed by German mathematician Karl Friedrich Gauss (1777–1855), known as the "prince of mathematics." He presented the theory of congruences, a beautiful arm of divisibility theory, in his outstanding work, *Disquistiones Arithmeticae*, published in 1801 when he was only twenty-four. Gauss is believed to have submitted a major portion of the book for publication by the French Academy, who rejected it.

The congruence relation shares many interesting properties with the equality relation, so it is no accident that the congruence symbol $\equiv$, invented by Gauss around 1800, parallels the equality symbol $=$. The congruence symbol facilitates the study of divisibility theory and has many fascinating applications.

Let us begin our discussion with a definition.

Congruence Modulo m

Let m be a positive integer. Then an integer a is *congruent* to an integer b modulo m if $m|(a - b)$. In symbols, we then write $a \equiv b \pmod{m}$; m is the *modulus* of the congruence relation.

If a is not congruent to b modulo m, then a is *incongruent* to b modulo m; we then write $a \not\equiv b \pmod{m}$.

For example, since $5|(43 - 3), 43 \equiv 3 \pmod 5$; likewise, $6|(54 - 12)$, so $54 \equiv 12 \pmod 6$; also $29 \equiv -3 \pmod{16}$. But $25 \not\equiv 3 \pmod 4$, since $4 \nmid (25 - 3)$; likewise, $23 \not\equiv -4 \pmod 7$.

Note that we use congruences in everyday life, often without realizing it. We use congruence modulo 12 to tell the time of the day, and congruence modulo 7 to tell the day of the week. Odometers in automobiles use 100,000 as the modulus.

A useful observation: It follows from the definition that $a \equiv 0 \pmod{m}$ if and only if $m|a$; that is, an integer is congruent to 0 if and only if it is divisible by m.

Thus $a \equiv 0 \pmod{m}$ and $m|a$ mean exactly the same thing. For instance, $28 \equiv 0 \pmod{7}$ and $7|28$.

Next, we enumerate several interesting and useful properties of the congruence relation, where $a, b, c,$ and d are any integers, and m is a positive integer. In the interest of brevity, we omit their properties:

- $a \equiv b \pmod{m}$ if and only if $a = b + km$ for some integer k.
- $a \equiv a \pmod{m}$ (Reflexive property).
- If $a \equiv b \pmod{m}$, then $b \equiv a \pmod{m}$ (Symmetric property).
- If $a \equiv b \pmod{m}$ and $b \equiv c \pmod{m}$, then $a \equiv c \pmod{m}$ (Transitive property).
- $a \equiv b \pmod{m}$ if and only if a and b leave the same remainder when divided by m.
- The integer r is the remainder when a is divided by m if and only if $a \equiv r \pmod{m}$, where $0 \leq r < m$.
- Let $a \equiv b \pmod{m}$ and $c \equiv d \pmod{m}$. Then $a + c \equiv b + d \pmod{m}$ and $ac \equiv bd \pmod{m}$.

A.6   Mathematical Induction

The principle of mathematical induction[†] (PMI) is a powerful proof technique used often in mathematics.

The principle of induction has two versions, the *weak version* and the *strong version*. It is based on the *well-ordering principle*: Every nonempty set of positive integers has a least element. We now state both versions and omit their proofs for convenience:

Weak Version

Let $P(n)$ be a statement satisfying the following conditions, where n is an integer:

- $P(n_0)$ is true for some integer n_0.
- If $P(k)$ is true for an arbitrary integer $\geq n_0$, then $P(k + 1)$ is also true.

Then $P(n)$ is true for every integer $n \geq n_0$.

[†] The term *mathematical induction* was coined by Augustus DeMorgan (1806–1871), although Venetian scientist Francesco Maurocylus (1491–1575) applied it much earlier, in proofs in a book he wrote in 1575.

Strong Version

Let $P(n)$ be a statement satisfying the following conditions, where n is an integer:

- $P(n_0)$ is true for some integer n_0.
- If k is an arbitrary integer $\geq n_0$ such that $P(n_0), P(n_0 + 1), \ldots, P(k)$ are true, then $P(k + 1)$ is also true.

Then $P(n)$ is true for every integer $n \geq n_0$.

A.7   Extended Binomial Coefficients

First, we extend the definition of $\binom{n}{r}$ to negative integers n.

A Definition of $\binom{-n}{r}$

Although $\binom{-n}{r}$ has no combinatorial interpretation, it can still be assigned a meaningful value. To this end, first we return to the definition of $\binom{n}{r}$:

$$\binom{n}{r} = \frac{n(n-1)\cdots(n-r+1)}{r!}$$

Now change n to $-n$:

$$\binom{-n}{r} = \frac{-n(-n-1)\cdots(-n-r+1)}{r!}$$

$$= (-1)^r \frac{n(n+1)\cdots(n+r-1)}{r!}$$

Thus

$$\binom{-n}{r} = (-1)^r \binom{n+r-1}{r} \tag{A.1}$$

For example, $\binom{-5}{3} = (-1)^3 \binom{7}{3} = -35$ and $\binom{-3}{5} = (-1)^5 \binom{7}{5} = -21$.

Using definition (A.1), Pascal's triangle can be extended in the opposite direction, as in Figure A.1.* Just as row n in Pascal's triangle can be used to expand $(1+x)^n$, row $-n$ in the extended triangle can be used to expand $(1+x)^{-n}$.

For example, using row -3,

$$(1+x)^{-3} = 1 - 3x + 6x^2 - 10x^3 + 15x^4 - \cdots$$

---

* An extended Pascal's triangle can be found in E. Lucas's *Théorie des Nombres*.

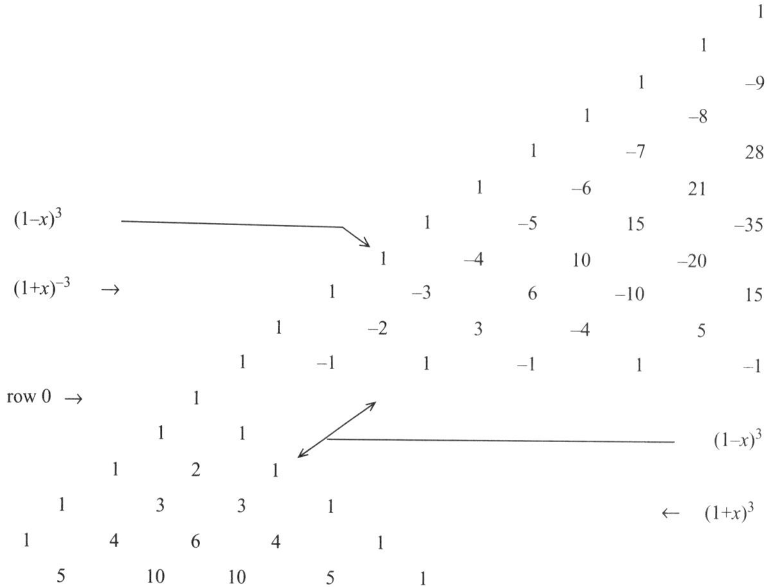

Figure A.1 Extended Pascal's Triangle

The southeast diagonals in the upper half can be used to expand $(1 - x)^n$; for example,

$$(1 - x)^3 = 1 - 3x + 3x^2 - x^3$$

Newton, using definition (A.1), extended the binomial theorem to negative integral exponents in 1664 or 1665.

A.8   Maclaurin's Series

Interestingly enough, $\binom{n+r-1}{r}$ is the coefficient of x^n in the Maclaurin's series for $(1 - x)^{-n}$:

$$(1 - x)^{-n} = \sum_{r=0}^{\infty} \binom{-n}{r}(-x)^r$$

$$= \sum_{r=0}^{\infty} \binom{n+r-1}{r} x^r$$

$$= 1 + \binom{n}{1}x + \binom{n+1}{2}x^2 + \binom{n+2}{3}x^3 + \cdots$$

Consequently, row $-n$ in the extended Pascal's triangle can be used to expand $(1-x)^{-n}$ by ignoring the negative signs in the row. For example,

$$
\begin{aligned}
(1-x)^{-3} &= \sum_{r \geq 0} \binom{2+r}{r} x^r \\
&= 1 + 3x + 6x^2 + 10x^3 + 15x^4 + \cdots \\
&= \sum_{r=0}^{\infty} t_{n+1} x^n
\end{aligned}
$$

where t_k denotes the kth triangular number.

A.9   A Definition of $\binom{p/q}{r}$

The definition of the binomial coefficient

$$
\binom{n}{r} = \frac{n(n-1)\cdots(n-r+1)}{r!}
$$

can formally be extended to fractional numbers $n = \dfrac{p}{q}$:

$$
\begin{aligned}
\binom{p/q}{r} &= \frac{\frac{p}{q}\left(\frac{p}{q}-1\right)\cdots\left(\frac{p}{q}-r+1\right)}{r!} \\
&= \frac{p(p-q)(p-2q)\cdots[p-(r-1)q]}{q^r\, r!}
\end{aligned}
\tag{A.2}
$$

As expected, this formula yields the familiar binomial coefficient when $q = 1$. For example,

$$
\begin{aligned}
\binom{2/3}{4} &= \frac{2(2-1\cdot 3)(2-2\cdot 3)(2-3\cdot 3)}{3^4\, 4!} \\
&= -\frac{7}{243}
\end{aligned}
$$

and

$$
\begin{aligned}
\binom{1/2}{n} &= \frac{\frac{1}{2}\left(\frac{1}{2}-1\right)\left(\frac{1}{2}-2\right)\cdots\left(\frac{1}{2}-n+1\right)}{n!} \\
&= \frac{(-1)^{n-1}\, 1(2\cdot 1 - 1)(2\cdot 2 - 1)\cdot[2(n-1)-1]}{2^n\, n!} \\
&= (-1)^{n-1}\frac{1\cdot 3\cdot 5\cdots(2n-3)}{2^n\, n!}
\end{aligned}
$$

$$= (-1)^{n-1} \frac{(2n-2)!}{2^n\, n![2 \cdot 4 \cdot 6 \cdots (2n-2)]}$$

$$= (-1)^{n-1} \frac{(2n-2)!}{2^n\, n!\, 2^{n-1}\,(n-1)!}$$

$$= (-1)^{n-1} \frac{(2n-2)!}{2^{2n-1}\, n!\,(n-1)!}$$

$$= \frac{(-1)^{n-1}}{2^{2n-1}\, n} \binom{2n-2}{n-1}$$

$$= \frac{(-1)^{n-1}}{2^{2n-1}} C_{n-1}$$

where C_k denotes the kth Catalan number.

So

$$(1+x)^{1/2} = \sum_{n=0}^{\infty} \binom{1/2}{n} x^n$$

$$= 1 + \sum_{n=1}^{\infty} \binom{1/2}{n} x^n$$

$$= 1 + \sum_{n=1}^{\infty} \frac{(-1)^{n-1}}{n \cdot 2^{2n-1}} \binom{2n-2}{n-1} x^n$$

$$= 1 + \frac{x}{2} - \frac{x^2}{8} + \frac{x^3}{16} - \cdots$$

$$= 1 + \left(\frac{x}{2}\right) - \frac{1}{2}\left(\frac{x}{2}\right)^2 + \frac{1}{2}\left(\frac{x}{2}\right)^3 - \cdots$$

and

$$(1-x)^{1/2} = \sum_{n=0}^{\infty} (-1)^n \binom{1/2}{n} x^n$$

$$= 1 + \sum_{n=1}^{\infty} (-1)^n \binom{1/2}{n} x^n$$

$$= 1 + \sum_{n=1}^{\infty} \frac{(-1)^{2n-1}}{n \cdot 2^{2n-1}} \binom{2n-2}{n-1} x^n$$

$$= 1 - \sum_{n=1}^{\infty} \frac{1}{n \cdot 2^{2n-1}} \binom{2n-2}{n-1} x^n$$

$$= 1 - \frac{x}{2} - \frac{x^2}{8} - \frac{x^3}{16} - \cdots$$

$$= 1 - \left(\frac{x}{2}\right) - \frac{1}{2}\left(\frac{x}{2}\right)^2 - \frac{1}{2}\left(\frac{x}{2}\right)^3 - \cdots$$

It follows from this that

$$(1 - 4x)^{1/2} = 1 - 2\sum_{n=1}^{\infty} \frac{1}{n}\binom{2n-2}{n-1}x^n$$

$$= 1 - 2\sum_{n=1}^{\infty} C_{n-1}x^n$$

By changing p to $-p$ in formula (A.2), $\binom{-p/q}{r}$ also can be defined:

$$\binom{-p/q}{r} = \frac{-p(-p-q)(-p-2q)\cdots[-p-(r-1)q]}{q^r\, r!}$$

$$= (-1)^r \frac{p(p+q)(p+2q)\cdots[p+(r-1)q]}{q^r\, r!} \tag{A.3}$$

For example,

$$\binom{-2/3}{4} = (-1)^4 \frac{2(2+1\cdot3)(2+2\cdot3)(2+3\cdot3)}{3^4\,4!}$$

$$= \frac{110}{243}$$

and

$$\binom{-1/2}{n} = (-1)^n \frac{1\cdot(1+2\cdot1)(1+2\cdot2)(1+2\cdot3)\cdots[1+2(n-1)]}{2^n\,n!}$$

$$= (-1)^n \frac{1\cdot3\cdot5\cdot7\cdots(2n-1)}{2^n\,n!}$$

$$= (-1)^n \frac{(2n-1)!}{2^n\,n!\,[2\cdot4\cdot6\cdots(2n-2)]}$$

$$= (-1)^n \frac{(2n-1)!}{2^{2n-1}\,n!(n-1)!}$$

$$= \frac{(-1)^n}{2^{2n}}\binom{2n}{n}$$

$$= \left(-\frac{1}{4}\right)^n \binom{2n}{n}$$

Using formula (A.3), Maclaurin's series can be extended to negative fractional exponents:

$$(1-x)^{-p/q} = \sum_{r=0}^{\infty} (-1)^r \binom{-p/q}{r} x^r$$

$$= \sum_{r=0}^{\infty} \frac{p(p+q)(p+2q)\cdots[p+(r-1)q]}{r!} \left(\frac{x}{q}\right)^r \qquad \text{(A.4)}$$

For example,

$$(1-x)^{-1/2} = \sum_{r=0}^{\infty} \frac{1(1+1\cdot 2)(1+2\cdot 2)\cdots[1+2(n-1)]}{n!} \left(\frac{x}{2}\right)^n$$

$$= 1 + \frac{1}{1!}\left(\frac{x}{2}\right) + \frac{1\cdot 3}{2!}\left(\frac{x}{2}\right)^2 + \frac{1\cdot 3\cdot 5}{3!}\left(\frac{x}{2}\right)^3 + \cdots$$

Appendix B

The First 100 Catalan Numbers

n	C_n
1	1
2	2
3	5
4	14
5	42
6	132
7	429
8	1,430
9	4,862
10	16,796
11	58,786
12	208,012
13	742,900
14	2,674,440
15	9,694,845
16	35,357,670
17	129,644,790
18	477,638,700
19	1,767,263,190
20	6,564,120,420
21	24,466,267,020
22	91,482,563,640
23	343,059,613,650
24	1,289,904,147,324
25	4,861,946,401,452
26	18,367,353,072,152

(continued)

(Continued)

n	C_n
27	69,533,550,916,004
28	263,747,951,750,360
29	1,002,242,216,651,368
30	3,814,986,502,092,304
31	14,544,636,039,226,909
32	55,534,064,877,048,198
33	212,336,130,412,243,110
34	812,944,042,149,730,764
35	3,116,285,494,907,301,262
36	11,959,798,385,860,453,492
37	45,950,804,324,621,742,364
38	176,733,862,787,006,701,400
39	680,425,371,729,975,800,390
40	2,622,127,042,276,492,108,820
41	10,113,918,591,637,898,134,020
42	39,044,429,911,904,443,959,240
43	150,853,479,205,085,351,660,700
44	583,300,119,592,996,693,088,040
45	2,257,117,854,077,248,073,253,720
46	8,740,328,711,533,173,390,046,320
47	33,868,773,757,191,046,886,429,490
48	131,327,898,242,169,365,477,991,900
49	509,552,245,179,617,138,054,608,572
50	1,978,261,657,756,160,653,623,774,456
51	7,684,785,670,514,316,385,230,816,156
52	29,869,166,945,772,625,950,142,417,512
53	116,157,871,455,782,434,250,553,845,880
54	451,959,718,027,953,471,447,609,509,424
55	1,759,414,616,608,818,870,992,479,875,972
56	6,852,456,927,844,873,497,549,658,464,312
57	26,700,952,856,774,851,904,245,220,912,664
58	104,088,460,289,122,304,033,498,318,812,080
59	405,944,995,127,576,985,730,643,443,367,112
60	1,583,850,964,596,120,042,686,772,779,038,896
61	6,182,127,958,584,855,650,487,080,847,216,336
62	24,139,737,743,045,626,825,711,458,546,273,312
63	94,295,850,558,771,979,787,935,384,946,380,125
64	368,479,169,875,816,659,479,009,042,713,546,950
65	1,440,418,573,150,919,668,872,489,894,243,865,350
66	5,632,681,584,560,312,734,993,915,705,849,145,100
67	22,033,725,021,956,517,463,358,552,614,056,949,950
68	86,218,923,998,960,285,726,185,640,663,701,108,500
69	337,485,502,510,215,975,556,783,793,455,058,624,700
70	1,321,422,108,420,282,270,489,942,177,190,229,544,600
71	5,175,569,924,646,105,559,418,940,193,995,065,716,350

(continued)

n	C_n
72	20,276,890,389,709,399,862,928,998,568,254,641,025,700
73	79,463,489,365,077,377,841,208,237,632,349,268,884,500
74	311,496,878,311,103,321,137,536,291,518,809,134,027,240
75	1,221,395,654,430,378,811,828,760,722,007,962,130,791,020
76	4,790,408,930,363,303,911,328,386,208,394,864,461,024,520
77	18,793,142,726,809,884,575,211,361,279,087,545,193,250,040
78	73,745,243,611,532,458,459,690,151,854,647,329,239,335,600
79	289,450,081,175,264,899,454,283,846,029,490,767,264,392,230
80	1,136,359,577,947,336,271,931,632,877,004,667,456,667,613,940
81	4,462,290,049,988,320,482,463,241,297,506,133,183,499,654,740
82	17,526,585,015,616,776,834,735,140,517,915,655,636,396,234,280
83	68,854,441,132,780,194,707,888,052,034,668,647,142,985,206,100
84	270,557,451,039,395,118,028,642,463,289,168,566,420,671,280,440
85	1,063,353,702,922,273,835,973,036,658,043,476,458,723,103,404,520
86	4,180,080,073,556,524,734,514,695,828,170,907,458,428,751,314,320
87	16,435,314,834,665,426,797,069,144,960,762,886,143,367,590,394,940
88	64,633,260,585,762,914,370,496,637,486,146,181,462,681,535,261,000
89	254,224,158,304,000,796,523,953,440,778,841,647,086,547,372,026,600
90	1,000,134,600,800,354,781,929,399,250,536,541,864,362,461,089,950,800
91	3,935,312,233,584,004,685,417,853,572,763,349,509,774,031,680,023,800
92	15,487,357,822,491,889,407,128,326,963,778,343,232,013,931,127,835,600
93	60,960,876,535,340,415,751,462,563,580,829,648,891,969,728,907,438,000
94	239,993,345,518,077,005,168,915,776,623,476,723,006,280,827,488,229,600
95	944,973,797,977,428,207,852,605,870,454,939,596,837,230,758,234,904,050
96	3,721,443,204,405,954,385,563,870,541,379,246,659,709,506,697,378,694,300
97	14,657,929,356,129,575,437,016,877,846,657,032,761,712,954,950,899,755,100
98	57,743,358,069,601,357,782,187,700,608,042,856,334,020,731,624,756,611,000
99	227,508,830,794,229,349,661,819,540,395,688,853,956,041,682,601,541,047,340
100	896,519,947,090,131,496,687,170,070,074,100,632,420,837,521,538,745,909,320

References

Abbott, H. L. Problem E 2799, *American Mathematical Monthly* 87 (Dec. 1980), 825.

Alagar, V. S. *Fundamentals of Computing: Theory and Practice*, Prentice-Hall, Englewood Cliffs, NJ, 1989, 621–622.

Alter, R. "Some Remarks and Results on Catalan Numbers," *Proceedings of the Louisiana Conference on Combinatorics, Graph Theory and Computing* (1971) 109–132.

Alter, R., and K. K. Kubota. "Prime and Prime Power Divisiblity of Catalan Numbers," *Notices of the AMS* 19 (1972), A-48.

Alter, R., and K. K. Kubota. "Prime and Prime Power Divisiblity of Catalan Numbers," *Journal of Combinatorial Theory*, Series A, 15 (1973), 243–256.

Alter, R., and T. B. Curtz. "On Binary Nonassociative Products and Their Relation to a Classical Problem of Euler," *Roczniki Polskiego Towarzystwa Mathematics* 17 (1973), 1–8.

Andersen, E. Solution to Problem E 2083, *American Mathematical Monthly* 76 (April 1969), 419–420.

Andreoli, M. Problem 630, *College Mathematics Journal* 30 (May 1999), 234.

Ayers, F. Jr. Problem 3803, *American Mathematical Monthly* 43 (Nov. 1936), 580.

Bager, A. Solution to Problem E 2659, *American Mathematical Monthly* 85 (Nov. 1978), 767–768.

Bailey, D. F. "Counting Arrangements of 1s and -1s," *Mathematics Magazine* 69 (1996), 128–131.

Balakrishnan, V. K. *Theory and Problems of Combinatorics,* Schaum's Outline Series, McGraw-Hill, New York, 1995, 27–33.

Balakram, H. "On the Values of n Which Make $\frac{(2n)!}{(n+1)!(n+1)!}$ an Integer," *Journal of the Indian Mathematical Society* 18 (Oct. 1929), 97–100.

Barlaz, J. Problem 4742, *American Mathematical Monthly* 54 (May 1957), 370.

Bassali, W. A. Letter to the Editor, *Mathematics Magazine* 62 (April 1989), 148.

Becker, H. W. "Rooks and Rhymes," *Bulletin of the AMS* 52 (1946), 1016.

Becker, H. W. "Books and Rhymes," *Mathematics Magazine* 22 (1948), 23–26.

Becker, H. W. Solution to Problem 4277, *American Mathematical Monthly*, 56 (Dec. 1949), 697–698.

Becker, H. W. "The Closed Diagonal Binomial Coefficient Expansion," *Bulletin of the AMS* 57 (1951), 298.

Becker, H. W. "Planar Rhyme Schemes," *Bulletin of the AMS* 58 (1952), 39.

Beineke, L. W., and R. E. Pippert. "A Census of Ball and Disk Dissections," *Lecture Notes in Mathematics* (by A. Dold), Springer-Verlag, New York, 1964.

Beineke, L. W., and R. E. Pippert. "Enumerating Dissectible Polyhedra by Their Automorphism Groups," *Canadian Journal of Mathematics* 1 (1974), 50–67.

Berman, G., and K. D. Fryer. *Introduction to Combinatorics*, Academic Press, New York, 1972.

Bernstein, K. L. Solution to Problem 1213, *Mathematics Magazine* 59 (April 1986), 116–117.

Beumer, M. G. Solution to Problem E 2054, *American Mathematical Monthly* 76 (Feb. 1969), 193.

Beyler, R. H. http://muse.jhu.edu/journals/technology-and culture/v047/47.1beyler.html.

Biasotti, J. et al. Solution to Problem 1213, *Mathematics Magazine* 59 (April 1986), 116.

Bird, M. T. Solution to Problem 5178, *American Mathematical Monthly* 73 (Feb. 1965), 203–204.

Birkhoff, G. Problem 3674, *American Mathematical Monthly* 41 (April 1934), 269.

Bivens, I. C., and B. G. Klein. Solution to Problem 1269, *Mathematics Magazine* 61 (June 1988), 200–201.

Bizley, M. T. L. Solution to Problem 5178, *American Mathematical Monthly* 72 (Feb. 1965), 203.

Blackwell, D., and J. L. Hodges Jr. "Elementary Path Counts," *American Mathematical Monthly* 74 (1967), 801–804.

Borwein, J. M., and P. B. Borwein. *Pi and the AGM: A Study in Analytic Number Theory and Computational Complexity*, Wiley, New York, 1987.

Boucher, C. L. "Path Representation of a Free Throw Shooter's Progress," *Mathematics Magazine* 79 (June 2006), 213–217.

Boyd, A. V. Problem E 2409, *American Mathematical Monthly* 80 (April 1973), 434.

Boyer, R. H. Problem 4510, *American Mathematical Monthly* 61 (Feb. 1954), 129.

Brawley, J. V. "Dedicated to Leonard Carlitz: The Man and His Work," *Finite Fields and Their Applications* 1 (1995), 135–151.

Brenner, J. L., and R. J. Evans. Letter to the Editor, *Mathematics Magazine* 61 (June 1988), 207.

Breusch, R. Problem 4742, *American Mathematical Monthly* 55 (April 1958), 293.

Breusch, R. Problem 5178, *American Mathematical Monthly* 71 (Feb. 1964), 217.

Bristol, W. A., and W. R. Church. Solution to Problem 3421, *American Mathematical Monthly* 38 (Jan. 1931), 54–55.

Brown, J. L., and V. E. Hoggatt Jr. "A Primer for the Fibonacci Numbers, Part XVI: The Central Column Sequence," *Fibonacci Quarterly* 16 (Feb. 1978), 41–46.

Brown, J. R. Solution to Problem 4931, *American Mathematical Monthly* 68 (Dec. 1961), 1012.

Brown, S. H. "An Interview with H. W. Gould," *College Mathematics Journal,* 37 (Nov. 2006), 370–379.

Brown, W. G. "Historical Note on a Recurrent Combinatorial Problem," *American Mathematical Monthly* 72 (1965), 973–977.

Bruckman, P. S. Problem H-616, *Fibonacci Quarterly* 42 (Nov. 2004), 377.

Bruckman, P. S. Problem 1134, *Pi Mu Epsilon Journal* 12 (Spring 2007), 365.

Callan, D. "Certificates of Integrality for Linear Binomials," *Fibonacci Quarterly* 38 (Aug. 2000), 317–325.

Callan, D. Problem 11091, *American Mathematical Monthly* 111 (June-July 2004), 534.

Campbell, D. M. "The Computation of Catalan Numbers," *Mathematics Magazine* 57 (Sept. 1984), 195–208.

Cantor, D. G. Problem 4764, *American Mathematical Monthly* 65 (Oct. 1958), 637.

Carlitz, L. Solution to Problem E 1854, *American Mathematical Monthly* 74 (May 1967), 596.

Carlitz, L. "Enumeration of Certain Sequences," *Acta Arithmetica* 18 (1971), 221–232.

Carlitz, L. "Sequences, Paths, Ballot Numbers," *Fibonacci Quarterly* 10 (Nov. 1972), 531–549.

Carlitz, L. Problem H-209, *Fibonacci Quarterly* 11 (Feb. 1973), 73.

Carlitz, L. Solution to Problem H-209, *Fibonacci Quarterly* 12 (Dec. 1974), 400.

Carlitz, L. et al, "Some Remarks on Ballot-Type Sequences of Positive Integers," *Combinatorial Structures and Their Applications*, Gordon and Breach, New York, 1970.

Carlitz, L. et al, "Some Remarks on Ballot-Type Sequences of Positive Integers," *Journal of Combinatorial Theory,* Series A, 11 (1971), 258–271.

Carlitz, L., and J. Riordan. "Two Element Lattice Permuation Numbers and Their q-Generalization," *Duke Mathematical Journal* 31 (1964), 371–388.

Carlitz, L., and D. P. Roselle. "Triangular Arrays Subject to MacMahon's Conditions," *Fibonacci Quarterly* 10 (Dec. 1972), 591–597, 658.

Carlitz, L., and R. A. Scoville, Problem E 2054, *American Mathematical Monthly* 75 (Jan. 1968), 77.

Carlitz, L., and R. A. Scoville. "A Generating Function for Triangular Partitions," *Mathematics of Computation* 29 (Jan. 1975), 67–77.

Carlitz, L., and R. A. Scoville. "Note on Weighted Sequences," *Fibonacci Quarterly* 13 (Dec. 1975), 303–306.

Castellanos, D. "The Ubiquitous π," *Mathematics Magazine* 61 (April 1988), 67–98, 148–163.

Cayley, A. "On the Partitions of a Polygon," *Proceedings of the London Mathematical Society* 22 (1891), 237–262.

Chamberlain, M. Solution to Problem 290, *College Mathematics Journal* 17 (Sept. 1986), 364–365.

Chandrasekharan, K. *Introduction to Analytic Number Theory*, Springer-Verlag, New York, 1968.

Chang, G. Problem 290, *College Mathematics Journal* 15 (Nov. 1984), 446.

Chapman, R. Solution to Problem 11165, *American Mathematical Monthly* 114 (May 2007), 454–455.

Chernoff, P. R. Solution to Problem E 1896, *American Mathematical Monthly* 74 (Oct. 1967), 1007.

Christophe, L. M. Jr. Problem 1269, *Mathematics Magazine* 60 (June 1987), 178.

Chu, W. "A New Combinatorial Interpretation for Generalized Catalan Number," *Discrete Mathematics* 65 (1987), 91–94.

Church, C. A. Jr. Solution to Problem 5343, *American Mathematical Monthly* 73 (Dec. 1966), 1129–1130.

Clarke, R. J. Letter to the Editor, *Mathematics Magazine* 61 (Oct. 1988), 269.

Cofman, J. "Catalan Numbers for the Classroom?," *Elemente der Mathematik* 52 (1997), 108–117.

Cohn. D. I. A. *Basic Techniques of Combinatorial Theory*, Wiley, New York, 1978.

Comtet, L. *Advanced Combinatorics*, Dreidel, Boston, Massachusetts, 1974.

Conway, J. H., and H. S. M. Coexter. "Triangulated Polygons and Frieze Patterns," *Mathematical Gazette* 57 (1973), 87–94; 175–183.

Conway, J. H., and R. K. Guy. *The Book of Numbers,* Copernicus, New York, 1996, 97–106.

Cook, R. "A Mathematical Centenary: J. J. Sylvester 1814–1897," *Mathematical Spectrum* 30 (1997–1998), 1-2.

Corey, S. A. Problem 270, *American Mathematical Monthly* 16 (Jan. 1909), 40.

Corey, S. A. Solution to Problem 270, *American Mathematical Monthly* 17 (June–July 1909), 121.

Currier, A. E. Problem 4764, *American Mathematical Monthly* 64 (Dec. 1957), 737.

Dean, R. A., and Keller, G. "Natural Partial Orders," *Canadian Journal of Mathematics* 20 (1968), 535–554.

de Bruijn, N. G. and Morselt, B. J. M. "A Note on Plane Trees," *Journal of Combinatorial Theory* 2 (1967), 27–34.

Debus, A. G. (ed.) *World Who's Who in Science*, Marquis—Who's Who, Chicago, Illinois.

Deutsch, E. "An Involution on Dyck Paths and its Consequences," *Discrete Mathematics*, 204 (1999), 163–166.

Deutsch, E. Problem 11071, *American Mathematical Monthly* 111 (March 2004), 259.

Deutsch, E. Problem 11150, *American Mathematical Monthly* 112 (April 2005), 367.

Deutsch, E. Problem 11170, *American Mathematical Monthly* 112 (Aug.–Sept. 2005), 655.

Deutsch, E. Problem 1739, *Mathematics Magazine* 79 (Feb. 2006), 67–68.

Deutsch, E., and I. M. Gessel. Problem 1494, *Mathematics Magazine* 69 (April 1996), 143.

Deutsch, E., and W. P. Johnson. "Create Your Own Permutation Statistics," *Mathematics Magazine* 77 (April 2004), 130–134.

Deustch, E., and B. E. Sagan. "Congruences for Catalan and Motzkin Numbers and Related Sequences," *Journal of Number Theory*, 117 (2006), 191–215.

de Vries, H. L. "On Property B and on Steiner Systems," *Mathematische Zeitschrift* 153 (1977), 155–159.

Došlić, T. "Perfect Matchings, Catalan Numbers, and Pascal's Triangle," *Mathematics Magazine* 80 (June 2007), 219–226.

Dunkel, O. Problem 3421, *American Mathematical Monthly* 38 (1930), 196.

Dvoretzky, A., and T. Motzkin. "A Problem of Arrangements," *Duke Mathematical Journal* 14 (1947), 305–313.

Dymàček, W. M., and H. Sharp. *Introduction to Discrete Mathematics*, W. C. Brown/ McGraw-Hill, Boston, Massachusetts, 1998.

Eggleton, R. B., and R. K. Guy. "Catalan Strikes Again! How Likely Is a Function to be Convex?," *Mathematics Magazine* 61 (Oct. 1988), 211–219.

Eldredge, G. Problem E 1903, *American Mathematical Monthly* 73 (1966), 666.

Elsner, C. "On Recurrence Formulae for Sums Involving Binomial Coefficients," *Fibonacci Quarterly* 43 (Feb. 2005), 31–45.

Eğecioğlu, O. "The Parity of the Catalan Numbers via Lattice Paths," *Fibonacci Quarterly,* 21 (Feb. 1983), 65–66.

Erdös, P. Problem 4252, *American Mathematical Monthly* 54 (May 1947), 287.

Erdös, P. "On Some Divisibility Properties of $\binom{2n}{n}$," *Canadian Mathematics Bulletin* 7 (1964), 513–518.

Erdös, P. et al. "On the Prime Factors of $\binom{2n}{n}$," *Mathematics of Computation* 29 (Jan. 1975), 83–92.

Erdös, P., and I. Kaplansky. "Sequences of Plus and Minus," *Scripta and Mathematics* 12 (1946), 73–75.

Etherington, I. M. H. "Non-Associate Powers and a Functional Equation," *Mathematical Gazette* 21 (1937), 36–39, 153.

Etherington, I. M. H. "On Non-Associative Combinations," *Proceedings of the Royal Society of Edinburgh* 59 (1939), 153–162.

Etherington, I. M. H. "Some Problems of Nonassociative Combinations," *Edinburgh Mathematical Notes* 32 (1940), 1–6.

Even, S. *Graph Algorithms*, Computer Science Press, Rockville, Maryland, 1979.

Fechter, M. A. Solution to Problem E 1896, *American Mathematical Monthly* 74 (Oct. 1967), 1007–1008.

Feemster, H. C. Problem 172, *American Mathematical Monthly* 17 (March 1910), 75.

Feemster, H. C. Problem 175, *American Mathematical Monthly* 17 (June-July 1910), 150.

Feldman, H. M. Solution to Problem 230, *American Mathematical Monthly* 29 (Nov.–Dec. 1955), 109.

Feller, W. *An Introduction to Probability Theory and Its Applications*, vol. 1, Wiley, New York, 1957, 66–73.

Fine, N. J. "Binomial Coefficients Modulo a Prime," *American Mathematical Monthly* 54 (1947), 589–592.

Finkel, B. F. (ed.). Problem 2, *American Mathematical Monthly* 1 (Feb. 1894), 50.

Forder, H. G. "Some Problems in Combinatorics," *Mathematical Gazette* 45 (1961), 199–201.

Fraliegh, J. B. *A First Course in Abstract Algebra*, 7th ed., Addison-Wesley, Reading, Massachusetts, 2004.

Frame, J. S. Problem 5113, *American Mathematical Monthly* 70 (June–July 1963), 672.

Franklin, P. Problem 2681, *American Mathematical Monthly* 25 (March 1918), 118.

Fray, R. D. Solution to Problem E 2253, *American Mathematical Monthly* 78 (Sept. 1971), 797.

Fray, R. D., and D. P. Roselle. "Weighted Lattice paths," *Pacific Journal of Mathematics* 37 (1971), 85–96.

Fryer, W. D. Solution to Problem E 1700, *American Mathematical Monthly* 72 (May 1965), 553.

Gardner, M. "Mathematical Games," *Scientific American* 215 (Dec. 1966), 131.

Gardner, M. "Catalan Numbers: An Integer Sequence that Materializes in Unexpected Places," *Scientific American* 234 (June 1976), 120–125.

Gandhi, J. M. Problem E 1346, *American Mathematical Monthly* 66 (Jan. 1959), 61.

Gandhi, J. M. Problem E 5343, *American Mathematical Monthly* 72 (Dec. 1965), 1135.

GCHQ Problem Solving Group. Solution to Problem 11091, *American Mathematical Monthly* 113 (May 2006), 462–463.

Gehami, L. A. Problem E 2253, *American Mathematical Monthly* 77 (Oct. 1970), 882.

Gillispie, C. C. (ed.). *Dictionary of Scientific Biography*, Charles Scribner's Sons, New York, 1973.

Goheen, C. C. Problem 4931, *American Mathematical Monthly* 67 (Nov. 1960), 926.

Goldstone, L. D. Solution to Problem E 1346, *American Mathematical Monthly* 66 (Sept. 1959), 591.

Gould, H. W. Solution to Problem 676, *American Mathematical Monthly* 41 (May–June 1968), 166.

Gould, H. W. "Equal Products of Generalized Binomial Coeffceints," *Fibonacci Quarterly* 9 (Oct. 1971), 337–346.

Gould, H. W. "Fibonomial Catalan Numbers: Arithmetic Properties and a Table of the First Fifty Numbers," *Notices of the AMS* 18 (1971), 938.

Gould, H. W. "A New Primality Criterion of Mann and Shanks and its Relation to a Theorem of Hermite," *Fibonacci Quarterly* 10 (1972), 355–364, 372.

Gould, H. W. "Generalization of a Formula of Touchard for Catalan Numbers," *Journal of Combinatorial Theory*, Series A, 23 (1977), 351–353.

Gould, H. W. *Bell and Catalan Numbers: Research Bibliography of Two Special Number Sequences,* rev. ed., Combinatorial Research Institute, Morgantown, West Verginia, 1978.

Graham, R. L. et al. *Concrete Mathematics*, 2nd ed., Addison-Wesley, Reading, Massachusetts, 1994.

Griffiths, R. L. "Catch-up Numbers," *Mathematical Gazette* 91 (Nov. 2007), 500–509.

Grimaldi, R. P. "The Catalan Numbers," MAA Minicourse, Orlando, Florida (Jan. 1996).

Grimaldi, R. P. *Discrete and Combinatorial Mathematics,* 5th ed., Pearson/Addison-Wesley, 2004.

Grossman, H. D. "Fun with Lattice Points," *Scripta Mathematica* 12 (1946), 223–225.

Grossman, H. D. "More about Lattice-Paths," *Scripta Mathematica* 14 (1948), 160–162.

Grossman, H. D. "Fun with Lattice Points," *Scripta Mathematica* 15 (1949), 79–81.

Grossman, H. D. "Fun with Lattice Points," *Scripta Mathematica* 16 (1950), 207–212.

Grosswald, E. Solution to Problem 6019, *American Mathematical Monthly* 84 (Jan. 1977), 63–65.

Growney, J. A. S. *Finitely Generated Free Groups,* Ph.D. dissertation, University of Oklahoma, Norman, 1970.

Gummer, C. F. Solution to Problem 2681, *American Mathematical Monthly* 27 (March 1919), 128.

Gupta, H. "On a Problem in Parity," *Indian Journal of Mathematics* 11 (1969), 157–163.

Gupta, H. "On the Parity of $\frac{(n+m-1)(n,n)}{n!m!}$," *Research Bulletin, Punjab University* 20 (1969), 571–575.

Gupta, H. "Meanings of Non-Associative Expressions," *Research Bulletin, Punjab University* 21 (1970), 225–226.

Guy, R. K. "Dissecting a Polygon into Triangles," Research Paper no.9, Department of Mathematics, University of Calgary, Alberta, Canada (Jan. 1967).

Guy, R. K. "Dissecting a Polygon into Triangles," *Bulletin of Malayan Mathematical Society* 5 (1968), 57–60.

Guy, R. K. "A Medley of Malayan Mathematical Memories and Mantissae," *Mathematical Medley* 12 (1984), 9–17.

Guy, R. K. Letter to the Editor, *Mathematics Magazine* 61 (Oct. 1988), 269.

Guy, R. K. "The Second Strong Law of Small Numbers," *Mathematics Magazine* 63 (Feb 1990), 3–20.

Guy, R. K. "Parker's Permutation Problem Involves the Catalan Numbers," *American Mathematical Monthly* 100 (March 1993), 287–289.

Guy, R. K., and J. L. Selfridge. "The Nesting and Roosting Habits of the Laddered Parenthesis," Research Paper no. 127, Department of Mathematics, University of Calgary, Alberta, Canada (June 1971).

Guy, R. K., and J. L. Selfridge. "The Nesting and Roosting Habits of the Laddered Parenthesis," *American Mathematical Monthly* 80 (Oct. 1973), 868–876.

Handelsman, M. B. Problem 436, *College Mathematics Journal* 21 (Nov. 1990), 423.

Harary, F. *Applied Combinatorial Mathematics*, Wiley, New York, 1964.

Harary, F. et al. "The Number of Plane Trees," *Indagationes Mathematicae* 26 (1964), 319–329.

Hayes, D. R. "Leonard Carlitz (1907–1999)," *Notices of the AMS* 48 (Dec. 2001), 1322–1323.

Herzog, F. Solution to Problem 4252, *American Mathematical Monthly* 56 (Jan. 1949), 42–43.

Heuer, G. A. Solution to Problem E 2054, *American Mathematical Monthly* 76 (Feb. 1969), 192–193.

Hilton, P., and J. J. Pedersen. Letter to the Editor, *Mathematics Magazine* 61 (June 1988), 207.

Hilton, P., and J. Pedersen. "The Ballot Problem and Catalan Numbers," *Nieuw Archief Voor Wiskunde* 7–8 (1990), 209–216.

Hilton, P., and J. Pedersen. "Catalan Numbers, Their Generalizations, and Their Uses," *Mathematics Intelligencer* 13 (1991), 64–75.

Hodges, J. H. Solution to Problem E 1346, *American Mathematical Monthly* 66 (Sept. 1959), 590.

Hoggatt, V. E. Jr., and M. Bicknell. "Pascal, Catalan, and General Sequence Convolution Arrays in a Matrix," *Fibonacci Quarterly* 14 (1976), 135–143.

Hoggatt, V. E. Jr., and M. Bicknell. "Catalan and Related Sequences Arising from Inverses of Pascal's Triangle Matrices," *Fibonacci Quarterly* 14 (Oct. 1976), 395–405.

Hoggatt, V. E. Jr., and M. Bicknell. "Sequences of Matrix Inverses from Pascal, Catalan, and Related Convolution Arrays," *Fibonacci Quarterly* 14 (Dec. 1976), 224–232.

Hoggatt, V. E. Jr., and M. Bicknell. "Numerator Polynomial Coefficient Arrays for Catalan and Related Sequence Convolution Triangles," *Fibonacci Quarterly* 15 (1977), 30–34.

Holshouser, A. L. Problem E 2659, *American Mathematical Monthly* 84 (June–July 1977), 486.

Howard, F. T. "In Memoriam—Leonard Carlitz," *Fibonacci Quarterly* 38 (Aug. 2000), 316.

Howard, H. R. Problem 2714, *American Mathematical Monthly* 25 (June 1918), 259.

Howard, H. R. Solution to Problem 2714, *American Mathematical Monthly* 26 (Sept. 1919), 318.

Howell, J. M. Problem 230, *Mathematics Magazine* 28 (March 1955), 233.

Hsu, L. C. Problem E 995, *American Mathematical Monthly* 58 (Dec. 1951), 700.

Huff, G. B. Problem E 2520, *American Mathematical Monthly* 82 (Feb. 1975), 169.

Irwin, F. Problem 2688, *American Mathematical Monthly* 25 (March 1918), 119.

Irwin, F. Solution to Problem 2688, *American Mathematical Monthly* 33 (Feb. 1926), 105.

Jackson, D. E., and R. C. Entringer. "Enumeration of Certain Binary Matrices," *Journal of Combinatorial Theory* 8 (1970), 291–298.

Jarvis, F. "Catalan Numbers," *Mathematical Spectrum* 36 (2003–2004), 9–12.

Just, E. Problem E 2083, *American Mathematical Monthly* 75 (April 1968), 404.

Kapur, J. N. "Combinatorial Analysis and School Mathematics," *Educational Studies in Mathematics* 3 (1970), 111–127.

Kelly, J. B. Solution to Problem 4983, *American Mathematical Monthly* 69 (Nov. 1962), 931.

Kemeny, J. G., and J. L. Snell. "A Combinatorial Formula," *Mathematical Models in the Social Sciences*, MIT Press, Cambridge, Massachusetts, 1975.

Kim, K. H., and F. W. Roush. "Enumeration of Semiorders and Transitive Relations," *Notices of the AMS* 24 (1977), 1-10.

King, J., and P. King. Solution to Problem 87.F, *Mathematical Gazette* 90 (March 2006), 163–164.

Klambauer, G. *Problems and Propositions in Analysis*, Marcel Dekker, New York, 1979.

Klamkin, M. S. Problem 4983, *American Mathematical Monthly* 68 (Oct. 1961), 807.

Klamkin, M. S. Problem 488, *College Mathematics Journal* 23 (Nov. 1992), 435–436.

Klarner, D. A. "A Correspondence between Two Sets of Trees," *Indagationes Mathematicae* 31 (1969), 292–296.

Klarner, D. A. "Correspondence between Plane Trees and Binary Sequences," *Journal of Combinatorial Theory* 9 (1970), 401–411.

Klazar, M. "On *abab*-Free and *abba*-Free Set Partitions," *European Journal of Mathematics* 1 (1996), 53–68.

Klazar, M. "Counting Even and Odd Partitions," *American Mathematical Monthly* 110 (June–July 2003), 527–531.

Knott, G. D. "A Numbering System for Binary Trees," *Communications of the ACM* 20 (Feb. 1977), 113 115.

Koehler, J. E. "Folding a Strip of Stamps," *Journal of Combinatorial Theory* 5 (1968), 135–152.

Koshy, T. *Fibonacci and Lucas Numbers with Applications*, Wiley, New York, 2001.

Koshy, T. *Discrete Mathematics with Applications*, Elsevier/Academic Press, Boston, Massachusets, 2004.

Koshy, T. "The Ubiquitous Catalan Numbers," *Mathematics Teacher* 100 (Oct. 2006), 184–188.

Koshy, T. *Elementary Number Theory with Applications*, 2nd ed., Academic Press, Burlington, Massachusetts, 2007.

Koshy, T., and M. Salmassi. "Parity and Primality of Catalan Numbers," *College Mathematics Journal* 37 (Jan, 2006), 52–53.

Kotlarski, I. I. Solution to Problem E 2409, *American Mathematical Monthly* 81 (April 1974), 408–409.

Krier, N. K., and F. R. Bernhart. Problem 1213, *Mathematics Magazine* 58 (March 1985), 111.

Kruse, R. L. *Data Structures and Program Design*, 2nd ed., Prentice Hall, Englewood cliffs, New Jersey, 1987.

Kuchinski, M. J. "Catalan Structures and Correspondences," Master's thesis, West Virginia University, Morgantown, West Virginia, 1977.

Lafer, P., and Long, C. T. "A Combinatorial Problem," *American Mathematical Monthly* 69 (Nov. 1962), 876–883.

Larcombe, P. J. "On Pre-Catalan Catalan Numbers: Kotelnikow (1766)," *Mathematics Today,* 35 (Feb. 1999), 25.

Larcombe, P. J. "On the History of the Catalan Numbers: A First Record in China," *Mathematics Today* 35 (June 1999), 89.

Larcombe, P. J. "The 18th Century Chinese Discovery of the Catalan Numbers," *Mathematical Spectrum* 32 (Nov. 1999), 5–7.

Larcombe, P. J., and P. D. C. Wilson. "On the Trail of the Catalan Sequence," *Mathematics Today* 34 (Aug. 1998), 114–117.

Lass, H. Solution to Problem E 2174. *American Mathematical Monthly* 77 (Feb. 1970), 195–196.

Lehmer, D. H. Solution to Problem E 760, *American Mathematical Monthly* 54 (Oct. 1947), 478.

Lehmer, D. H. "Interesting Series Involving the Central Binomial Coefficients," *American Mathematical Monthly* 92 (Aug.–Sept. 1985), 449–457.

Lehmer, E. T. Solution to Problem 3414, *American Mathematical Monthly* 37 (Dec. 1930), 558.

Lehmer, E. T. Solution to Problem 3803, *American Mathematical Monthly* 45 (Nov. 1938), 633–634.

Leighton, F. T., and M. Newman. "Positive Definite Matrices and Catalan Numbers," *Proceedings of the AMS* 79 (June 1980), 177–181.

Levine, J. "Note on the Number of Pairs of Nonintersecting Routes," *Scripta Mathematica*, 24 (1959), 335–338.

Linis, V. Solution to Problem E 2409, *American Mathematical Monthly* 81 (April 1974), 408.

Little, T. M. Solution to Problem E 1346, *American Mathematical Monthly* 66 (Sept. 1959), 590.

Liu, C. L. *Introduction to Combinatorial Mathematics*, McGraw-Hill, New York, 1968.

Liulevicius, A. "Homotopy Rigidity of Linear Actions: Characters Tell All," *Bulletin of the AMS* 84 (March 1978), 213–221.

Lobb. A. "Deriving the nth Catalan Number," *Mathematical Gazette 83* (March 1999), 109–110.

Loève, M. "Random Walk," *Probability Theory*, D. van Nostrand, Princeton, New Jersey, 1963, 47–48.

Loney, S. L. *Plane Trigonometry,* Cambridge University Press, New York, 1933.

Lossers, O. P. Solution to Problem E 2520, *American Mathematical Monthly* 83 (May 1976), 383–384.

Lossers, O. P. Solution to Problem 11150, *American Mathematical Monthly* 114 (March 2007), 264–265.

Lovit, D. Solution to Problem 1739, *Mathematics Magazine* 80 (Feb. 2007), 80–81.

Mattson, H. F. Jr. *Discrete Mathematics with Applications,* Wiley, New York, 1993, 535–538.

Mays, M. E., and J. Wojciechowski. "A Determinant Property of Catalan Numbers," *Discrete Mathematics* 211 (2000), 125–133.

McCammond. J. "Noncrossing Partitions in Surprising Locations," *American Mathematical Monthly* 113 (Aug.–Sept. 2006), 598–610.

McKay, J. H. "The William Lowell Putnam Mathematical Competition," *American Mathematical Monthly* 73 (Aug.–Sept. 1966), 726–732.

Merlini, D. et al. "The Method of Coefficients," *American Mathematical Monthly* 114 (Jan. 2007), 40–57.

Miksa, F. L. et al. "Restricted Partitions of Finite Sets," *Canadian Mathematics Bulletin* 1 (May 1958), 87–96.

Moen, C. "Infinite Series with Binomial Coeffceints," *Mathematics Magazine* 64 (Feb. 1991), 53–55.

Mohanty, S. G. "Some Properties of Compositions and Their Application to the Ballot Problem," *Canadian Mathematics Bulletin* 8 (1956), 359–372.

Moon, J. W. *Topics on Tournaments,* Holt, Rinehart and Winston, New York, 1968, 68–70.

Moon, J. W., and L. Moser. "Triangular Dissections of n-gons," *Canadian Mathematics Bulletin* 6 (1963), 175–178.

Moore, R. E. Problem 2752, *American Mathematical Monthly* 26 (Feb. 1919), 72.

More, Y. Problem 11165, *American Mathematical Monthly* 112 (June–July 2005), 568.

Moser, L. Solution to Problem E 1346, *American Mathematical Monthly* 66 (Sept. 1959), 591.

Moser, L. "Insolvability of $\binom{2n}{n} = \binom{2a}{a}\binom{2b}{b}$," *Canadian Mathematics Bulletin* 6 (1963), 167–169.

Moser, L., and W. Zayachkowski. "Lattice Paths with Diagonal Steps," *Scripta Mathematica* 26 (1963), 223–229.

Motzkin, T. "Relations between Hypersurface Cross Ratios, and a Combinatorial Formula for Partitions of a Polygon, for Permanent Preponderance, and for

Nonassociative Products," *Bulletin of the American Mathematical Society* 54 (April 1948), 352–360.

Mullin, R. C. "On Counting Rooted Triangular Maps," *Canadian Journal of Mathematics* 17 (1965), 373–382.

Mullin, R. C., and R. G. Stanton. "A Map-Theoretic Approach to Davenport-Schinzel Sequences," *Pacific Journal of Mathematics* 40 (1972), 167–172.

Nederpelt, R. P. Solution to Problem E 1903, *American Mathematical Monthly* 79 (April 1972), 396.

Nelsen, R. B. Solution to Problem 1049, *Mathematics Magazine* 52 (Nov. 1979), 321–322.

Niven, I. "Interpretations of a Non-Associative Product," *Mathematics of Choice*, Random House, New York, 1965, 140–152.

Niven, I. "Formal Power Series," *American Mathematical Monthly* 75 (1969), 871–889.

Obituary. "H. Balakram," *Journal of the Indian Mathematical Society* 18 (June 1929), 71–72.

Odlyzko, A. M., and H. S. Wilf. "The Editor's Corner: n Coins in a Fountain," *American Mathematical Monthly* 95 (Nov. 1988), 840–843.

Olds, C. D. Problem E 760, *American Mathematical Monthly* 54 (Feb. 1947), 108.

Olds, C. D. Problem 4277, *American Mathematical Monthly* 55 (Jan. 1948), 34.

Ore, O. Problem 3954, *American Mathematical Monthly* 48 (April 1940), 245.

Otter, R. "The Number of Trees," *Annals of Mathematics* 49 (July 1948), 583–599.

Pennisi, L. L. "Expressions for π and π^2," *American Mathematical Monthly* 62 (Nov. 1955), 653–654.

Pennisi, L. L. "A Method for Solving $\int \sin^{2n} \alpha x \, dx$ and $\int \cos^{2n} \alpha x \, dx$," *Mathematics Magazine* 29 (May–June 1956), 271–272.

Petersen, J. Problem E 1700, *American Mathematical Monthly* 71 (May 1964), 555.

Philbrick, P. H. Solution to Problem 2, *American Mathematical Monthly* 1 (March 1894), 82.

Pollak, H. O. Solution to Problem 420, *College Mathematics Journal* 22 (Jan. 1991), 76.

Pólya, G. *Mathematics and Plausible Reasoning*, vol. 1, Princeton University Press, New Jersey, 1954.

Pólya, G. "On Picture-Writing," *American Mathematical Monthly* 63 (Dec. 1956), 689–697.

Poonen, B. Solution to Problem 1251, *Mathematics Magazine* 60 (Nov. 1987), 248.

Rainville, E. D. Solution to Problem 3674, *American Mathematical Monthly* 43 (Oct. 1935), 519.

Rainville, E. D. Solution to Problem 230, *American Mathematical Monthly* 29 (Nov.–Dec. 1955), 109–110.

Raney, G. N. "Functional Composition Patterns and Power Series Reversion," *Transactions of the AMS* 94 (1960), 441–451.

Renault, M. Solution to Problem 11071, *American Mathematical Monthly* 113 (May 2004), 460–461.

Renault, M. "Four proofs of the Ballot Theorem," *Mathematics Magazine* (Dec. 2007), 345–352.

Renault, M. "Lost (and Found) in Translation: Andreé's *Actual Method* and its Application to the Generalized Ballot Problem," *American Mathematical Monthly* 115 (April 2008), 358–363.

Rey, J. G. Problem 2070, *Crux Mathematicorum* 21 (1995), 236.

Rey, J. G. Problem 2070, *Crux Mathematicorum* 22 (1996), 279–281.

Richards, D. "Ballot Sequences and Restricted Permutations," *Ars Combinatoria* 25 (1988), 83–86.

Riddle, L. H. "An Occurrence of the Ballot Numbers in Operator Theory," *American Mathematical Monthly* 98 (Aug. 1991), 613–617.

Rinaldi, S., and Rogers, D. G. "Indecomposability: Polyominoes and Polyomino Tilings," *Mathematics Gazette* 92 (2008).

Riordan, J. *An Introduction to Combinatorial Analysis*, Princeton University Press, New Jersey, 1980, 125–139.

Riordan, J. *Combinatorial Identities,* Wiley, New York, 1968.

Riordan, J. "Ballots and Trees," *Journal of Combinatorial Theory* 6 (1969), 408–411.

Riordan, J. "A Note on Catalan Parentheses," *American Mathematical Monthly* 81 (Oct. 1973), 904–906.

Riordan, J. "The Distribution of Crossings of Chords Joining Pairs of $2n$ Points on a Circle," *Mathematics of Computation* 29 (1975), 215–222.

Rogers, D. G. "A Schröder Triangle: Three Combinatorial Problems," *Australian Conference on Combinatorial Mathematics*, Lecture Notes in Mathematics 622, Springer, Berlin, 1977, 175–196.

Rogers, D. G. "Pascal Triangles, Catalan Numbers, and Renewal Arrays," *Discrete Mathematics* 22 (1978), 301–310.

Rohatgi, V. K. "On Lattice Paths with Diagonal Steps," *Canadian Mathematics Bulletin* 7 (1964), 470–472.

Rohatgi, V. K. "Some Combinatorial Identities Involving Lattice Paths," *American Mathematical Monthly* 73 (1966), 507–508, 751.

Roselle, D. P. "An Algorithmic Approach to Davenport-Schinzel Sequences," *Utilitas Mathematica* 6 (1974), 91–93.

Roselle, D. P. "A Combinatorial Problem Involving q-Catalan Numbers," *Notices of the AMS* 21 (1974), A-609.

Rota, G. C. "Nomination for Ronald L. Graham for AMS President," *Notices of the AMS* 38 (Sep. 1991), 757–758.

Rousseau, C. C. Solution to Problem 1134, *Pi Mu Epsilon Journal* 12 (Spring 2007), 365–367.

Ruderman, H. D. Problem Q699, *Mathematics Magazine* 58 (Sept. 1985), 238, 245.

Ruderman, H. D. Problem 1251, *Mathematics Magazine* 59 (Oct. 1986), 240.

Sandham, H. F. Problem 4519, *American Mathematical Monthly* 60 (Jan. 1953), 47.

Sandham, H. F. Solution to Problem 4519, *American Mathematical Monthly* 61 (Jan. 1953), 265.

Sands, A. D. "On Generalized Catalan Numbers," *Discrete Mathematics* 21 (1978), 219–221.

Schneider, J. E. Problem E 1600, *American Mathematical Monthly* 70 (May 1963), 569.

Schneider, R. Problem E 1854, *The American Mathematical Monthly* 73 (Jan. 1966), 83.

Schwatt, I. J. *An Introduction to the Operators with Series,* Chelsea, New York, 1924.

Schwenk, A. J. Solution to Problem 1213, *American Mathematical Monthly* 59 (April 1986), 115–116.

Scoville, R. "Some Properties of the Catalan Numbers," *Notices of the AMS* 21 (1974), A-602.

Seiffert, H. Problem 11172, *American Mathematical Monthly* 112 (Aug.–Sept. 2005), 655.

Shafer, R. E. Problem 6019, *American Mathematical Monthly* 82 (March 1975), 307.

Shapiro, L. W. "Rings and Catalan Numbers," *American Mathematical Monthly* 81 (1974), 940.

Shapiro, L. W. "Upper Triangular Rings, Ideals, and Catalan Numbers," *American Mathematical Monthly* 82 (1975), 634–637.

Shapiro, L. W. "Catalan Numbers and "Total Information" Numbers," *Proceedings of the Sixth Southwestern Conference on Combinatorial Graph Theory and Computing*, Florida Atlantic University, 1975, 531–539.

Shapiro, L. W. "A Catalan Triangle," *Discrete Mathematics* 14 (1976), 83–90.

Shapiro, L. W. "A Short Proof of an Identity of Touchard's Concerning Catalan Numbers," *Journal of Combinatorial Theory*, Series A, 20 (1976), 375–376.

Shapiro, L. W. "Positive Definite Matrices and Catalan Numbers, Revisited," *Proceedings of the AMS* 90 (March 1984), 488–496.

Shapiro, L. W., and W. Hamilton. "The Catalan Numbers Visit the World Series," *Mathematics Magazine* 66 (Feb. 1993), 20–22.

Shen, Y., and J. T. Shult. "Sailing around Binary Trees and Krusemeyer's Bijection," *Mathematics Magazine* 73 (June 2000), 216–220.

Sholander, M., and E. B. Leach. Problem E 2799, *American Mathematical Monthly* 86 (Nov. 1979), 785.

Silberger, D. M. "The Parity of the Integer $\frac{(2n-2)!}{n!(n-1)!}$," *Notices of the AMS* 15 (1968), 615.

Silberger, D. M. "Occurrences of the Integer $\frac{(2n-2)!}{n!(n-1)!}$," *Roczniki Polskiego Towarzystwa Mathematics* 13 (1969), 91–96.

Simion, R., and F. W. Schmidt. "Restricted Permutations," *European Journal of Combinatorics* 6 (1985), 383–406.

Simion, R., and D. Ullman. "On the Structure of the Lattice of Noncrossing Partitions," *Discrete Mathematics*, 98 (1991), 193–206.

Singmaster, D. Problem E 1896, *American Mathematical Monthly* 73 (June–July 1966), 665.

Singmaster, D. "An Elementary Evaluation of the Catalan Numbers," *American Mathematical Monthly* 85 (May 1978), 366–368.

Sloane, N. J. A. *A Handbook of Integer Sequences*, Academic Press, New York, 1973, 18–22.

Speigel, M. R. Problem E 995, *American Mathematical Monthly* 59 (June 1952), 411.

Squire, W. Problem 676, *Mathematics Magazine* 40 (Nov. 1967), 279.

Stanley, R. P. *Enumerative Combinatorics*, vol. 1, Wadsworth & Brooks/Cole, Monterey, California, 1986, 9–10, 52–63.

Stanley, R. P. *Enumerative Combinatorics*, vol. 2, Cambridge University Press, New York, 1999.

Stanley, R. P. "Catalan Addendum," http://www.math.mit.edu/rstan/ec/catadd.pdf.

Steelman, H. Solution to Problem 630, *College Mathematics Journal* 30 (May 1999), 234–235.

Steffen, K. Problem E 2174, *American Mathematical Monthly* 76 (May 1969), 553.

Stenger, A. Solution to Problem E 2520, *American Mathematical Monthly* 83 (May 1976), 383.

Strehl, V. "Enumeration of Alternating Permutations According to Peak Sets," *Journal of Combinatorial Theory*, Series A, 20 (1978) 238–240.

Sumner, J. S., and K. L. Dove. Solution to Problem 488, *College Mathematics Journal* 24 (Nov. 1993), 477.

Swylan, E. "Proving Numbers Catalan," *Mathematical Gazette* 86 (Nov. 2002), 460–463.

Szekeres, G. Solution to Problem 3954, *American Mathematical Monthly* 41 (Oct. 1941), 565.

Takács, L. Letter to the Editor, *Mathematics Magazine* 61 (Oct. 1988), 269.

Thielman, H. P. Problem 4510, *American Mathematical Monthly* 59 (Nov. 1952), 640.

van Hamme, L. Problem Q745, *Mathematics Magazine* 62 (April 1989), 137, 143.

van Lint, J. H. *Lecture Notes in Mathematics*, Springer-Verlag, New York, 1974.

Vaughn, H. E. Solution to Problem 3954, *American Mathematical Monthly* 41 (Oct. 1941), 564–565.

Vun, I., and P. Belcher. "Catalan Numbers," *Mathematical Spectrum* 29 (1996–1997), 3–5.

Wagner, R. J. Solution to Problem 436, *College Mathematics Journal* 22 (Nov. 1991), 444–445.

Wahlin, G. E. Solution to Problem 175, *American Mathematical Monthly* 18 (Feb. 1971), 41–43.

Walkup, D. W. "The Number of Plane Trees," *Mathematika* 19 (1972), 200–204.

Wang, E. T. H. Problem 420, *The College Mathematics Journal* 21 (Jan. 1990), 66.

Wang, E. T. H. Problem 1049, *Mathematics Magazine* 51 (Sept. 1978), 245.

Weinmann, A. Solution to Problem 5113, *American Mathematical Monthly* 71 (June–July, 1964), 691–692.

Weisstein, E. W. (ed.). *CRC Concise Encyclopedia of Mathematics,* Boca Raton, Florida.

Wells, M. B. *Elements of Combinatorial Computing,* Pergamon, New York, 1971.

West, J. "Generating Trees and the Catalan and Schröder Numbers," *Discrete Mathematics* 146 (1995), 247–262.

Wigner, E. P. "Characteristic Vectors of Bordered Matrices with Infinite Dimensions," *Annals of Mathematics* 62 (Nov. 1955), 548–564.

Woan, W. et al. "The Catalan Numbers, the Lebesgue Integral, and 4^{n-2}," *The American Mathematical Monthly* 104 (Dec. 1997), 926–931.

Wong, B. C. Problem 3414, *The American Mathematical Monthly* 37 (March 1930), 157.

Yaglom, A. M., and I. M. Yaglom. *Challenging Mathematical Problems with Elementary Solutions*, vol. 1, Holden-Day, San Francisco, CA, 1964, 117–121.

Young, J. W. Problem 106, *American Mathematical Monthly* 9 (March 1902), 86.

D. Zeitlin. Solution to Problem E 1346, *American Mathematical Monthly* 66 (Sept. 1959), 590.

Zerr, G. B. M. Solution to Problem 106, *American Mathematical Monthly* 9 (June–July 1902), 86.

Zerr, G. B. M. Solution to Problem 172, *American Mathematical Monthly* 17 (June–July 1910), 147.

Zucker, I. J. "On the Series $\sum_{k=1}^{\infty} \binom{2k}{k} k^{-n}$ and Related Series," *Journal of Number Theory* 20 (1985), 92–102.

Index of Mathematicians

Index